STUDENT STUDY GUIDE

Richard N. Aufmann

Vernon C. Barker

Richard D. Nation

Sandy Doerfel

Palomar College

COLLEGE TRIGONOMETRY

FOURTH EDITION

Aufmann/Barker/Nation

HOUGHTON MIFFLIN COMPANY   BOSTON NEW YORK

Senior Sponsoring Editor: Lynn Cox
Senior Development Editor: Dawn Nuttall
Editorial Assistant: Melissa Parkin
Senior Manufacturing Coordinator: Jane Spelman
Executive Marketing Manager: Michael Busnach

Printed in the U.S.A.

ISBN: 0-618-13088-8

1 2 3 4 5 6 7 8 9-CRS-05 04 03 02 01

CONTENTS

Preface

Study Tips 1

Chapter Tests 7

Solutions to the Chapter Tests 17

Solutions to the Odd-Numbered Exercises in the Text 37

PREFACE

The *Student Study Guide* for the Aufmann/Barker/Nation *College Trigonometry*, Fourth Edition, text contains study tips, additional chapter tests (with solutions), and solutions to the odd-numbered exercises in the text.

The *Study Tips* explain how to best utilize the text and your time in order to succeed in this course.

The *Chapter Tests* contain one test for each chapter in the text. They are modeled after the chapter tests found at the end of each chapter in the text and can be used to provide additional practice for an in-class chapter test. *Solutions* are provided in the section following the tests.

The *Solutions to the Odd-Numbered Exercises in the Text* provide complete, worked-out solutions to all the odd-numbered section and Chapter Review exercises and to **all** the Chapter Test exercises in the text.

Study Tips

STUDY TIPS

The skills you will learn in any mathematics course will be important in your future career—no matter what career you choose. In your textbook, we have provided you with the tools to master these skills. There's no mystery to success in this course; a little hard work and attention to your instructor will pay off. Here are a few tips to help ensure your success in this class.

Know Your Instructor's Requirements

To do your best in this course, you must know exactly what your instructor requires. If you don't, you probably will not meet his or her expectations and are not likely to earn a good grade in the course.

Instructors ordinarily explain course requirements during the first few days of class. Course requirements may be stated in a *syllabus*, which is a printed outline of the main topics of the course, or they may be presented orally. When they are listed in a syllabus or on other printed pages, keep them in a safe place. When they are presented orally, make sure to take complete notes. In either case, understand them completely and follow them exactly.

Attend Every Class

Attending class is vital if you are to succeed in this course. Your instructor will provide not only information but also practice in the skills you are learning. Be sure to arrive on time. You are responsible for everything that happens in class, even if you are absent. If you *must* be absent from a class session:

1. Deliver due assignments to the instructor as soon as possible.
2. Contact a classmate to learn about assignments or tests announced in your absence.
3. Hand copy or photocopy notes taken by a classmate while you were absent.

Take Careful Notes in Class

You need a notebook in which to keep class notes and records about assignments and tests. Make sure to take complete and well-organized notes. Your instructor will explain text material that may be difficult for you to understand on your own and may supply important information that is not provided in the textbook. Be sure to include in your notes everything that is written on the chalkboard.

Information recorded in your notes about assignments should explain exactly what they are, how they are to be done, and when they are due. Information about tests should include exactly what text material and topics will be covered on each test and the dates on which the tests will be given.

Survey the Chapter

Before you begin reading a chapter, take a few minutes to survey it. Glancing through the chapter will give you an overview of its contents and help you see how the pieces fit together as you read.

Begin by reading the chapter title. The title summarizes what the chapter is about. Next, read the section headings. The section headings summarize the major topics presented in the chapter. Then read the topic headings that appear in green within each section. The topic headings describe the concepts for that section. Keep these headings in mind as you work through the material. They provide direction as you study.

Use the Textbook to learn the Material

For each concept studied, read very carefully all of the material from the topic heading to the examples provided for that concept. As you read, note carefully the formulas and words printed in **boldface** type. It is important for you to know these formulas and the definitions of these words.

You will note that each example references an exercise. The example is worked out for you; the exercise, which is highlighted in red in the section's exercise set, is left for you to do. After studying the example, do the exercise. Immediately look up the answer to this exercise in the Solutions section at the back of the text. If your answer is correct, continue. If your answer is incorrect, check your solution against the one given in the Solutions section. It may be helpful to review the worked-out example also. Determine where you made your mistakes.

Next, do the other problems in the exercise set that correspond to the concept just studied. The answers to all the odd-numbered exercises appear in the answer section in the back of the text, and the solutions appear in this Study Guide. Check your answers to the exercises against these.

If you have difficulty solving problems in the exercise set, review the material in the text. Many examples are solved within the text material. Review the solutions to these problems. Reread the examples provided for the concept. If, after checking these sources and trying to find your mistakes, you are still unable to solve a problem correctly, make a note of the exercise number so that you can ask someone for help with that problem.

Review Material

Reviewing material is the repetition that is essential for learning. Much of what we learn is soon forgotten unless we review it. If you find that you do not remember information that you studied previously, you probably have not reviewed it sufficiently. *You will remember best what you review most.*

One method of reviewing material is to begin a study session by reviewing a concept you have studied previously. For example, before trying to solve a new type of problem, spend a few minutes solving a kind of problem you already know how to solve. Not only will you provide yourself with the review practice you need, but you are also likely to put yourself in the right frame of mind for learning how to solve the new type of problem.

Use the End-of-Chapter Material

To help you review the material presented within a chapter, a Chapter Review appears at the end of each chapter. In the Chapter Review, the main concepts of each section are summarized. Included are important definitions and formulas. After completing a chapter, be sure to read the Chapter Review. Use it to check your understanding of the material presented and to determine what concepts you need to review. Return to any section that contains a concept you need to study again.

Each chapter ends with Chapter Review Exercises and a Chapter Test. The problems these contain summarize what you should have learned when you have finished the chapter. Do these exercises as you prepare for an examination. Check your answers against those in the back of the text. Answers to the odd-numbered Chapter Review Exercises and all the Chapter Test exercises are provided there. For any problem you answer incorrectly, review the material corresponding to that concept in the textbook. Determine *why* your answer was wrong.

Finding Good Study Areas

Find a place to study where you are comfortable and can concentrate well. Many students find the campus library to be a good place. You might select two or three places at the college library where you like to study. Or there may be a small, quiet lounge on the third floor of a building where you find you can study well. Take the time to find places that promote good study habits.

Determining When to Study

Spaced practice is generally superior to massed practice. For example, four half-hour study periods will produce more learning than one two-hour study session. The following suggestions may help you decide when to study.

1. A free period immediately before class is the best time to study about the lecture topic for the class.
2. A free period immediately after class is the best time to review notes taken during the class.
3. A brief period of time is good for reciting or reviewing information.

4. A long period of an hour or more is good for doing challenging activities such as learning to solve a new type of problem.
5. Free periods just before you go to sleep are good times for learning information. (There is evidence that information learned just before sleep is remembered longer than information learned at other times.)

Determining How Much to Study

Instructors often advise students to spend twice as much time outside of class studying as they spend in the classroom. For example, if a course meets for three hours each week, instructors customarily advise students to study for six hours each week outside of class.

Any mathematics course requires the learning of skills, which are abilities acquired through practice. It is often necessary to practice a skill more than a teacher requires. For example, this textbook may provide 50 practice problems on a specific concept, and the instructor may assign only 25 of them. However, some students may need to do 30, 40, or all 50 problems.

If you are an accomplished athlete, musician, or dancer, you know that long hours of practice are necessary to acquire a skill. Do not cheat yourself of the practice you need to develop the abilities taught in this course.

Study followed by reward is usually productive. Schedule something enjoyable to do following study sessions. If you know that you only have two hours to study because you have scheduled an enjoyable activity for yourself, you may be inspired to make the best use of the two hours that you have set aside for studying.

Keep Up to Date with Course Work

College terms start out slowly. Then they gradually get busier and busier, reaching a peak of activity at final examination time. If you fall behind in the work for a course, you will find yourself trying to catch up at a time when you are very busy with all of your other courses. Don't fall behind—keep up to date with course work.

Keeping up with course work is doubly important for a course in which information and skills learned early in the course are needed to learn information and skills later in the course. Any mathematics course falls into this category. Skills must be learned immediately and reviewed often.

Your instructor gives assignments to help you acquire a skill or understand a concept. Do each assignment as it is assigned, or you may well fall behind and have great difficulty catching up. Keeping up with course work also makes it easier to prepare for each exam.

Be Prepared for Tests

The Chapter Test at the end of a chapter should be used to prepare for an examination. Additional Chapter Tests are also provided in this Study Guide. We suggest that you try a Chapter Test a few days before your actual exam. Do these exercises in a quiet place, and try to complete the exercises in the same amount of time as you will be allowed for your exam. When completing the exercises, practice the strategies of successful test takers: 1) look over the entire test before you begin to solve any problem; 2) write down any rules or formulas you may need so they are readily available; 3) read the directions carefully; 4) work the problems that are easiest for you first; 5) check your work, looking particularly for careless errors.

When you have completed the exercises in the Chapter Test, check your answers. If you missed a question, review the material in the appropriate section and then rework some of the exercises from that concept. This will strengthen your ability to perform the skills in that concept.

Get Help for Academic Difficulties

If you do have trouble in this course, teachers, counselors, and advisers can help. They usually know of study groups, tutors, or other sources of help that are available. They may suggest visiting an office of academic skills, a learning center, a tutorial service or some other department or service on campus.

Students who have already taken the course and who have done well in it may be a source of assistance. If they have a good understanding of the material, they may be able to help by explaining it to you.

Chapter Tests

CHAPTER 1 of *College Trigonometry* Chapter Test

Functions and Graphs

1. Solve $4(3x-2)-3(2x-5)=25$

2. Solve by factoring and apply the zero product property.
$12x^2-17x+6=0$

3. Use interval notation to express the solution set.
$|x-3| \le 2$

4. Find the midpoint and the length of the line segment with endpoints $(-3, 4)$ and $(-7, -5)$.

5. Determine the x- and y-intercepts, and then graph the equation $x+y^2=6$.

6. Graph the equation $y=-2|x+3|+4$.

7. Find the center and radius of the circle that has the general form $x^2+6x+y^2-8y-11=0$.

8. Determine the domain of the function $f(x)=\sqrt{25-x^2}$.

9. Use the formula $d=\dfrac{|mx_1+b-y_1|}{\sqrt{1+m^2}}$ to find the distance from the point $(-2, 5)$ to the line given by the equation $y=4x-3$.

10. Graph $f(x)=\begin{cases} x^2 & \text{if } x \le -1 \\ |x| & \text{if } -1<x<1 \\ 1 & \text{if } x \ge 1 \end{cases}$.

    Identify the intervals over which the function is
    a. increasing
    b. constant
    c. decreasing

11. Graph the function $f(x)=(x-2)^2-1$. From the graph, find the domain and range of the function.

12. Use the graph of $f(x)=x^2$ to graph $y=3f(x+1)+2$.

13. Classify each of the following as either an even function, an odd function, or neither an even nor an odd function.
    a. $f(x)=x^2+3$
    b. $f(x)=x^2+2x$
    c. $f(x)=4x$

14. Let $f(x)=4x^2-3$ and $g(x)=x+6$. Find $(f-g)$ and (fg).

15. Find the difference quotient of the function
$f(x)=x^2-x-4$.

16. Evaluate $(f \circ g)$, where $f(x)=\sqrt{2x}$ and $g(x)=\frac{1}{2}x^2$.

17. Find the inverse of $f(x)=\frac{2}{3}x-4$. Graph f and f^{-1} on the same coordinate axes.

18. Find the inverse of $f(x)=\dfrac{x-3}{3x+6}$. State the domain and the range of f^{-1}.

19. The distance traveled by a ball rolling down a ramp is given by $s(t)=4t^2$, where t is the time in seconds after the ball is released and $s(t)$ is measured in feet. Evaluate the average velocity of the ball for each of the following time intervals.
    a. $[1, 2]$ b. $[1, 1.5]$ c. $[1, 1.01]$

20. a. Determine the linear regression equation for the given set.

x	2	3	4	6	10
y	3	4	6	10	16

    b. Using the linear model from part **a**, find the expected y-value when the x-value is 8.

Trigonometric Functions

In Problems 1 and 2, convert the measure of the given angle from degrees to radians.

1. $-135°$

2. $240°$

In Problems 3 and 4, convert the measure of the given angle from radians to degrees.

3. $\dfrac{9\pi}{2}$

4. $\dfrac{16\pi}{3}$

5. Determine the measure in radians and in degrees of the central angle that cuts an arc of length 12π centimeters in a circle with radius 8 centimeters.

In Problems 6-9, determine the given trigonometric function value.

6. $\sin\dfrac{5\pi}{4}$

7. $\cos\left(-\dfrac{4\pi}{3}\right)$

8. $\sec\dfrac{19\pi}{2}$

9. $\tan\left(-\dfrac{17\pi}{3}\right)$

In Problems 10 and 11, the given point is located on the terminal side of angle θ. Find $\sin\theta$ and $\tan\theta$.

10. $(-10, 24)$

11. $(-12, -9)$

In Problems 12 and 13, evaluate each quantity.

12. $\cos\left(-270°\right)$

13. $\sin 180°$

In Problems 14-17, sketch the graph on the interval $\left[0, 2\pi\right]$.

14. $y = -2 + 3\cos(-x)$

15. $y = -1 + \sin(2x - \pi)$

16. $y = \sec\dfrac{x}{2}$

17. $y = \cot(2x + \pi)$

18. Determine the amplitude and period and then sketch the graph of the function $y = 3\cos\dfrac{1}{2}x$.

19. Determine the amplitude, period, and phase shift and then sketch the graph of the function $y = 14\sin(2x + \pi)$.

20. Sketch the graph of the function $y = -3\sin\dfrac{1}{2}x + x$.

21. What is the angle of elevation of the sun at the time that a flagpole 68 feet tall casts a shadow 25 feet long?

22. A ladder 25 feet long is leaning against a building. If the foot of the ladder is 6 feet from the base of the building (on ground level), what acute angle does the ladder make with the ground?

23. The equation $x = 4\sin\left(\dfrac{t}{2} - 2\pi\right)$ represents simple harmonic motion. Determine the amplitude, period, and frequency. Sketch the graph.

Trigonometric Identities and Equations

In Problems 1 and 2, verify the identity.

1. $\dfrac{1-\tan x}{\sec x} + \sin x = \cos x$

2. $\dfrac{\tan^2 x + 1}{\tan^2 x + 2} = \dfrac{1}{1 + \cos^2 x}$

3. Evaluate $\cot \dfrac{5\pi}{12}$ without using a calculator.

In Problems 4 and 5, verify the identity.

4. $\dfrac{\sin(a+b)}{\cos a \cos b} = \tan a + \tan b$

5. $\dfrac{\tan a - \tan b}{\tan a + \tan b} = \dfrac{\sin(a-b)}{\sin(a+b)}$

6. Given that $\alpha = 315°$, find $\sin 2\alpha$, $\cos 2\alpha$, and $\tan 2\alpha$, using the double-angle identities.

In Problems 7 and 8, from the given information, determine $\cos \dfrac{\alpha}{2}$.

7. $\sin \alpha = -\dfrac{2}{3}, \alpha$ in Quadrant III

8. $\tan \alpha = -\dfrac{3}{2}, \alpha$ in Quadrant II

In Problems 9 and 10, write the given expression as a single trigonometric function.

9. $\dfrac{\cot \varphi - \tan \varphi}{\cot \varphi + \tan \varphi}$

10. $\dfrac{\sin 2\alpha}{1 - \cos 2\alpha}$

In Problems 11 and 12, verify the identity.

11. $\dfrac{1}{\sin x} - \sin x = \dfrac{\sin x}{\tan^2 x}$

12. $\sin\left(\dfrac{\pi}{4} - \alpha\right) = \cos\left(\alpha + \dfrac{\pi}{4}\right)$

In Problems 13 and 14, solve for x. Express the answer in radians.

13. $2\cos^2 x - \cos x = 1$

14. $\sin^2 x + 3\sin x = 4$

In Problems 15 and 16, given that $\sin 72° = 0.95$ and $\cos 72° = 0.31$, find the function value. Round your answer to two decimal places.

15. $\sec 18°$

16. $\cot 18°$

In Problems 17 to 20, find the exact value of each expression.

17. $\sin^{-1}\left(-\dfrac{\sqrt{2}}{2}\right)$

18. $\cos^{-1}\left(-\dfrac{1}{2}\right)$

19. $\cos^{-1}\left(\cos\left(-\dfrac{\pi}{4}\right)\right)$

20. $\sin^{-1}\left(\cos \dfrac{\pi}{6}\right)$

Applications of Trigonometry

In Problems 1 - 9, solve triangle *ABC*.

1. $a = 17$, $B = 46°$, $C = 81°$

2. $a = 11$, $b = 16$, $A = 40°$

3. $a = 14$, $A = 38°$, $B = 73°$

4. $a = 18$, $b = 7.0$, $B = 23°$

5. $A = 41°$, $B = 83°$, $c = 44.6$

6. $A = 120°$, $b = 20$, $c = 16$

7. $a = 78$, $B = 128°$, $c = 125$

8. $a = 14$, $b = 19$, $c = 13$

9. $a = 10$, $b = 13$, $C = 100°$

In Problems 10 - 12, find the area of triangle *ABC*.

10. $B = 30°$, $C = 135°$, $b = 4\sqrt{2}$

11. $a = 7.0$, $b = 9.0$, $C = 26°$

12. $a = 12$, $b = 15$, $c = 23$

13. The vector **u** has initial point (0, 0) and terminal point (−3, 3). The vector **v** has initial point (0, 0) and terminal point (1, 1).
    a. Determine the magnitude and direction of vector **u**.
    b. Draw $\mathbf{u} - 2\mathbf{v}$.

In Problems 14 and 15, determine the dot product of the given vectors. Find the angle between the vectors to the nearest degree.

14. $\mathbf{u} = \langle 3, -1 \rangle$, $\mathbf{v} = \langle -2, -2 \rangle$

15. $\mathbf{u} = \langle 4, 6 \rangle$, $\mathbf{v} = \langle -3, 2 \rangle$

In Problems 16 and 17, use the vectors $\mathbf{u} = \left\langle -\frac{1}{2}, \frac{7}{2} \right\rangle$ and $\mathbf{v} = \langle 6, -1 \rangle$ to determine the following.

16. $2\mathbf{u} - \dfrac{1}{2}\mathbf{v}$

17. $\mathbf{u} - 4\mathbf{v}$

18. A plane's air speed is 500 mph and its bearing is N 75° E. An 6-mph wind is blowing in the direction of N 79° W. Determine the ground speed and true bearing of the plane.

19. Find the force required to hold a 120-pound crate stationary on a ramp inclined at 25°.

Complex Numbers

In Problems 1 – 14, perform the indicated operation; then write the answer in standard form.

1. i^{51}

2. $(-i)^{35}$

3. $(3i)^4 \cdot i^{44}$

4. $(1-3i)+(-2+i)$

5. $(4-3i)-(6-2i)$

6. $5i(4-2i)$

7. $2i-i(8+3i)$

8. $(1-4i)(3+2i)$

9. $(3+5i)(7-4i)$

10. $(2-9i)(2+9i)$

11. $(1+i\sqrt{5})(1-i\sqrt{5})$

12. $\dfrac{4}{3i}$

13. $\dfrac{1}{3+11i}$

14. $\dfrac{6-2i}{2+i}$

In Problems 15 and 16, express the complex number in trigonometric form.

15. $3-i\sqrt{3}$

16. $-4-4i$

17. Write $z = 4$ cis 45° in standard form.

18. Multiply: $(-2$ cis 225°$)(\sqrt{3}$ cis 75°$)$. Write the answer in standard form.

19. Divide: $\dfrac{6\sqrt{3}+6i}{1-i\sqrt{3}}$. Write the answer in trigonometric form.

20. Find $\left(\dfrac{-\sqrt{3}}{2}+\dfrac{1}{2}i\right)^5$. Write the answer in standard form.

21. Find the fourth roots of $-1+i\sqrt{3}$. Write the answer in trigonometric form.

Topics in Analytic Geometry

In Problems 1 – 4, write the equation of each conic in standard form and state the type of conic, the foci, and the vertices of each.

1. $x^2 + 4y^2 - 36 = 0$ **2.** $y^2 + 2y - x + 3 = 0$ **3.** $x^2 - 4y^2 - 36 = 0$ **4.** $y^2 - 6y + x = 0$

In Problems 5 – 8, find the foci and vertices of each conic. If the conic is a hyperbola, find the asymptotes. Graph each equation.

5. $y^2 - 6y - x + 5 = 0$

6. $\dfrac{(x-5)^2}{25} + \dfrac{y^2}{4} = 1$

7. $\dfrac{(y+2)^2}{24} - \dfrac{(x-1)^2}{4} = 1$

8. $4x^2 - 9y^2 - 8x + 18y = 149$

In Problems 9 and 10, the rectangular coordinates of a point are given. Determine two polar representations (r, θ) for the point with $0 \le \theta < 2\pi$.

9. $(-1, 1)$

10. $(-\sqrt{3}, -1)$

In Problems 11 – 14, convert the given polar equation into rectangular form.

11. $r = 3\sin\theta$ **12.** $r = 2\cos\theta$ **13.** $\theta = \pi/6$ **14.** $r^2 + \sin 2\theta = 0$

In Problems 15 and 16, graph each polar equation.

15. $r = 2 + \sin\theta$

16. $r = 3 - 2\cos\theta$

In Problems 17 and 18, graph the conic given by each polar equation.

17. $r = \dfrac{5}{1 + \sin\theta}$

18. $r = \dfrac{18}{2 + 3\cos\theta}$

19. Write the equation $13x^2 + 6\sqrt{3}xy + 7y^2 - 32 = 0$ without an xy term. Name the graph of the equation.

20. Eliminate the parameter and graph the curve given by the parametric equations

Exponential and Logarithmic Functions

1. Graph : $f(x) = 2^{-x}$

2. Graph : $f(x) = e^{2x}$

3. Write $\log_4 (3x + 2) = d$ in exponential form.

4. Write $4^{x+3} = y$ in logarithmic form.

5. Write $\log_b \dfrac{\sqrt{y}}{x^3 z^2}$ in terms of logarithms x, y, and z.

6. Write $\log_b (x-1) + \frac{1}{2}\log_b (y+2) - 4\log_b (z+3)$ as a single logarithm with a coefficient of 1.

7. Use the change-of-base formula and a calculator to approximate $\log_5 15$. Round your result to the nearest 0.0001.

8. Graph: $f(x) = \ln (x + 2)$

9. Solve: $6^x = 60$. Round your solution to the nearest 0.0001.

10. Find the *exact* solution(s) of

$$\frac{10^x - 10^{-x}}{10^x + 10^{-x}} = \frac{1}{4}$$

11. Solve: $\log (2x + 1) - 2 = \log (3x - 1)$

12. Solve: $\ln (x+1) + \ln(x-2) = \ln (4x - 8)$

13. Find the balance on $12,000 invested at an annual interest rate of 6.7% for 3 years:
 a. compounded quarterly.
 b. compounded continuously

14. a. What, to the nearest 0.1, will an earthquake measure on the Richter scale if it has an intensity of $I = 35{,}792{,}000 I_0$?

 b. Compare the intensity of an earthquake that measures 6.4 on the Richter scale to the intensity of an earthquake that measures 5.1 on the Richter scale by finding the ratio of the larger intensity to the smaller intensity. Round to the nearest whole number.

15. a. Find the exponential growth function for a city whose population was 51,300 in 1980 and 240,500 in 2000. Use $t = 0$ to represent the year 1980.
 b. Use the growth function to predict the population of the city in 2010. Round to the nearest 100.

16. Determine, to the nearest 10 years, the age of a bone if it now contains 89% of its original amount of carbon-14. The half-life of carbon-14 is 5730 years.

17. a. Use a graphing utility to find the exponential regression function for the following data:
 $\{(1.4, 3), (2.6, 7), (3.9, 15), (4.3, 20), (5.2, 26)\}$
 b. Use the function to predict, to the nearest whole number, the y-value associated with $x = 8.2$.

18. An altimeter is used to determine the height of an airplane above sea level. The following table shows the values for the pressure p and altitude h of an altimeter.

Pressure p in pounds per square inch	13.1	12.7	12.3	11.4	10.9
Altitude h in feet	3000	3900	4800	6800	8000

 a. Find the logarithmic regression function for the data and use the function to estimate, to the nearest 100 feet, the height of the airplane when the pressure is 12.0 pounds per square inch.
 b. According to your function, what will the pressure be, to the nearest 0.1 per square inch, when the airplane is 6200 feet above sea level?

Solutions to Chapter Tests

Functions and Graphs

1.
$$4(3x - 2) - 3(2x - 5) = 25$$
$$12x - 8 - 6x + 15 = 25$$
$$6x + 7 = 25$$
$$6x = 18$$
$$x = 3$$

2.
$$12x^2 - 17x + 6 = 0$$
$$(3x - 2)(4x - 3) = 0$$
$$3x - 2 = 0 \quad \text{or} \quad 4x - 3 = 0$$
$$3x = 2 \qquad\qquad 4x = 3$$
$$x = \frac{2}{3} \qquad\qquad x = \frac{3}{4}$$

3.
$$|x - 3| \le 2$$
$$-2 \le x - 3 \le 2$$
$$1 \le x \le 5$$
$$[1, 5]$$

4. midpoint $= \left(\dfrac{x_1 + x_2}{2}, \ \dfrac{y_1 + y_2}{2} \right) = \left(\dfrac{-3 + (-7)}{2}, \ \dfrac{4 + (-5)}{2} \right) = \left(\dfrac{-3 - 7}{2}, \ \dfrac{4 - 5}{2} \right) = \left(\dfrac{-10}{2}, \ \dfrac{-1}{2} \right) = \left(-5, \ -\dfrac{1}{2} \right)$

length $= d = \sqrt{(x_1 - x_2)^2 + (y_1 - y_2)^2} = \sqrt{[-3 - (-7)]^2 + [4 - (-5)]^2} = \sqrt{(-3 + 7)^2 + (4 + 5)^2} = \sqrt{(4)^2 + (9)^2} = \sqrt{16 + 81} = \sqrt{97}$

5. $x + y^2 = 6$

$y = 0 \Rightarrow x + (0)^2 = 6 \Rightarrow x = 6$
Thus the *x*-intercept is (6, 0).

$x = 0 \Rightarrow 0 + y^2 = 6$
$$y^2 = 6$$
$$y = \pm\sqrt{6}$$
Thus the *y*-intercepts are $\left(0, \ -\sqrt{6} \right)$ and $\left(0, \ \sqrt{6} \right)$.

6. $y = -2|x + 3| + 4$

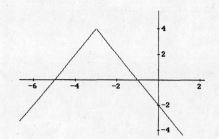

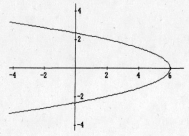

7.
$$x^2 + 6x + y^2 - 8y - 11 = 0.$$
$$(x^2 + 6x) + (y^2 - 8y) = 11$$
$$(x^2 + 6x + 9) + (y^2 - 8y + 16) = 11 + 9 + 16$$
$$(x + 3)^2 + (y - 4)^2 = 36$$
$$(x + 3)^2 + (y - 4)^2 = 6^2$$
center (−3, 4), radius 6

8.
$$25 - x^2 \ge 0$$
$$(5 - x)(5 + x) \ge 0$$
The product is positive or zero.
The critical values are 5 and −5.

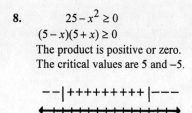

The domain is $\left\{ x \mid -5 \le x \le 5 \right\}$.

9. $y = 4x - 3, \ (x_1, y_1) = (-2, 5)$

$$d = \frac{|mx_1 + b - y_1|}{\sqrt{1 + m^2}} = \frac{|4(-2) + (-3) - 5|}{\sqrt{1 + 4^2}} = \frac{|-8 - 3 - 5|}{\sqrt{1 + 16}} = \frac{|-16|}{\sqrt{17}} = \frac{16}{\sqrt{17}} = \frac{16\sqrt{17}}{17}$$

20   **Solutions to Chapter Tests**

10.

$$f(x) = \begin{cases} x^2 & \text{if } x \le -1 \\ |x| & \text{if } -1 < x < 1 \\ 1 & \text{if } x \ge 1 \end{cases}$$

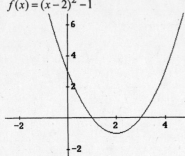

a. increasing on $[0, 1]$
b. constant on $[1, \infty)$
c. decreasing on $(-\infty, 0]$

11. $f(x) = (x-2)^2 - 1$

domain: all real numbers
range: $\{y \mid y \ge -1\}$

12. The graph is shifted 1 unit to the left and 2 units up, and is stretched vertically by multiplying each y-coordinate by 3.

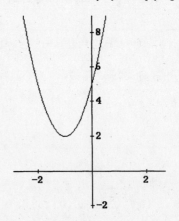

13. a. $f(x) = x^2 + 3$

$f(-x) = (-x)^2 + 3 = x^2 + 3 = f(x)$

$f(x)$ is an even function.

b. $f(x) = x^2 + 2x$

$f(-x) = (-x)^2 + 2(-x) = x^2 - 2x$

$f(-x) = x^2 - 2x \neq f(x)$ not an even function

$f(-x) = x^2 - 2x \neq -f(x)$ not an odd function

$f(x)$ is neither an even function nor an odd function.

c. $f(x) = 4x$

$f(-x) = 4(-x) = -4x = -f(x)$

$f(x)$ is an odd function.

$f(3) = -(3)^2 + 6(3) + 7$

$= -9 + 18 + 7$

$= 16$

The maximum value of the function is 16.

14. $(f - g)(x) = f(x) - g(x)$

$= (4x^2 - 3) - (x + 6)$

$= 4x^2 - 3 - x - 6$

$= 4x^2 - x - 9$

$(fg) = f(x) \cdot g(x)$

$= (4x^2 - 3)(x + 6)$

$= 4x^3 + 24x^2 - 3x - 18$

15. $f(x) = x^2 - x - 4$

$\dfrac{f(x+h) - f(x)}{h} = \dfrac{(x+h)^2 - (x+h) - 4 - (x^2 - x - 4)}{h}$

$= \dfrac{x^2 + 2xh + h^2 - x - h - 4 - x^2 + x + 4}{h}$

$= \dfrac{2xh + h^2 - h}{h}$

$= \dfrac{h(2x + h - 1)}{h}$

$= 2x + h - 1$

16. $f(x) = \sqrt{2x}$ and $g(x) = \frac{1}{2}x^2$

$(f \circ g)(x) = f[g(x)] = f\left(\frac{1}{2}x^2\right)$

$= \left(\sqrt{2\left(\frac{1}{2}x^2\right)}\right)$

$= \sqrt{x^2}$

$= x$

17. $y = \frac{2}{3}x - 4$

Interchange x and y, then solve for y.

$$x = \frac{2}{3}y - 4$$

$$3(x) = 3\left(\frac{2}{3}y - 4\right)$$

$$3x = 2y - 12$$

$$3x + 12 = 2y$$

$$\frac{1}{2}(3x + 12) = \frac{1}{2}(2y)$$

$$\frac{3}{2}x + 6 = y$$

$$f^{-1}(x) = \frac{3}{2}x + 6$$

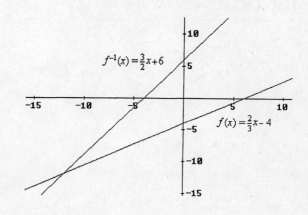

18. $y = \dfrac{x-3}{3x+6}$

Interchange x and y, then solve for y.

$$x = \frac{y-3}{3y+6}$$

$$x(3y+6) = y-3$$

$$3xy + 6x = y - 3$$

$$3xy - y = -6x - 3$$

$$y(3x - 1) = -6x - 3$$

$$y = \frac{-6x - 3}{3x - 1}$$

$$f^{-1}(x) = \frac{-6x - 3}{3x - 1}$$

Domain $f^{-1}(x)$: all real numbers except $\frac{1}{3}$

Range $f^{-1}(x)$: all real numbers except -2

19. $s(t) = 4t^2$

a. Average velocity $= \dfrac{4(2)^2 - 4(1)^2}{2-1}$

$$= \frac{4(4) - 4(1)}{1}$$

$$= 16 - 4$$

$$= 12\ \text{ft/sec}$$

b. Average velocity $= \dfrac{4(1.5)^2 - 4(1)^2}{1.5-1}$

$$= \frac{4(2.25) - 4(1)}{1.5 - 1}$$

$$= \frac{9 - 4}{0.5}$$

$$= \frac{5}{0.5}$$

$$= 10\ \text{ft/sec}$$

c. Average velocity $= \dfrac{4(1.01)^2 - 4(1)^2}{1.01-1}$

$$= \frac{4(1.0201) - 4(1)}{0.01}$$

$$= \frac{4.0804 - 4}{0.01}$$

$$= \frac{0.0804}{0.01}$$

$$= 8.04\ \text{ft/sec}$$

20. a. Enter the data on your calculator. The technique for a TI-83 calculator is illustrated here.

```
EDIT CALC TESTS
1:Edit
2:SortA(
3:SortD(
4:ClrList
5:SetUpEditor
```

```
L1    L2    L3    2
2     3     ------
3     4
4     6
6     10
10    16
------
L2(6)=
```

```
EDIT CALC TEST
1:1-Var Stats
2:2-Var Stats
3:Med-Med
4:LinReg (ax+b)
5:QuadReg
6:CubicReg
7↓QuartReg
```

```
LinReg
y=ax+b
a=1.675
b=-.575
r2=.9949024823
r=.9974479847
```

$y = 1.675x - 0.575$

b. $y = 1.675(8) - 0.575$

$= 12.825$

Trigonometric Functions

1. $\alpha = \dfrac{d}{180°} \cdot \pi = -\dfrac{135°}{180°} \cdot \pi = -\dfrac{3\pi}{4}$

2. $\alpha = \dfrac{d}{180°} \cdot \pi = \dfrac{240°}{180°} \cdot \pi = \dfrac{4\pi}{3}$

3. $d = \dfrac{\alpha}{\pi} \cdot 180° = \dfrac{\frac{9\pi}{2}}{\pi} \cdot 180° = 810°$

4. $d = \dfrac{\alpha}{\pi} \cdot 180° = \dfrac{\frac{16\pi}{3}}{\pi} \cdot 180° = 960°$

5. $\theta = \dfrac{s}{r} = \dfrac{12\pi}{8} = \dfrac{3\pi}{2}$

    The angle measures $\dfrac{3\pi}{2}$ radians, or $270°$.

6. $\sin \dfrac{5\pi}{4} = -\sin\left(\dfrac{5\pi}{4} - \pi\right) = -\sin\dfrac{\pi}{4} = -\dfrac{\sqrt{2}}{2}$

7. $\cos\left(-\dfrac{4\pi}{3}\right) = \cos\left(2\pi - \dfrac{4\pi}{3}\right) = \cos\left(\dfrac{2\pi}{3}\right) = -\cos\left(\pi - \dfrac{2\pi}{3}\right) = -\cos\dfrac{\pi}{3} = -\dfrac{1}{2}$

8. $\sec\left(\dfrac{19\pi}{2}\right) = \sec\left(5 \cdot 2\pi - \dfrac{19}{2}\pi\right) = \sec\dfrac{\pi}{2}$: undefined

9. $\tan\left(-\dfrac{17\pi}{6}\right) = \tan\left(2 \cdot 2\pi - \dfrac{17\pi}{6}\right) = \tan\dfrac{7\pi}{6} = \tan\left(\dfrac{7\pi}{6} - \pi\right) = \tan\dfrac{\pi}{6} = \dfrac{\sqrt{3}}{3}$

10. $r = \sqrt{x^2 + y^2} = \sqrt{(-10)^2 + 24^2}$
    $= \sqrt{676} = 26$

    $\sin\theta = \dfrac{y}{r} = \dfrac{24}{26} = \dfrac{12}{13}$

    $\tan\theta = \dfrac{y}{x} = \dfrac{24}{-10} = -\dfrac{12}{5}$

11. $r = \sqrt{x^2 + y^2} = \sqrt{(-12)^2 + (-9)^2}$
    $= \sqrt{225} = 15$

    $\sin\theta = \dfrac{y}{r} = \dfrac{-9}{15} = -\dfrac{3}{5}$

    $\tan\theta = \dfrac{y}{x} = \dfrac{-9}{-12} = \dfrac{3}{4}$

12. For the angle $-270°$, use the point $(0, 1)$.

    $\cos(-270°) = \dfrac{x}{r} = \dfrac{0}{1} = 0$

13. For the angle $180°$, use the point $(-1, 0)$.

    $\sin 180° = \dfrac{y}{r} = \dfrac{0}{1} = 0$

14.

15.

16.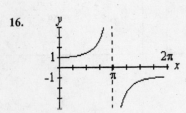

17.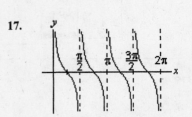

18. Amplitude 3, period 4π

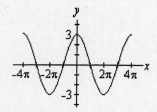

19. Amplitude 4, period π, phase shift $-\frac{\pi}{2}$

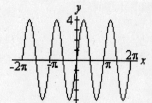

20.

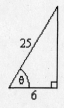

21.

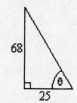

$\tan \theta = \dfrac{68}{25}$

$\tan \theta = 2.72$

$\theta \approx 70°$

The angle of elevation is approximately $70°$.

22.

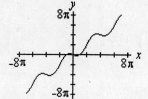

$\cos \theta = \dfrac{6}{25}$

$\cos \theta = 0.24$

$\theta \approx 76°$

The ladder makes an angle of approximately $76°$ with the ground.

23. Amplitude 4, period 4π, frequency $\frac{1}{4\pi}$

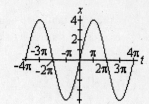

Trigonometric Identities and Equations

1. $\dfrac{1-\tan x}{\sec x}+\sin x = \dfrac{1-\dfrac{\sin x}{\cos x}}{\dfrac{1}{\cos x}}+\sin x = \cos x - \sin x + \sin x = \cos x$

2. $\dfrac{\tan^2 x+1}{\tan^2 x+2} = \dfrac{\sec^2 x}{\left(\tan^2 x+1\right)+1} = \dfrac{\sec^2 x}{\sec^2 x+1} = \dfrac{\dfrac{1}{\cos^2 x}}{\dfrac{1}{\cos^2 x}+1}\cdot\dfrac{\cos^2 x}{\cos^2 x} = \dfrac{1}{1+\cos^2 x}$

3. $\cot\dfrac{5\pi}{12} = \cot\left(\dfrac{\pi}{4}+\dfrac{\pi}{6}\right) = \dfrac{1}{\tan\left(\dfrac{\pi}{4}+\dfrac{\pi}{6}\right)} = \dfrac{1}{\dfrac{\tan\frac{\pi}{4}+\tan\frac{\pi}{6}}{1-\tan\frac{\pi}{4}\tan\frac{\pi}{6}}} = \dfrac{1-\tan\frac{\pi}{4}\tan\frac{\pi}{6}}{\tan\frac{\pi}{4}+\tan\frac{\pi}{6}} = \dfrac{1-(1)\left(\frac{\sqrt{3}}{3}\right)}{1+\frac{\sqrt{3}}{3}} = \dfrac{3-\sqrt{3}}{3+\sqrt{3}}$

4. $\dfrac{\sin(a+b)}{\cos a\cos b} = \dfrac{\sin a\cos b+\cos a\sin b}{\cos a\cos b} = \dfrac{\sin a\cos b}{\cos a\cos b}+\dfrac{\cos a\sin b}{\cos a\cos b} = \dfrac{\sin a}{\cos a}+\dfrac{\sin b}{\cos b} = \tan a+\tan b$

5. $\dfrac{\tan a-\tan b}{\tan a+\tan b} = \dfrac{\frac{\sin a}{\cos a}-\frac{\sin b}{\cos b}}{\frac{\sin a}{\cos a}+\frac{\sin b}{\cos b}}\cdot\dfrac{\cos a\cos b}{\cos a\cos b} = \dfrac{\sin a\cos b-\sin b\cos a}{\sin a\cos b+\sin b\cos a} = \dfrac{\sin(a-b)}{\sin(a+b)}$

6. $\sin(2\cdot 315°) = 2\sin 315°\cos 315° = 2\left(-\dfrac{\sqrt{2}}{2}\right)\left(\dfrac{\sqrt{2}}{2}\right) = -1$

   $\cos(2\cdot 315°) = 2\cos^2 315° - 1 = 2\left(\dfrac{\sqrt{2}}{2}\right)^2 - 1 = 0$

   $\tan(2\cdot 315°) = \dfrac{2\tan 315°}{1-\tan^2 315°} = \dfrac{2(-1)}{1-1^2} = -\dfrac{2}{0}$ Undefined

7. $\sin\alpha = -\dfrac{2}{3}$

   $\dfrac{y}{r} = \dfrac{-2}{3}$

   $x^2 = r^2-y^2 = 3^2-(-2)^2 = 5$

   $x = -\sqrt{5}$

   $\cos\alpha = \dfrac{x}{r} = -\dfrac{\sqrt{5}}{3}$

   $\cos\dfrac{\alpha}{2} = -\sqrt{\dfrac{1+\cos\alpha}{2}} = -\sqrt{\dfrac{1-\frac{\sqrt{5}}{3}}{2}}$

   $\phantom{\cos\dfrac{\alpha}{2}} = -\sqrt{\dfrac{\frac{3-\sqrt{5}}{3}}{2}} = -\sqrt{\dfrac{3-\sqrt{5}}{6}}$

8. $\tan\alpha = -\dfrac{3}{2}$

   $\dfrac{y}{x} = \dfrac{3}{-2}$

   $r = \sqrt{x^2+y^2} = \sqrt{(-2)^2+3^2} = \sqrt{13}$

   $\cos\alpha = \dfrac{x}{r} = \dfrac{-2}{\sqrt{13}} = -\dfrac{2\sqrt{13}}{13}$

   $\cos\dfrac{\alpha}{2} = \sqrt{\dfrac{1+\cos\alpha}{2}} = \sqrt{\dfrac{1-\frac{2\sqrt{13}}{13}}{2}}$

   $\phantom{\cos\dfrac{\alpha}{2}} = \sqrt{\dfrac{\frac{13-2\sqrt{13}}{13}}{2}} = \sqrt{\dfrac{13-2\sqrt{13}}{26}}$

9. $\dfrac{\cot\varphi-\tan\varphi}{\cot\varphi+\tan\varphi} = \dfrac{\frac{\cos\varphi}{\sin\varphi}-\frac{\sin\varphi}{\cos\varphi}}{\frac{\cos\varphi}{\sin\varphi}+\frac{\sin\varphi}{\cos\varphi}}\cdot\dfrac{\sin\varphi\cos\varphi}{\sin\varphi\cos\varphi} = \dfrac{\cos^2\varphi-\sin^2\varphi}{\cos^2\varphi+\sin^2\varphi} = \dfrac{\cos^2\varphi}{1} = \cos 2\varphi$

10. $\dfrac{\sin 2\alpha}{1-\cos 2\alpha} = \dfrac{2\sin \alpha \cos \alpha}{1-(1-2\sin^2 \alpha)} = \dfrac{2\sin \alpha \cos \alpha}{2\sin^2 \alpha} = \dfrac{\cos \alpha}{\sin \alpha} = \cot \alpha$

11. $\dfrac{1}{\sin x} - \sin x = \dfrac{1-\sin^2 x}{\sin x} = \dfrac{\cos^2 x}{\sin x} \cdot \dfrac{\dfrac{\sin x}{\cos^2 x}}{\dfrac{\sin x}{\cos^2 x}} = \dfrac{\sin x}{\dfrac{\sin^2 x}{\cos^2 x}} = \dfrac{\sin x}{\tan^2 x}$

12. $\sin\left(\dfrac{\pi}{4}-\alpha\right) = \sin\dfrac{\pi}{4}\cos\alpha - \cos\dfrac{\pi}{4}\sin\alpha = \dfrac{\sqrt{2}}{2}\cos\alpha - \dfrac{\sqrt{2}}{2}\sin\alpha = \cos\dfrac{\pi}{4}\cos\alpha - \sin\dfrac{\pi}{4}\sin\alpha = \cos\left(\dfrac{\pi}{4}+\alpha\right)$

13.
$$2\cos^2 x - \cos x = 1$$
$$2\cos^2 x - \cos x - 1 = 0$$
$$(2\cos x + 1)(\cos x - 1) = 0$$

$2\cos x + 1 = 0$	$\cos x - 1 = 0$
$\cos x = -\dfrac{1}{2}$	$\cos x = 1$
$x = \dfrac{2\pi}{3}$ or $\dfrac{4\pi}{3}$	$x = 0$

The solutions are $\dfrac{2\pi}{3}+2n\pi$, $\dfrac{4\pi}{3}+2n\pi$, and $0+2n\pi$.

14.
$$\sin^2 x + 3\sin x = 4$$
$$\sin^2 x + 3\sin x - 4 = 0$$
$$(\sin x - 1)(\sin x + 4) = 0$$

$\sin x - 1 = 0$	$\sin x + 4 = 0$
$\sin x = 1$	$\sin x = -4$
$x = \dfrac{\pi}{2}$	no solutions

The solution is $\dfrac{\pi}{2}+2n\pi$.

15. $\sec 18° = \dfrac{1}{\cos 18°} = \dfrac{1}{\sin(90°-18°)} = \dfrac{1}{\sin 72°} = \dfrac{1}{0.95} \approx 1.05$

16. $\cot 18° = \tan(90°-18°) = \tan 72° = \dfrac{\sin 72°}{\cos 72°} = \dfrac{0.95}{0.31} \approx 3.06$

17. $\sin^{-1}\left(-\dfrac{\sqrt{2}}{2}\right) = -\dfrac{\pi}{4}$

18. $\cos^{-1}\left(-\dfrac{1}{2}\right) = \dfrac{2\pi}{3}$

19. $\cos^{-1}\left(\cos\left(-\dfrac{\pi}{4}\right)\right) = \cos^{-1}\left(\dfrac{\sqrt{2}}{2}\right) = \dfrac{\pi}{4}$

20. $\sin^{-1}\left(\cos\dfrac{\pi}{6}\right) = \sin^{-1}\left(\dfrac{\sqrt{3}}{2}\right) = \dfrac{\pi}{3}$

Applications of Trigonometry

1. $A = 180° - (46° + 81°) = 53°$

$$\frac{17}{\sin 53°} = \frac{b}{\sin 46°}$$

$$\frac{17}{0.7986} = \frac{b}{0.7193}$$

$$b \approx 15$$

$$\frac{17}{\sin 53°} = \frac{c}{\sin 81°}$$

$$\frac{17}{0.7986} = \frac{c}{0.9877}$$

$$c \approx 21$$

2. $$\frac{11}{\sin 40°} = \frac{16}{\sin B}$$

$$\frac{11}{0.6428} = \frac{16}{\sin B}$$

$$\sin B \approx 0.9350$$

$$B \approx 69° \text{ or } 111°$$

Case 1:

$B \approx 69°$

$C = 180° - (40° + 69°) = 71°$

$$\frac{11}{\sin 40°} = \frac{c}{\sin 71°}$$

$$\frac{11}{0.6428} = \frac{c}{0.9455}$$

$$c \approx 16$$

Case 2:

$B \approx 111°$

$C = 180° - (40° + 111°) = 29°$

$$\frac{11}{\sin 40°} = \frac{c}{\sin 29°}$$

$$\frac{11}{0.6428} = \frac{c}{0.4848}$$

$$c \approx 8.3$$

Thus, there are two possible triangles: one with $B \approx 69°, C \approx 71°$, and $c \approx 16$, and one with $B \approx 111°, C \approx 29°$, and $c \approx 8.3$.

3. $C = 180° - (38° + 73°) = 69°$

$$\frac{14}{\sin 38°} = \frac{b}{\sin 73°}$$

$$\frac{14}{0.6157} = \frac{b}{0.9563}$$

$$b \approx 22$$

$$\frac{14}{\sin 38°} = \frac{c}{\sin 69°}$$

$$\frac{14}{0.6157} = \frac{c}{0.9336}$$

$$c \approx 21$$

4. $$\frac{18}{\sin A} = \frac{7.0}{\sin 23°}$$

$$\frac{18}{\sin A} = \frac{7.0}{0.3907}$$

$$\sin A \approx 1.0047$$

Since there is no value A such that $\sin A > 1$, no triangle is possible.

5. $C = 180° - (41° + 83°) = 56°$

$$\frac{a}{\sin 41°} = \frac{44.6}{\sin 56°}$$

$$\frac{a}{0.6561} = \frac{44.6}{0.8290}$$

$$a \approx 35$$

$$\frac{b}{\sin 83°} = \frac{44.6}{\sin 56°}$$

$$\frac{b}{0.9925} = \frac{44.6}{0.8290}$$

$$b \approx 53$$

6. $$a^2 = b^2 + c^2 - 2bc \cos A$$

$$a^2 = 20^2 + 16^2 - 2(20)(16)\cos 120°$$

$$a^2 = 656 - 640(-0.5)$$

$$a^2 \approx 976$$

$$a \approx 31$$

$$b^2 = a^2 + c^2 - 2ac \cos B$$

$$20^2 = 31^2 + 16^2 - 2(31)(16)\cos B$$

$$\cos B \approx 0.8236$$

$$B \approx 35°$$

$$C \approx 180° - (120° + 35°) = 25°$$

7. $b^2 = a^2 + c^2 - 2ac \cos B$

$b^2 = 78^2 + 125^2 - 2(78)(125) \cos 128°$

$b^2 \approx 21,709 - 19,500(-0.6157)$

$b^2 \approx 33,715.15$

$b \approx 184$

$a^2 = b^2 + c^2 - 2bc \cos A$

$78^2 = 184^2 + 125^2 - 2(184)(125) \cos A$

$\cos A \approx 0.9434$

$A \approx 19°$

$C \approx 180^0 - (19° + 128°) = 33°$

8. $a^2 = b^2 + c^2 - 2bc \cos A$

$14^2 = 19^2 + 13^2 - 2(19)(13) \cos A$

$\cos A \approx 0.6791$

$A \approx 47°$

$b^2 = a^2 + c^2 - 2ac \cos B$

$19^2 = 14^2 + 13^2 - 2(14)(13) \cos B$

$\cos B \approx 0.0110$

$B \approx 89°$

$C \approx 180° - (47° + 89°) = 44°$

9. $c^2 = a^2 + b^2 - 2ab \cos C$

$c^2 = 10^2 + 13^2 - 2(10)(13) \cos 100°$

$c^2 \approx 269 - 260(-0.1736)$

$c^2 \approx 314.1$

$c \approx 18$

$a^2 = b^2 + c^2 - 2bc \cos A$

$10^2 = 13^2 + 18^2 - 2(13)(18) \cos A$

$\cos A \approx 0.8397$

$A \approx 33°$

$B \approx 180° - (100° + 33°) = 47°$

10. $\dfrac{4\sqrt{2}}{\sin 30°} = \dfrac{c}{\sin 135°}$

$\dfrac{4\sqrt{2}}{\frac{1}{2}} = \dfrac{c}{\frac{\sqrt{2}}{2}}$

$c = 8.0$

$A = 180° - (30° + 135°) = 15°$

$K = \dfrac{1}{2} bc \sin A$

$= \dfrac{1}{2}(4\sqrt{2})(8.0) \sin 15°$

≈ 6 square units

11. $K = \dfrac{1}{2} ab \sin C$

$= \dfrac{1}{2}(7.0)(9.0) \sin 26°$

≈ 14 square units

12. $s = \dfrac{1}{2}(a + b + c)$

$= \dfrac{1}{2}(12 + 15 + 23)$

$= 25$

$K = \sqrt{s(s-a)(s-b)(s-c)}$

$= \sqrt{25(25-12)(25-15)(25-23)}$

$= \sqrt{25(13)(10)(2)}$

$= \sqrt{6500}$

≈ 81 square units

13. a. $d = \sqrt{(-3-0)^2 + (3-0)^2} = \sqrt{18} = 3\sqrt{2}$

$\tan \theta = \dfrac{y}{x}$

$\tan \theta = \dfrac{3}{-3}$

$\tan \theta = -1$

$\theta = 135°$

b.

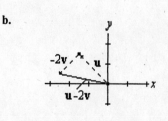

14. $\langle 3,-1\rangle \cdot \langle -2,-2\rangle = 3(-2)+(-1)(-2)$
$$= -4$$
$$\cos\theta = \frac{\mathbf{u}\cdot\mathbf{v}}{\|\mathbf{u}\|\cdot\|\mathbf{v}\|}$$
$$\cos\theta = \frac{-\sqrt{4}}{(\sqrt{3^2+(-1)^2})(\sqrt{(-2)^2+(-2)^2})}$$
$$\cos\theta \approx -0.4472$$
$$\theta \approx 117°$$

15. $\langle 4,6\rangle \cdot \langle -3,2\rangle = 4(-3)+6(2)$
$$= 0$$
$$\cos\theta = \frac{\mathbf{u}\cdot\mathbf{v}}{\|\mathbf{u}\|\cdot\|\mathbf{v}\|}$$
$$\cos\theta = \frac{0}{(\sqrt{4^2+6^2})(\sqrt{(-3)^2+2^2})}$$
$$\cos\theta = 0$$
$$\theta = 90°$$

16. $2\mathbf{u}-\frac{1}{2}\mathbf{v} = 2\left\langle -\frac{1}{2},\frac{7}{2}\right\rangle - \frac{1}{2}\langle 6,-1\rangle$
$$= \langle -1,7\rangle + \left\langle -3,\frac{1}{2}\right\rangle$$
$$= \left\langle -4,\frac{15}{2}\right\rangle$$

17. $\mathbf{u}-4\mathbf{v} = \langle 6,-1\rangle - 4\left\langle -\frac{1}{2},\frac{7}{2}\right\rangle$
$$= \langle 6,-1\rangle + \langle 2,-14\rangle$$
$$= \langle 8,-15\rangle$$

18. $\|\mathbf{v}\|^2 = 500^2 + 8^2 - 2(500)(8)\cos 35°$
$$\|\mathbf{v}\|^2 = 250{,}064 - 8000(0.8192)$$
$$\|\mathbf{v}\|^2 = 243{,}510.4$$
$$\|\mathbf{v}\| = 493$$

$$\frac{\sin\theta}{8} = \frac{\sin 35°}{493}$$
$$\frac{\sin\theta}{8} = \frac{0.5736}{493}$$
$$\sin\theta \approx 0.0093$$
$$\theta \approx 0°32'$$
$$75° - \theta \approx 75° - 0°32' = 74°28'$$

The jet's ground speed is approximately 493 mph, and its true bearing is approximately N 74°28' E.

19. $\|\mathbf{F}\| = (\sin 25°)(120) = (0.4226)(120) \approx 51$
The required force is approximately 51 pounds.

Complex Numbers

1. $i^{51} = i^{48}i^3 = (1)(-i) = -i$

2. $(-i)^{35} = (-1)^{35}(i)^{35} = (-1)i^{32}i^3 = (-1)(1)(-i) = -i$

3. $(3i)^4 \cdot i^{44} = 3^4 i^4 i^{44} = 81(1)(1) = 81$

4. $(1-3i) + (-2+i) = (1-2) + (-3+1)i = -1-2i$

5. $(4-3i) - (6-2i) = (4-6) + (-3+2)i = -2-i$

6. $5i(4-2i) = 20i - 10i^2 = 20i - 10(-1) = 10 + 20i$

7. $2i - i(8+3i) = 2i - 8i - 3i^2$
   $= -6i - 3(-1)$
   $= 3 - 6i$

8. $(1-4i)(3+2i) = 3 + 2i - 12i - 8i^2$
   $= 3 - 10i - 8(-1)$
   $= 3 - 10i + 8$
   $= 11 - 10i$

9. $(3+5i)(7-4i) = 21 - 12i + 35i - 20i$
   $= 21 + 23i - 20(-1)$
   $= 21 + 23i + 20$
   $= 41 + 23i$

10. $(2-9i)(2+9i) = 4 - 81i^2$
    $= 4 - 81(-1)$
    $= 4 + 81$
    $= 85$

11. $(1+i\sqrt{5})(1-i\sqrt{5}) = 1 - 5i^2 = 1 - 5(-1)$
    $= 1 + 5$
    $= 6$

12. $\dfrac{4}{3i} = \dfrac{4}{3i} \cdot \dfrac{i}{i} = \dfrac{4i}{3i^2} = \dfrac{4i}{3(-1)} = \dfrac{4i}{-3}$
    $= -\dfrac{4}{3}i$

13. $\dfrac{1}{3+11i} = \dfrac{1}{3+11i} \cdot \dfrac{3-11i}{3-11i} = \dfrac{3-11i}{9-121i^2} = \dfrac{3-11i}{9-121(-1)}$
    $= \dfrac{3-11i}{9+121} = \dfrac{3-11i}{130}$
    $= \dfrac{3}{130} - \dfrac{11}{130}i$

14. $\dfrac{6-2i}{2+i} = \dfrac{6-2i}{2+i} \cdot \dfrac{2-i}{2-i} = \dfrac{12-6i-4i+2i^2}{4-i^2}$
    $= \dfrac{12-10i+2(-1)}{4-(-1)} = \dfrac{12-10i-2}{4+1}$
    $= \dfrac{10-10i}{5} = \dfrac{10}{5} - \dfrac{10}{5}i$
    $= 2 - 2i$

In Problems 15 and 16, express the complex number in trigonometric form.

15. $3 - i\sqrt{3} \Rightarrow a = 3,\ b = -\sqrt{3}$
    $r = \sqrt{3^2 + (-\sqrt{3})^2} = \sqrt{12} = 2\sqrt{3}$
    $\tan\theta = -\dfrac{\sqrt{3}}{3}$
    $\theta = \dfrac{11\pi}{6}$
    $3 - i\sqrt{3} = 2\sqrt{3}\left(\cos\dfrac{11\pi}{6} + i\sin\dfrac{11\pi}{6}\right)$
    $= 2\sqrt{3}\ \text{cis}\ \dfrac{11\pi}{6},\quad \text{or}\quad 2\sqrt{3}\ \text{cis}\ 330°$

16. $-4 - 4i \Rightarrow a = -4,\ b = -4$
    $r = \sqrt{(-4)^2 + (-4)^2} = \sqrt{32} = 4\sqrt{2}$
    $\tan\theta = \dfrac{-4}{-4}$
    $\tan\theta = 1$
    $\theta = \dfrac{5\pi}{4}$
    $-4 - 4i = 4\sqrt{2}\left(\cos\dfrac{5\pi}{4} + i\sin\dfrac{5\pi}{4}\right)$
    $= 4\sqrt{2}\ \text{cis}\ \dfrac{5\pi}{4},\quad \text{or}\quad 4\sqrt{2}\ \text{cis}\ 225°$

17. $z = 4 \text{ cis } 45°$

$z = 4(\cos 45° + i \sin 45°)$

$z = 4\left(\dfrac{\sqrt{2}}{2} + \dfrac{i\sqrt{2}}{2}\right)$

$z = 2\sqrt{2} + 2i\sqrt{2}$

18. $(-2 \text{ cis } 225°)(\sqrt{3} \text{ cis } 75°) = -2\sqrt{3} \text{ cis } (225° + 75°)$

$= -2\sqrt{3} \text{ cis } (300°)$

$= -2\sqrt{3}(\cos 300° + i \sin 300°)$

$= -2\sqrt{3}\left(\dfrac{\sqrt{3}}{2} - \dfrac{i}{2}\right)$

$= -3 + i\sqrt{3}$

19.

$\tan \theta = \dfrac{6}{6\sqrt{3}} = \dfrac{1}{\sqrt{3}} = \dfrac{\sqrt{3}}{3}$

$\theta = 30°$

$r = \sqrt{\left(6\sqrt{3}\right)^2 + (6)^2}$

$= \sqrt{36(3) + 36}$

$= \sqrt{144}$

$= 12$

$\tan \alpha = \dfrac{-\sqrt{3}}{1} = -\sqrt{3}$

$\alpha = -60°$

$\beta = 360° - 60°$

$\beta = 300°$

$r = \sqrt{1^2 + \left(-\sqrt{3}\right)^2}$

$= \sqrt{1 + 3}$

$= \sqrt{4}$

$= 2$

$6\sqrt{3} + 6i = 12 \text{ cis } 30°$ $1 - i\sqrt{3} = 2 \text{ cis } 300°$

$\dfrac{6\sqrt{3} + 6i}{1 - i\sqrt{3}} = \dfrac{12 \text{ cis } 30°}{2 \text{ cis } 300°}$

$= \dfrac{12}{2} \text{ cis } (30° - 300°)$

$= 6 \text{ cis } (-270°)$

$= 6 \text{ cis } 90°$

20. $\left(\dfrac{-\sqrt{3}}{2} + \dfrac{1}{2}i\right)^5 = (\text{cis } 150°)^5$

$= \text{cis } 5(150°)$

$= \text{cis } 750°$

$= (\cos 750° + i \sin 750°)$

$= (\cos 30° + i \sin 30°)$

$= \dfrac{\sqrt{3}}{2} + \dfrac{1}{2}i$

21. $-1 + i\sqrt{3} = 2 \text{ cis } 120°$

$w_k = 2^{1/4} \text{ cis } \dfrac{120° + 360°k}{4}$ $k = 0, 1, 2, 3$

$w_0 \approx 1.19 \text{ cis } \dfrac{120°}{4}$

$w_0 \approx 1.19 \text{ cis } 30°$

$w_1 \approx 1.19 \text{ cis } \dfrac{120° + 360°}{4}$

$w_1 \approx 1.19 \text{ cis } 120°$

$w_2 \approx 1.19 \text{ cis } \dfrac{120° + 720°}{4}$

$w_2 \approx 1.19 \text{ cis } 210°$

$w_3 \approx 1.19 \text{ cis } \dfrac{120° + 1080°}{4}$

$w_3 \approx 1.19 \text{ cis } 300°$

Topics in Analytic Geometry

1. $x^2 + 4y^2 - 36 = 0$

    $x^2 + 4y^2 = 36$

    $\dfrac{x^2}{36} + \dfrac{4y^2}{36} = \dfrac{36}{36}$

    $\dfrac{x^2}{36} + \dfrac{y^2}{9} = 1$

    The graph is an ellipse.

    vertices: $(-6, 0)$ and $(6, 0)$
    foci: $(-3\sqrt{3}, 0)$ and $(3\sqrt{3}, 0)$

2. $y^2 + 2y - x + 3 = 0$

    $y^2 + 2y = x - 3$

    $y^2 + 2y + 1 = x - 3 + 1$

    $(y + 1)^2 = x - 2$

    The graph is a parabola.

    $4p = 1$

    $p = \dfrac{1}{4}$

    vertex: $(2, -1)$
    focus: $(2 + \frac{1}{4}, -1) = \left(\frac{9}{4}, -1\right)$

3. $x^2 - 4y^2 - 36 = 0$

    $x^2 - 4y^2 = 36$

    $\dfrac{x^2}{36} - \dfrac{4y^2}{36} = \dfrac{36}{36}$

    $\dfrac{x^2}{36} - \dfrac{y^2}{9} = 1$

    The graph is a hyperbola.

    vertices: $(-6, 0)$ and $(6, 0)$

    $c^2 = a^2 + b^2$

    $c^2 = 36 + 9$

    $c^2 = 45$

    $c = \sqrt{45} = 3\sqrt{5}$

    foci: $(-3\sqrt{5}, 0)$ and $(3\sqrt{5}, 0)$

4. $y^2 - 6y + x = 0$

    $y^2 - 6y = -x$

    $y^2 - 6y + 9 = -x + 9$

    $(y - 3)^2 = -(x - 9)$

    The graph is a parabola

    vertex: $(9, 3)$

    $4p = -1$

    $p = -\dfrac{1}{4}$

    focus: $\left(9 - \frac{1}{4}, 3\right) = \left(\frac{35}{4}, 3\right)$

5. $y^2 - 6y - x + 5 = 0$

    $y^2 - 6y = x - 5$

    $y^2 - 6y + 9 = x - 5 + 9$

    $(y - 3)^2 = x + 4$

    $4p = 1$

    $p = \dfrac{1}{4}$

    vertex: $(-4, 3)$
    focus: $\left(-4 + \frac{1}{4}, 3\right) = \left(-\frac{15}{4}, 3\right)$

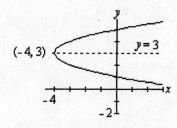

6. $\dfrac{(x-5)^2}{25} + \dfrac{y^2}{4} = 1$

    ellipse with center at $(5, 0)$

    $c^2 = a^2 - b^2$

    $c^2 = 25 - 4$

    $c^2 = 21$

    $c = \sqrt{21}$

    vertices: $(0, 0)$ and $(10, 0)$
    foci: $(5 - \sqrt{21}, 0)$ and $(5 + \sqrt{21}, 0)$

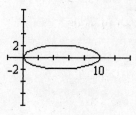

7. $\dfrac{(y+2)^2}{24} - \dfrac{(x-1)^2}{4} = 1$

hyperbola with vertical transverse axis, center at $(1, -2)$

$$a^2 = 24 \qquad b^2 = 4 \qquad c^2 = a^2 + b^2 \qquad \dfrac{a}{b} = \dfrac{2\sqrt{6}}{2}$$
$$a = 2\sqrt{6} \qquad b = 2 \qquad c^2 = 28 \qquad\qquad = \sqrt{6}$$
$$c = 2\sqrt{7}$$

vertices: $(1, -2 - 2\sqrt{6})$ and $(1, -2 + 2\sqrt{6})$

foci: $(1, -2 - 2\sqrt{7})$ and $(1, -2 + 2\sqrt{7})$

asymptotes: $y + 2 = \pm\sqrt{6}(x-1)$

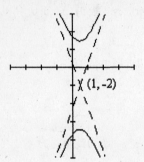

8.
$$4x^2 - 9y^2 - 8x + 18y = 149$$
$$4(x^2 - 2x + 1) - 9(y^2 - 2y + 1) = 149 + 4 - 9$$
$$4(x-1)^2 - 9(y-1)^2 = 144$$
$$\dfrac{4(x-1)^2}{144} - \dfrac{9(y-1)^2}{144} = \dfrac{144}{144}$$
$$\dfrac{(x-1)^2}{36} - \dfrac{(y-1)^2}{16} = 1$$

hyperbola with horizontal transverse axis, center at $(1, 1)$ a

$$a^2 = 36 \qquad b^2 = 16 \qquad c^2 = a^2 + b^2 \qquad \dfrac{b}{a} = \dfrac{4}{6}$$
$$a = 6 \qquad b = 4 \qquad c^2 = 36 + 16 \qquad\qquad = \dfrac{2}{3}$$
$$c^2 = 52$$
$$c = \sqrt{52}$$
$$= 2\sqrt{13}$$

vertices: $(-5, 1)$ and $(7, 1)$

foci: $(1 - 2\sqrt{13},\ 1)$ and $(1 + 2\sqrt{13},\ 1)$

asymptotes: $y - 1 = \pm\dfrac{2}{3}(x-1)$

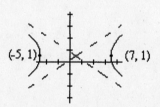

9. $r = \pm\sqrt{(-1)^2 + 1^2} = \pm\sqrt{2}$

$\tan\theta = \dfrac{y}{x} = \dfrac{1}{-1} = -1$

$\theta = \dfrac{3\pi}{4}$

Two polar representations are $\left(\sqrt{2},\ \dfrac{3\pi}{4}\right)$ and $\left(-\sqrt{2},\ \dfrac{7\pi}{4}\right)$.

10. $r = \pm\sqrt{(-\sqrt{3})^2 + (-1)^2} = \pm 2$

$\tan\theta = \dfrac{y}{x} = \dfrac{-1}{-\sqrt{3}} = \dfrac{\sqrt{3}}{3}$

$\theta = \dfrac{7\pi}{6}$

Two polar representations are $\left(2,\ \dfrac{7\pi}{6}\right)$ and $\left(-2,\ \dfrac{\pi}{6}\right)$.

11. $r = 3\sin\theta$

$r^2 = 3r\sin\theta$

$x^2 + y^2 = 3y$

12. $r = 2\cos\theta$

$r^2 = 2r\cos\theta$

$x^2 + y^2 = 2x$

13. $\theta = \pi/6$

$\tan\theta = \tan\pi/6$

$\dfrac{y}{x} = \dfrac{\sqrt{3}}{3}$

$y = \dfrac{\sqrt{3}}{3}x$

14. $r^2 + \sin 2\theta = 0$

$r^2 + 2\sin\theta\cos\theta = 0$

$(r^2)(r^2) + r^2(2\sin\theta\cos\theta) = 0$

$(r^2)^2 + 2(r\sin\theta)(r\cos\theta) = 0$

$x^2 + y^2 + 2xy = 0$

15.

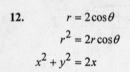

16.

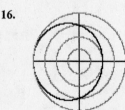

17.

18.

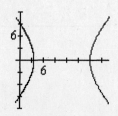

19. $13x^2 + 6\sqrt{3}xy + 7y^2 - 32 = 0$

$A = 13, \; B = 6\sqrt{3}, \; C = 7, \; D = 0, \; E = 0, \; F = -32$

$$\cot 2\alpha = \frac{13 - 7}{6\sqrt{3}} = \frac{6}{6\sqrt{3}} = \frac{1}{\sqrt{3}}$$

$$\csc^2 2\alpha = 1 + \cot^2 2\alpha = 1 + \frac{1}{3} = \frac{4}{3}$$

$\csc 2\alpha = \dfrac{2}{\sqrt{3}}$. Therefore, $\sin 2\alpha = \dfrac{\sqrt{3}}{2}$ and $\cos 2\alpha = \dfrac{1}{2}$.

$$\cos\alpha = \sqrt{\frac{1 + \frac{1}{2}}{2}} = \frac{\sqrt{3}}{2} \qquad\qquad \sin\alpha = \sqrt{\frac{1 - \frac{1}{2}}{2}} = \frac{1}{2}$$

$$A' = 13\left(\frac{\sqrt{3}}{2}\right)^2 + 6\sqrt{3}\left(\frac{\sqrt{3}}{2}\right)\left(\frac{1}{2}\right) + 7\left(\frac{1}{2}\right)^2 = 16$$

$B' = 0$

$$C' = 13\left(\frac{1}{2}\right)^2 - 6\sqrt{3}\left(\frac{\sqrt{3}}{2}\right)\left(\frac{1}{2}\right) + 7\left(\frac{\sqrt{3}}{2}\right)^2 = 4$$

$D' = 0$

$E' = 0$

$F' = -32$

$16x'^2 + 4y'^2 - 32 = 0$ or $\dfrac{x'^2}{2} + \dfrac{y'^2}{8} = 1$

The graph is an ellipse.

20. $x = 2t - 1 \qquad\qquad y = 2t^2 + t$

$t = \dfrac{x+1}{2} \qquad\qquad y = 2\left(\dfrac{x+1}{2}\right)^2 + \dfrac{x+1}{2}$

$2y = x^2 + 3x + 2$

$y = \dfrac{1}{2}x^2 + \dfrac{3}{2}x + 1$

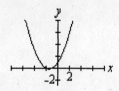

Exponential and Logarithmic Functions

1.

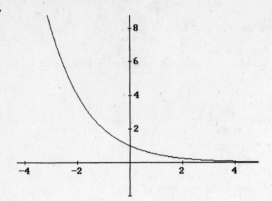

2.

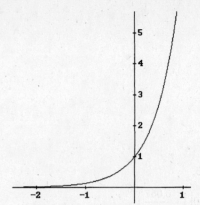

3. $4^d = 3x + 2$

4. $\log_4 y = x + 3$

5. $\frac{1}{2}\log_b y - 3\log_b x - 2\log_b z$

6. $\log_b \dfrac{(x-1)\sqrt{y+2}}{(z+3)^4}$

7. $\log_5 15 = \dfrac{\log 15}{\log 5} = \dfrac{1.176091259}{0.6989700043} \approx 1.6826$

Note that you could also have used natural logarithms.

$\log_5 15 = \dfrac{\ln 15}{\ln 5} = \dfrac{2.708050201}{1.609437912} \approx 1.6826$

8.

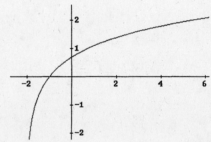

9. $6^x = 60$

$x = \log_6 60$

$x = \dfrac{\log 60}{\log 6} = \dfrac{1.77815125}{0.7781512504} \approx 2.2851$

Note that you could also have used natural logarithms.

$x = \log_6 60 = \dfrac{\ln 60}{\ln 6} = \dfrac{4.094344562}{1.791759469} \approx 2.2851$

10. $\left(\dfrac{10^x}{10^x}\right)\left(\dfrac{10^x - 10^{-x}}{10^x + 10^{-x}}\right) = \dfrac{1}{4}$

$\dfrac{10^{2x} - 1}{10^{2x} + 1} = \dfrac{1}{4}$

$4(10^{2x} - 1) = 10^{2x} + 1$

$4(10^{2x}) - 4 = 10^{2x} + 1$

$4(10^{2x}) - 10^{2x} = 1 + 4$

$10^{2x}(4 - 1) = 1 + 4$

$10^{2x}(3) = 5$

$10^{2x} = \dfrac{5}{3}$

$\log\left(10^{2x}\right) = \log\left(\dfrac{5}{3}\right)$

$2x = \log 5 - \log 3$

$x = \dfrac{1}{2}\log 5 - \dfrac{1}{2}\log 3$

11. $\log(2x+1) - \log(3x-1) = 2$

$$\log\frac{2x+1}{3x-1} = 2$$

$$10^2 = \frac{2x+1}{3x-1}$$

$$100(3x-1) = 2x+1$$

$$300x - 100 = 2x + 1$$

$$298x = 101$$

$$x = \frac{101}{298}$$

12. $\ln[(x+1)(x-2)] = \ln(4x-8)$

$$(x+1)(x-2) = 4x-8$$

$$x^2 - x - 2 = 4x - 8$$

$$x^2 - 5x + 6 = 0$$

$$(x-3)(x-2) = 0$$

$$x - 3 = 0 \Rightarrow x = 3$$

$x - 2 = 0 \Rightarrow x = 2$ [No; 2 is not in the domain of $\log(x-2)$

and not in the domain of $\log(4x-8)$.]

The solution is 3.

13. a. Use $A = P\left(1+\dfrac{r}{n}\right)^{nt}$ with $P = \$12,000$, $r = 0.067$,

$n = 4$, $t = 3$.

$$A = 12,000\left(1+\frac{0.067}{4}\right)^{4(3)}$$

$$= 12,000(1+0.01675)^{12}$$

$$= 12,000(1.01675)^{12}$$

$$= 12,000(1.220591027)$$

$$\approx \$14,647.09$$

b. Use $A = Pe^{rt}$ with $P = \$12,000$, $r = 0.067$,

$t = 3$.

$$A = 12,000e^{0.067(3)}$$

$$= 12,000e^{0.201}$$

$$= 12,000(1.222624772)$$

$$\approx \$14,671.50$$

14. a. $M = \log\left(\dfrac{I}{I_0}\right)$

$$= \log\left(\frac{35,792,000I_0}{I_0}\right)$$

$$= \log 35,792,000$$

$$\approx 7.6 \text{ on the Richter scale}$$

b. $\log\left(\dfrac{I_1}{I_0}\right) = 6.4$ and $\log\left(\dfrac{I_2}{I_0}\right) = 5.1$

$$\frac{I_1}{I_0} = 10^{6.4} \qquad\qquad \frac{I_2}{I_0} = 10^{5.1}$$

$$I_1 = 10^{6.4}I_0 \qquad\qquad I_2 = 10^{5.1}I_0$$

To compare the intensities, compute the ratio $\dfrac{I_1}{I_2}$.

$$\frac{I_1}{I_2} = \frac{10^{6.4}I_0}{10^{5.1}I_0} = \frac{10^{6.4}}{10^{5.1}} = 10^{6.4-5.1} = 10^{1.3} \approx 20$$

An earthquake that measures 6.4 on the Richter scale is approximately 20 times as intense as an earthquake that measures 5.1 on the Richter scale.

15. a. $N(t) = N_0e^{kt}$

$$N_0 = N(0) = 51,300$$

$$N(20) = 240,500$$

To determine k, substitute $t = 20$ and

$N_0 = 51,300$ into $N(t) = N_0e^{kt}$ to produce

$$N(20) = 51,300e^{k(20)}$$

$$240,500 = 51,300e^{20k}$$

$$\frac{240,500}{51,300} = e^{20k}$$

$$\ln\frac{240,500}{51,300} = \ln e^{20k}$$

$$\ln\frac{240,500}{51,300} = 20k$$

$$\frac{1}{20}\ln\frac{240,500}{51,300} = k$$

$$0.0773 \approx k$$

The exponential growth function is

$N(t) = 51,300e^{0.0773t}$.

b. Use $N(t) = 51,300e^{0.0773t}$ with $t = 30$.

$$N(30) = 51,300e^{0.0773(30)}$$

$$\approx 521,500$$

The exponential growth function yields 521,500 as the approximate population of the town in 2010.

16. The percent of carbon-14 present at time t is

$P(t) = 0.5^{t/5730}$.

$0.89 = 0.5^{t/5730}$

$\ln 0.89 = \ln 0.5^{t/5730}$

$\ln 0.89 = \left(\dfrac{t}{5730}\right)\ln 0.5$

$5730\left(\dfrac{\ln 0.89}{\ln 0.5}\right) = t$

$960 \approx t$

The bone is about 960 years old.

17. a.

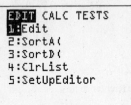

$y = 1.447348651(1.79279858)^x$

b. Use $y = 1.447348651(1.79279858)^x$ with $x = 8.2$.

$y = 1.447348651(1.79279858)^{8.2}$

≈ 174

18. a.

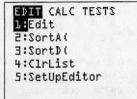

The logarithmic regression function is $h = 72610.5637 - 27040.63079\ln p$.

When $p = 12.0$, $h = 72610.5637 - 27040.63079\ln 12.0 \approx 5400$ feet.

b. $6200 = 72610.5637 - 27040.63079\ln p \Rightarrow \dfrac{6200 - 72610.5637}{-27040.63079} = \ln p$

$\ln p = \dfrac{-66410.5637}{-27040.63079} = 2.455954679 \Rightarrow e^{2.455954679} = e^{\ln p} \Rightarrow e^{2.455954679} = p \approx 11.7$

When the altitude is 6200 feet, the pressure is about 11.7 pounds per square inch.

**Solutions to the
Odd-Numbered Exercises
in the Text**

SECTION 1.1, Page 11

1.
$$2x + 10 = 40$$
$$2x + 10 = 40$$
$$2x = 40 - 10$$
$$2x = 30$$
$$x = 15$$

3.
$$5x + 2 = 2x - 10$$
$$5x - 2x = -10 - 2$$
$$3x = -12$$
$$x = -4$$

5.
$$2(x - 3) - 5 = 4(x - 5)$$
$$2x - 6 - 5 = 4x - 20$$
$$2x - 11 = 4x - 20$$
$$2x - 4x = -20 + 11$$
$$-2x = -9$$
$$x = \frac{9}{2}$$

7.
$$\frac{3}{4}x + \frac{1}{2} = \frac{2}{3}$$
$$12\left(\frac{3}{4}x + \frac{1}{2}\right) = 12\left(\frac{2}{3}\right)$$
$$9x + 6 = 8$$
$$9x = 8 - 6$$
$$9x = 2$$
$$x = \frac{2}{9}$$

9.
$$\frac{2}{3}x - 5 = \frac{1}{2}x - 3$$
$$6\left(\frac{2}{3}x - 5\right) = 6\left(\frac{1}{2}x - 3\right)$$
$$4x - 30 = 3x - 18$$
$$4x - 3x = -18 + 30$$
$$x = 12$$

11.
$$0.2x + 0.4 = 3.6$$
$$0.2x = 3.2$$
$$x = 16$$

13.
$$\frac{3}{5}(n + 5) - \frac{3}{4}(n - 11) = 0$$
$$20\left[\frac{3}{5}(n + 5) - \frac{3}{4}(n - 11)\right] = 20 - 0$$
$$12(n + 5) - 15(n - 11) = 0$$
$$12n + 60 - 15n + 165 = 0$$
$$-3n = -225$$
$$n = 75$$

15.
$$3(x + 5)(x - 1) = (3x + 4)(x - 2)$$
$$3x^2 + 12x - 15 = 3x^2 - 2x - 8$$
$$14x = 7$$
$$x = \frac{1}{2}$$

17.
$$0.08x + 0.12(4000 - x) = 432$$
$$0.08x + 480 - 0.12x = 432$$
$$-0.04x = -48$$
$$x = 1200$$

19.
$$x^2 - 2x - 15 = 0$$
$$a = 1 \quad b = -2 \quad c = -15$$
$$x = \frac{-(-2) \pm \sqrt{(-2)^2 - 4(1)(-15)}}{2(1)}$$
$$x = \frac{2 \pm \sqrt{4 + 60}}{2}$$
$$x = \frac{2 \pm \sqrt{64}}{2} = \frac{2 \pm 8}{2}$$
$$x = \frac{2 + 8}{2} = \frac{10}{2} = 5 \quad \text{or}$$
$$x = \frac{2 - 8}{2} = \frac{-6}{2} = -3$$

21.
$$x^2 + x - 1 = 0$$
$$a = 1 \quad b = 1 \quad c = -1$$
$$x = \frac{-1 \pm \sqrt{1^2 - 4(1)(-1)}}{2}$$
$$x = \frac{-1 \pm \sqrt{1 + 4}}{2}$$
$$x = \frac{-1 \pm \sqrt{5}}{2}$$

23. $2x^2 + 4x + 1 = 0$
$a = 2 \quad b = 4 \quad c = 1$

$x = \dfrac{-4 \pm \sqrt{4^2 - 4(2)(1)}}{2(2)}$

$x = \dfrac{-4 \pm \sqrt{16 - 8}}{4}$

$x = \dfrac{-4 \pm \sqrt{8}}{4} = \dfrac{-4 \pm 2\sqrt{2}}{4}$

$x = \dfrac{-2 \pm \sqrt{2}}{2}$

25. $3x^2 - 5x - 3 = 0$
$a = 3 \quad b = -5 \quad c = -3$

$x = \dfrac{-(-5) \pm \sqrt{(-5)^2 - 4(3)(-3)}}{2(3)}$

$x = \dfrac{5 \pm \sqrt{25 + 36}}{6}$

$x = \dfrac{5 \pm \sqrt{61}}{6}$

27. $\dfrac{1}{2}x^2 + \dfrac{3}{4}x - 1 = 0$
$a = \dfrac{1}{2} \quad b = \dfrac{3}{4} \quad c = -1$

$x = \dfrac{-\frac{3}{4} \pm \sqrt{\left(\frac{3}{4}\right)^2 - 4\left(\frac{1}{2}\right)(-1)}}{2\left(\frac{1}{2}\right)}$

$x = \dfrac{-\frac{3}{4} \pm \sqrt{\frac{9}{16} + 2}}{1}$

$x = -\dfrac{3}{4} \pm \sqrt{\dfrac{41}{16}}$

$x = -\dfrac{3}{4} \pm \dfrac{\sqrt{41}}{4}$

$x = \dfrac{-3 \pm \sqrt{41}}{4}$

29. $\sqrt{2}x^2 + 3x + \sqrt{2} = 0$
$a = \sqrt{2} \quad b = 3 \quad c = \sqrt{2}$

$x = \dfrac{-3 \pm \sqrt{3^2 - 4 \cdot \sqrt{2} \cdot \sqrt{2}}}{2\sqrt{2}}$

$x = \dfrac{-3 \pm \sqrt{9 - 8}}{2\sqrt{2}}$

$x = \dfrac{-3 + 1}{2\sqrt{2}} = \dfrac{-2}{2\sqrt{2}} = -\dfrac{\sqrt{2}}{2} \quad$ or

$x = \dfrac{-3 - 1}{2\sqrt{2}} = \dfrac{-4}{2\sqrt{2}} = -\sqrt{2}$

31. $x^2 = 3x + 5$
$x^2 - 3x - 5 = 0$
$a = 1 \quad b = -3 \quad c = -5$

$x = \dfrac{-(-3) \pm \sqrt{(-3)^2 - 4(1)(5)}}{2(1)}$

$x = \dfrac{3 \pm \sqrt{9 + 20}}{2}$

$x = \dfrac{3 \pm \sqrt{29}}{2}$

33. $x^2 - 2x - 15 = 0$
$(x + 3)(x - 5) = 0$
$x + 3 = 0 \quad$ or $\quad x - 5 = 0$
$\qquad x = -3 \qquad\qquad x = 5$

35. $8y^2 + 189y - 72 = 0$
$(8y - 3)(y + 24) = 0$
$8y - 3 = 0 \quad$ or $\quad y + 24 = 0$
$\qquad y = \dfrac{3}{8} \qquad\qquad y = -24$

37. $3x^2 - 7x = 0$
$x(3x - 7) = 0$
$x = 0 \quad$ or $\quad 3x - 7 = 0$
$\qquad\qquad x = \dfrac{7}{3}$

39. $\qquad\qquad (x - 5)^2 - 9 = 0$
$[(x - 5) - 3][(x - 5) + 3] = 0$
$x - 8 = 0 \quad$ or $\quad x - 2 = 0$
$\quad x = 8 \qquad\qquad x = 2$

41. $2x + 3 < 11$
$\quad 2x < 8$
$\quad x < 4$

43. $x + 4 > 3x + 16$
$\quad -2x > 12$
$\quad x < -6$

45. $-6x + 1 \geq 19$
$\quad -6x \geq 18$
$\quad x \leq -3$

47. $\quad -3(x + 2) \leq 5x + 7$
$\quad -3x - 6 \leq 5x + 7$
$\quad\quad -8 \leq 13$
$\qquad\quad x \geq -\dfrac{13}{8}$

49. $-4(3x - 5) > 2(x - 4)$
$\quad -12x + 20 > 2x - 8$
$\quad\quad -14x > -28$
$\quad\quad\quad x < 2$

51. $x^2 + 7x > 0$
$x(x + 7) > 0$

The product is positive.
The critical values are 0 and −7.

$x(x - 7) \qquad + + + + | - - - - - - - | + + + +$

$\qquad\qquad\qquad\qquad -7 \qquad\quad 0$

$(-\infty, -7) \cup (0, \infty)$

53. $x^2 + 7x + 10 < 0$
$(x + 5)(x + 2) < 0$

The product is negative.
The critical values are −5 and −2.

$(x + 5)(x + 2) \qquad + + + + | - - - | + + + + + +$

$\qquad\qquad\qquad\qquad -5 \quad -2 \; 0$

$(-5, -2)$

55. $x^2 - 3x \geq 28$

$x^2 - 3x - 28 \geq 0$

$(x+4)(x-7) \geq 0$

The product is positive or zero.
The critical values are –4 and 7.

$(x+4)(x-7)$

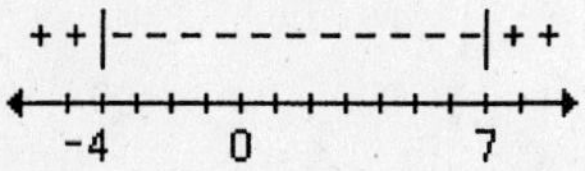

$(-\infty, -4] \cup [7, \infty)$

57. $6x^2 - 4 \leq 5x$

$6x^2 - 5x - 4 \leq 0$

$(3x-4)(2x+1) \leq 0$

The product is negative or zero.

The critical values are $\frac{4}{3}$ and $-\frac{1}{2}$.

$(3x-4)(2x+1)$

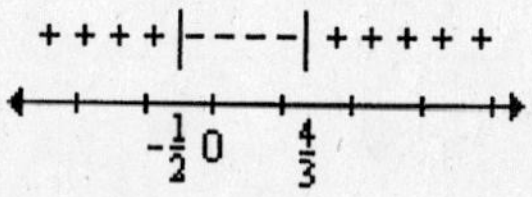

$\left[-\frac{1}{2}, \frac{4}{3}\right]$

59. $|x| < 4$

$-4 < x < 4$

$(-4, 4)$

61. $|x-1| < 9$

$-9 < x-1 < 9$

$-8 < x < 10$

$(-8, 10)$

63. $|x+3| > 30$

$x+3 < -30$ or $x+3 > 30$

$x < -33$ $\quad x > 27$

$(-\infty, -33) \cup (27, \infty)$

65. $|2x-1| > 4$

$2x-1 < -4$ or $2x-1 > 4$

$2x < -3$ $\quad 2x > 5$

$x < -\frac{3}{2}$ $\quad x > \frac{5}{2}$

$\left(-\infty, -\frac{3}{2}\right) \cup \left(\frac{5}{2}, \infty\right)$

67. $|x+3| \geq 5$

$x+3 \leq -5$ or $x+3 \geq 5$

$x \leq -8$ $\quad x \geq 2$

$(-\infty, -8] \cup [2, \infty)$

69. $|3x-10| \leq 14$

$-14 \leq 3x - 10 \leq 14$

$-4 \leq 3x \leq 24$

$-\frac{4}{3} \leq x \leq 8$

$\left[-\frac{4}{3}, 8\right]$

71. $|4-5x| \geq 24$

$4-5x \leq 24$ or $4-5x \geq 24$

$-5x \leq -28$ $\quad -5x \geq 20$

$x \geq \frac{28}{5}$ $\quad x \leq -4$

$(-\infty, -4] \cup \left[\frac{28}{5}, \infty\right)$

73. $|x-5| \geq 0$

Because an absolute value is always nonnegative, the inequality is always true. The solution set consists of all real numbers.

$(-\infty, \infty)$

75. $|x-4| \leq 0$

Because an absolute value is always nonnegative, the inequality $|x-4| < 0$ has no solution. Thus, the only solution of the inequality $|x-4| \leq 0$ is the solution of the equation $x - 4 = 0$.

$x = 4$

77. $A = 35$

$A = LW$

$LW = 35$

$L = \dfrac{35}{W}$

$P = 27$

$P = 2L + 2W$

$2L + 2W = 27$

$2\left(\dfrac{35}{W}\right) + 2W = 27$

$70 + 2W^2 = 27W$

$2W^2 - 27W + 70 = 0$

$(2W - 7)(W - 10) = 0$

$W = \dfrac{7}{2}$ or $W = 10$

$35 = \dfrac{7}{2}L$ $35 = LW$

$70 = 7L$ $35 = 10L$

$10 = L$ $3.5 = L$

The rectangle measures 3.5 cm by 10 cm.

81.   Plan A: $5 + 0.01x$

Plan B: $1 + 0.08x$

$5 + 0.01x < 1 + 0.08x$

$4 < .07x$

$57.1 < x$

Plan A is less expensive if you use at least 58 checks.

85. $68 \le F \le 104$

$68 \le \dfrac{9}{5}C + 32 \le 104$

$36 \le \dfrac{9}{5}C \le 72$

$20 \le C \le 40$

89. $R = 420x - 2x^2$

$420x - 2x^2 > 0$

$2x(210 - x) > 0$

The product is positive.
The critical values are 0 and 210.

$2x(210 - x)$ $----|+++++++++|---$

0 210

$(0, 210)$

79. $A = 1500 = lw$

$P = 600 - 2l + 3w$

$l = \dfrac{15000}{w}$

$2l + 3w = 600$

$2\left(\dfrac{15000}{w}\right) + 3w = 600$

$30,000 + 3w^2 = 600w$

$3w^2 - 600w + 30,000 = 0$

$3(w^2 - 200w + 10,000) = 0$

$3(w - 100)(w - 100) = 0$

$w = 100$ ft

$l = \dfrac{15000}{100} = 150$ ft

The dimensions are 100 feet by 150 feet.

83.   Plan A: $100 + 8x$

Plan B: $250 + 3.5x$

$100 + 8x > 250 + 3.5x$

$4.5x > 150$

$x > 33.3$

Plan A pays better if at least 34 sales are made.

87. $253 = \dfrac{1}{2}n(n + 1)$

$n^2 + n - 506 = 0$

$(n + 23)(n - 22) = 0$

$n = 22$

So $1 + 2 + 3 + \cdots + 21 + 22 = 253.$

91. $s = -16t^2 + v_0 t + s_0$ $s > 48,$ $v_0 = 64,$ $s_0 = 0$

$$-16t + 64t > 48$$
$$-16t^2 + 64t - 48 > 0$$
$$-16(t^2 - 4t + 3) > 0$$
$$-16(t - 3)(t - 1) > 0$$

The product is positive.
The critical values are $t = 3$ and $t = 1$.

$-16(t-3)(t-1)$ $|-|+\ +|- - - -$

$\overset{\longleftrightarrow}{\underset{0\ 1\quad 3}{+\!+\!+\!+\!+\!+\!+\!+\!+}}$

$1 \text{ second} < t < 3 \text{ seconds}$

The ball is higher than 48 ft between 1 and 3 seconds.

93. a. $|s - 4.25| \le 0.01$

b. $s - 4.25 = 0.01$ or $s - 4.25 = -0.01$
        $s = 4.26$ $s = 4.24$

critical values: 4.24 and 4.26

$4.24 \le s \le 4.26$

SECTION 1.2, Page 24

1.

3. a.

b. $\text{average} = \dfrac{(84-63)+(99-72)+(111-87)+(129-90)+(108-90)+(141-96)+(93-69)+(96-81)+(90-75)+(90-84)}{10}$

$= \dfrac{21+27+24+39+18+45+24+15+15+6}{10} = \dfrac{234}{10} = 23.4$

The average increase in heart rate is 23.4 beats per minute.

5. $d = \sqrt{(-8-6)^2 + (11-4)^2}$
$= \sqrt{(-14)^2 + (7)^2}$
$= \sqrt{196 + 49}$
$= \sqrt{245}$
$= 7\sqrt{5}$

7. $d = \sqrt{(-10-(-4))^2 + (15-(-20))^2}$
$= \sqrt{(-6)^2 + (35)^2}$
$= \sqrt{36 + 1225}$
$= \sqrt{1261}$

44   **Chapter 1/Functions and Graphs**

9. $d = \sqrt{(0-5)^2 + (0-(-8))^2}$

$= \sqrt{(-5)^2 + (8)^2}$

$= \sqrt{25 + 64}$

$= \sqrt{89}$

11. $d = \sqrt{(\sqrt{12} - \sqrt{3})^2 + (\sqrt{27} - \sqrt{8})^2}$

$= \sqrt{(2\sqrt{3} - \sqrt{3})^2 + (3\sqrt{3} - 2\sqrt{2})^2}$

$= \sqrt{(\sqrt{3})^2 + (3\sqrt{3} - 2\sqrt{2})^2}$

$= \sqrt{3 + (27 - 12\sqrt{6} + 8)}$

$= \sqrt{3 + 27 - 12\sqrt{6} + 8}$

$= \sqrt{38 - 12\sqrt{6}}$

13. $d = \sqrt{(-a-a)^2 + (-b-b)^2}$

$= \sqrt{(-2a)^2 + (-2b)^2}$

$= \sqrt{4a^2 + 4b^2}$

$= \sqrt{4(a^2 + b^2)}$

$= 2\sqrt{a^2 + b^2}$

15. $d = \sqrt{(-2x - x)^2 + (3x - 4x)^2}$ with $x < 0$

$= \sqrt{(-3x)^2 + (-x)^2}$

$= \sqrt{9x^2 + x^2}$

$= \sqrt{10x^2}$

$= -x\sqrt{10}$ (Note: since $x < 0, \sqrt{x^2} = -x$)

17. $\sqrt{(4-x)^2 + (6-0)^2} = 10$

$\left(\sqrt{(4-x)^2 + (6-0)^2}\right)^2 = 10^2$

$16 - 8x + x^2 + 36 = 100$

$x^2 - 8x - 48 = 0$

$(x - 12)(x + 4) = 0$

$x = 12$ or $x = -4$

The points are $(12, 0), (-4, 0)$.

19. $M = \left(\dfrac{x_1 + x_2}{2}, \dfrac{y_1 + y_2}{2}\right)$

$= \left(\dfrac{1+5}{2}, \dfrac{-1+5}{2}\right)$

$= \left(\dfrac{6}{2}, \dfrac{4}{2}\right)$

$= (3, 2)$

21. $M = \left(\dfrac{6+6}{2}, \dfrac{-3+11}{2}\right)$

$= \left(\dfrac{12}{2}, \dfrac{8}{2}\right)$

$= (6, 4)$

23. $M = \left(\dfrac{1.75 + (-3.5)}{2}, \dfrac{2.25 + 5.57}{2}\right)$

$= \left(-\dfrac{1.75}{2}, \dfrac{7.82}{2}\right)$

$= (-0.875, 3.91)$

25.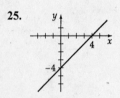

27.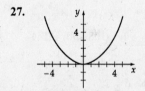

29.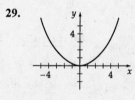

31.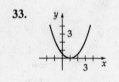

33.

35.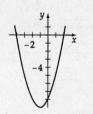

37.

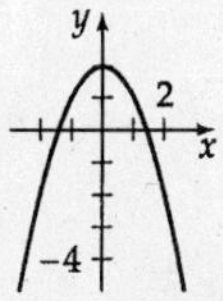

39. Intercepts: $\left(0, \frac{12}{5}\right)$, $(6, 0)$

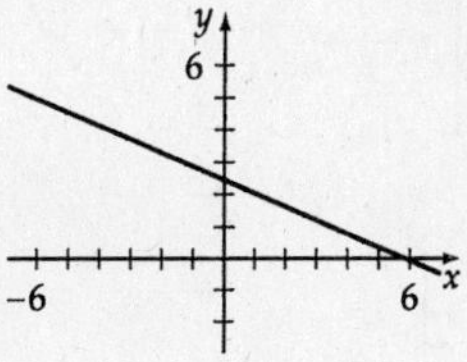

41. $(0, \sqrt{5})$, $(0, -\sqrt{5})$, $(5, 0)$

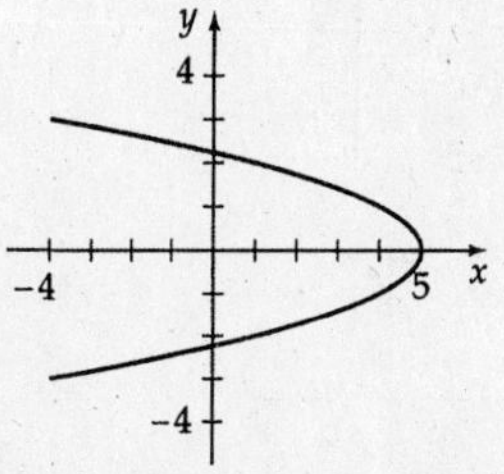

$x = -y^2 + 5$

43. $(0, 4)$, $(0, -4)$, $(-4, 0)$

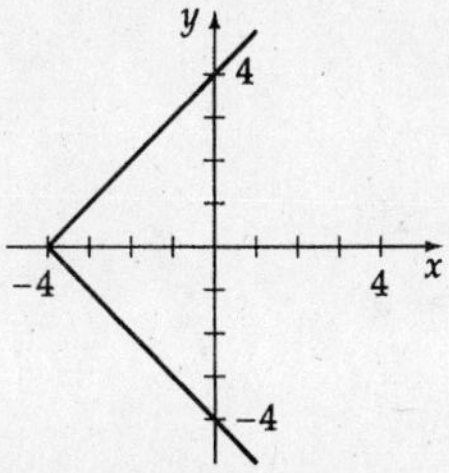

$x = |y| - 4$

45. $(0, \pm 2)$, $(\pm 2, 0)$

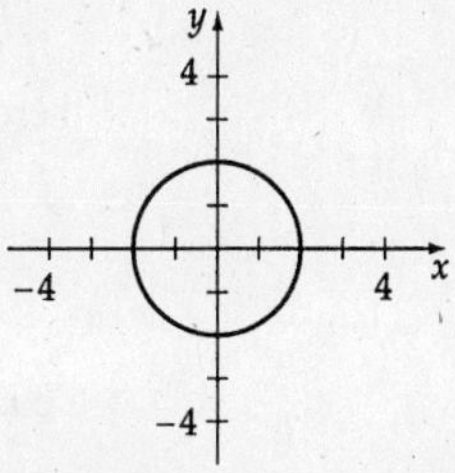

$x^2 + y^2 = 4$

47. $(0, \pm 4)$, $(\pm 4, 0)$

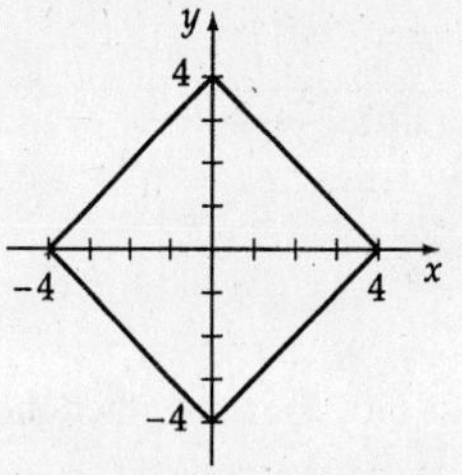

$|x| + |y| = 4$

49. center $(0, 0)$, radius 6

51. center $(0, 0)$, radius 10

53. center $(1, 3)$, radius 7

55. center $(-2, -5)$, radius 5

57. center $(8, 0)$, radius $\frac{1}{2}$

59. $(x-4)^2 + (y-1)^2 = 2^2$

61. $\left(x - \frac{1}{2}\right)^2 + \left(y - \frac{1}{4}\right)^2 = \left(\sqrt{5}\right)^2$

63.
$$
\begin{aligned}
(x-0)^2 + (y-0)^2 &= r^2 \\
(-3-0)^2 + (4-0)^2 &= r^2 \\
(-3)^2 + 4^2 &= r^2 \\
9 + 16 &= r^2 \\
25 = 5^2 &= r^2 \\
(x-0)^2 + (y-0)^2 &= 5^2
\end{aligned}
$$

65.
$$
\begin{aligned}
(x+2)^2 + (y-5)^2 &= r^2 \\
(x-1)^2 + (y-3)^2 &= r^2 \\
(4-1)^2 + (-1-3)^2 &= r^2 \\
3^2 + (-4)^2 &= r^2 \\
9 + 16 &= r^2 \\
25 = 5^2 &= r^2 \\
(x-1)^2 + (y-3)^2 &= 5^2
\end{aligned}
$$

67.
$$
\begin{aligned}
x^2 - 6x \quad + y^2 &= -5 \\
x^2 - 6x + 9 + y^2 &= -5 + 9 \\
(x-3)^2 + y^2 &= 2^2
\end{aligned}
$$
center $(3, 0)$, radius 2

69.
$$
\begin{aligned}
x^2 - 4x \quad + y^2 - 10y \quad &= -20 \\
x^2 - 4x + 4 + y^2 - 10y + 25 &= -20 + 4 + 25 \\
(x-2)^2 + (y-5)^2 &= 3^2
\end{aligned}
$$
center $(2, 5)$, radius 3

71.
$$
\begin{aligned}
x^2 - 14x \quad + y^2 + 8y \quad &= -56 \\
x^2 - 14x + 49 + y^2 + 8y + 16 &= -56 + 49 + 16 \\
(x-7)^2 + (y+4)^2 &= 3^2
\end{aligned}
$$
center $(7, -4)$, radius 3

73.
$$
\begin{aligned}
4x^2 + 4x \quad + 4y^2 &= 63 \\
x^2 + x \quad + y^2 &= \frac{63}{4} \\
x^2 + x + \frac{1}{4} + y^2 &= \frac{63}{4} + \frac{1}{4} \\
\left(x + \frac{1}{2}\right)^2 + y^2 &= 16 \\
\left(x + \frac{1}{2}\right)^2 + (y-0)^2 &= 4^2
\end{aligned}
$$
center $\left(-\frac{1}{2}, 0\right)$, radius 4

75.
$$
\begin{aligned}
x^2 - x \quad + y^2 + \frac{2}{3}y \quad &= -\frac{1}{3} \\
x^2 - x + \frac{1}{4} + y^2 + \frac{2}{3}y + \frac{1}{9} &= -\frac{1}{3} + \frac{1}{4} + \frac{1}{9} \\
\left(x - \frac{1}{2}\right)^2 + \left(y + \frac{1}{3}\right)^2 &= \left(\frac{1}{6}\right)^2
\end{aligned}
$$
center $\left(\frac{1}{2}, -\frac{1}{3}\right)$, radius $\frac{1}{6}$

77.

79.

81.

83.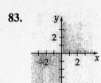

85.

87.

89. $\left(\dfrac{x+5}{2}, \dfrac{y+1}{2}\right) = (9, 3)$

therefore $\dfrac{x+5}{2} = 9$ and $\dfrac{y+1}{2} = 3$

$x + 5 = 18 \qquad y + 1 = 6$

$x = 13 \qquad y = 5$

Thus (13, 5) is the other endpoint.

91. $\left(\dfrac{x+(-3)}{2}, \dfrac{y+(-8)}{2}\right) = (2, -7)$

therefore $\dfrac{x-3}{2} = 2$ and $\dfrac{y-8}{2} = -7$

$x - 3 = 4 \qquad y - 8 = -14$

$x = 7 \qquad y = -6$

Thus (7, −6) is the other endpoint.

93. Let $P(x, y)$ be the point three-fourths the distance from $P_1(-8, 11)$ to $P_2(6, 20)$.

Then $P(x, y)$ divides the line segment from P_1 to P_2 into the ratio 3 to 1, or $\dfrac{d(P_1, P)}{d(P, P_2)} = \dfrac{3}{1} = 3$. Therefore $r = 3$.

$x = \dfrac{-8 + 3(6)}{1+3} = \dfrac{-8+18}{4} = \dfrac{10}{4} = \dfrac{5}{2}$ and $y = \dfrac{11 + 3(20)}{1+3} = \dfrac{11+60}{4} = \dfrac{71}{4}$

The coordinates of the point three-fourths the distance from P_1 to P_2 are $\left(\dfrac{5}{2}, \dfrac{71}{4}\right)$.

95. Let $P(x, y)$ be the point seven-eights the distance from $P_1(6, 10)$ to $P_2(2, 1)$.

Then $P(x, y)$ divides the line segment from P_1 to P_2 into the ratio 7 to 1, or $\dfrac{d(P_1, P)}{d(P, P_2)} = \dfrac{7}{1} = 7$. Therefore $r = 7$

$x = \dfrac{6 + 7(2)}{1+7} = \dfrac{6+14}{8} = \dfrac{20}{8} = \dfrac{5}{2}$ and $y = \dfrac{10 + 7(1)}{1+7} = \dfrac{10+7}{8} = \dfrac{17}{8}$

The coordinates of the point seven-eighths the distance from P_1 to P_2 are $\left(\dfrac{5}{2}, \dfrac{17}{8}\right)$.

97.
$d((1,-4),(3,2)) = \sqrt{(3-1)^2 + (2-(-4))^2} = \sqrt{2^2 + 6^2} = \sqrt{40}$

$d((3,2),(-3,4)) = \sqrt{(-3-3)^2 + (4-2)^2} = \sqrt{(-6)^2 + 2^2} = \sqrt{40}$

$d((-3,4),(-5,-2)) = \sqrt{(-5-(-3))^2 + (-2-4)^2} = \sqrt{(-2)^2 (-6)^2} = \sqrt{40}$

$d((-5,-2),(1,-4)) = \sqrt{(1-(-5))^2 + (-4-(-2))^2} = \sqrt{6^2 + (-2)^2} = \sqrt{40}$

The sides are all of equal length. Now we find the length of each diagonal:

$d((3,2),(-5,-2)) = \sqrt{(-5-3)^2 + (-2-2)^2} = \sqrt{(-8)^2 + (-4)^2} = \sqrt{80}$

$d((-3,4),(1,-4)) = \sqrt{(1-(-3))^2 + (-4-4)^2} = \sqrt{4^2 + (-8)^2} = \sqrt{80}$

Since the sides are of the same length and the diagonals are of equal length, we know these points are the vertices of a square.

99.
$$\sqrt{(3-x)^2+(4-y)^2}=5$$
$$(3-x)^2+(4-y)^2=5^2$$
$$9-6x+x^2+16-18y+y^2=25$$
$$x^2-6x+y^2-8y=0$$

101.
$$\sqrt{(4-x)^2+(0-y)^2}+\sqrt{(-4-x)^2+(0-y)^2}=10$$
$$(4-x)^2+(0-y)^2=100-20\sqrt{(-4-x)^2+(0-y)^2}+(-4-x)^2+(-y)^2$$
$$16-8x+x^2+y^2=100-20\sqrt{(-4-x)^2+(-y)^2}+16+8x+x^2+y^2$$
$$-16x-100=-20\sqrt{(-4-x)^2+(-y)^2}$$
$$4x+25=5\sqrt{(-4-x)^2+(-y)^2}$$
$$16x^2+200x+625=25\left[(-4-x)^2+(-y)^2\right]$$
$$16x^2+200x+625=25\left[16+8x+x^2+y^2\right]$$
$$16x^2+200x+625=400+200x+25x^2+25y^2$$

Simplifying yields $9x^2+25y^2=225$.

103.
$$d=\sqrt{(-4-2)^2+(11-3)^2}$$
$$=\sqrt{36+64}$$
$$=\sqrt{100}$$
$$=10$$
Since the diameter is 10, the radius is 5.
The center is the midpoint of the line segment from (2, 3)
to (−4, 11).
$$\left(\frac{2+(-4)}{2},\frac{3+11}{2}\right)=(-1,7)\text{ center}$$
$$(x+1)^2+(y-7)^2=5^2$$

105. Since it is tangent to the x-axis, its radius is 11.
$$(x-7)^2+(y-11)^2=11^2$$

107. The center is (−3,3). The radius is 3.
$$(x+3)^2+(y-3)^2=3^2$$

SECTION 1.3, Page 38

1. Given $f(x)=3x-1,$

   a. $f(2)=3(2)-1$
$$=6-1$$
$$=5$$

   b. $f(-1)=3(-1)-1$
$$=-3-1$$
$$=-4$$

   c. $f(0)=3(0)-1$
$$=0-1$$
$$=-1$$

   d. $f\left(\frac{2}{3}\right)=3\left(\frac{2}{3}\right)-1$
$$=2-1$$
$$=1$$

   e. $f(k)=3(k)-1$
$$=3k-1$$

   f. $f(k+2)=3(k+2)-1$
$$=3k+6-1$$
$$=3k+5$$

3. Given $A(w) = \sqrt{w^2 + 5}$,

    a. $A(0) = \sqrt{(0)^2 + 5}$

             $= \sqrt{5}$

    b. $A(2) = \sqrt{(2)^2 + 5}$

             $= \sqrt{9}$

             $= 3$

    c. $A(-2) = \sqrt{(-2)^2 + 5}$

             $= \sqrt{9}$

             $= 3$

    d. $A(4) = \sqrt{4^2 + 5}$

             $= \sqrt{21}$

    e. $A(r+1) = \sqrt{(r+1)^2 + 5}$

             $= \sqrt{r^2 + 2r + 1 + 5}$

             $= \sqrt{r^2 + 2r + 6}$

    f. $A(-c) = \sqrt{(-c)^2 + 5}$

             $= \sqrt{c^2 + 5}$

5. Given $f(x) = \dfrac{1}{|x|}$,

    a. $f(2) = \dfrac{1}{|2|}$

             $= \dfrac{1}{2}$

    b. $f(-2) = \dfrac{1}{|-2|}$

             $= \dfrac{1}{2}$

    c. $f\left(-\dfrac{3}{5}\right) = \dfrac{1}{\left|-\dfrac{3}{5}\right|}$

             $= \dfrac{1}{\dfrac{3}{5}}$

             $= 1 \div \dfrac{3}{5}$

             $= 1 \cdot \dfrac{5}{3}$

             $= \dfrac{5}{3}$

    d. $f(2) + f(-2) = \dfrac{1}{2} + \dfrac{1}{2}$

             $= 1$

    e. $f(c^2 + 4) = \dfrac{1}{\left|c^2 + 4\right|}$

             $= \dfrac{1}{c^2 + 4}$

    f. $f(2 + h) = \dfrac{1}{|2 + h|}$

7. Given $s(x) = \dfrac{x}{|x|}$,

    a. $s(4) = \dfrac{4}{|4|} = \dfrac{4}{4} = 1$

    b. $s(5) = \dfrac{5}{|5|} = \dfrac{5}{5} = 1$

    c. $s(-2) = \dfrac{-2}{|-2|} = \dfrac{-2}{2} = -1$

    d. $s(-3) = \dfrac{-3}{|-3|} = \dfrac{-3}{3} = -1$

    e.   Since $t > 0, |t| = t$.

             $s(t) = \dfrac{t}{|t|} = \dfrac{t}{t} = 1$

    f.   Since $t < 0, |t| = -t$.

             $s(t) = \dfrac{t}{|t|} = \dfrac{t}{-t} = -1$

9.   **a.**   Since $x = -4 < 2$, use $P(x) = 3x + 1$.

          $P(-4) = 3(-4) + 1 = -12 + 1 = -11$

    b.   Since $x = \sqrt{5} \geq 2$, use $P(x) = -x^2 + 11$.

          $P\left(\sqrt{5}\right) = -\left(\sqrt{5}\right)^2 + 11 = -5 + 11 = 6$

    c.   Since $x = c < 2$, use $P(x) = 3x + 1$.

          $P(c) = 3c + 1$

    d.   Since $k \geq 1$, then $x = k + 1 \geq 2$,

          so use $P(x) = -x^2 + 11$.

          $P(k+1) = -(k+1)^2 + 11 = -(k^2 + 2k + 1) + 11$

                          $= -k^2 - 2k - 1 + 11$

                          $= -k^2 - 2k + 10$

11. $2x + 3y = 7$

         $3y = -2x + 7$

         $y = -\dfrac{2}{3}x + \dfrac{7}{3}$, y is a function of x.

13. $-x + y^2 = 2$

         $y^2 = x + 2$

         $y = \pm\sqrt{x + 2}$, y is a not function of x.

15. $y = 4 \pm \sqrt{x}$, y is not a function of x since for each $x > 0$ there are two values of x.

17. $y = \sqrt[3]{x}$, y is a function of x.

19. $y^2 = x^2$

         $y = \pm\sqrt{x^2}$, y is a not function of x.

21. Function; each x is paired with exactly one y.

23. Function; each x is paired with exactly one y.

25. Function; each x is paired with exactly one y.

27. $f(x) = 3x - 4$ Domain is the set of all real numbers.

29. $f(x) = x^2 + 2$ Domain is the set of all real numbers.

31. $f(x) = \dfrac{4}{x + 2}$ Domain is $\{x \mid x \neq -2\}$.

33. $f(x) = \sqrt{7 + x}$ Domain is $\{x \mid x \geq -7\}$.

35. $f(x) = \sqrt{4 - x^2}$ Domain is $\{x \mid -2 \leq x \leq 2\}$.

37. $f(x) = \dfrac{1}{\sqrt{x + 4}}$ Domain is $\{x \mid x > -4\}$.

39.

Domain: the set of all real numbers

41.

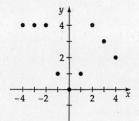

Domain: $\{-4, -3, -2, -1, 0, 1, 2, 3, 4\}$

43.

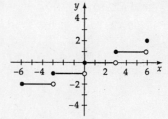

Domain: $\{x \mid -6 \leq x \leq 6\}$

45.

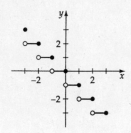

Domain: $\{x \mid -3 \leq x \leq 3\}$

47.   **a.** $C(3.97) = 0.29 - 0.29\,\text{int}(1 - 3.97)$

$\qquad\qquad\qquad = 0.29 - 0.29\,\text{int}(-2.97)$

$\qquad\qquad\qquad = 0.29 - 0.29(-3)$

$\qquad\qquad\qquad = 0.29 + 0.87$

$\qquad\qquad\qquad = \$1.16$

   b.

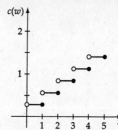

49.   **a.** Yes; every vertical line intersects the graph in one point.

   b. Yes; every vertical line intersects the graph in one point.

   c. No; some vertical lines intersect the graph at more than one point.

   d. Yes; every vertical line intersects the graph in at most one point.

51. Decreasing on $(-\infty, 0]$; increasing on $[0, \infty)$

53. Increasing on $(-\infty, \infty)$

55. Decreasing on $(-\infty, -3]$; increasing on $[-3, 0]$; decreasing on $[0, 3]$; increasing on $[3, \infty)$

57. Constant on $(-\infty, 0]$; increasing on $[0, \infty)$

59. Decreasing on $(-\infty, 0]$; constant on $[0, 1]$; increasing on $[1, \infty)$

61. g and F are one-to-one since every horizontal line intersects the graph at one point.
f, V, and p are not one-to-one since some horizontal lines intersect the graph at more than one point.

63. a.
$$P = 2l + 2w = 50$$
$$2w = 50 - 2l$$
$$w = 25 - l$$

b.
$$A = lw$$
$$A = l(25 - l)$$
$$A = 25l - l^2$$

65. $v(t) = 80{,}000 - 6500t, \quad 0 \le t \le 10$

67. a.
$$C(x) = 5(400) + 22.80x$$
$$= 2000 + 22.80x$$

b. $R(x) = 37.00x$

c.
$$P(x) = 37.00x - C(x)$$
$$= 37.00 - [2000 + 22.80x]$$
$$= 37.00x - 2000 - 22.80x$$
$$= 14.20x - 2000$$

Note x is a natural number.

69.
$$\frac{15}{3} = \frac{15 - h}{r}$$
$$5 = \frac{15 - h}{r}$$
$$5r = 15 - h$$
$$h = 15 - 5r$$
$$h(r) = 15 - 5r$$

71.
$$d = \sqrt{(3t)^2 + (50)^2}$$
$$d = \sqrt{9t^2 + 2500} \text{ meters}, \ 0 \le t \le 60$$

73. $d = \sqrt{(45 - 8t)^2 + (6t)^2}$ miles
where t is the number of hours after 12:00 noon

75.

x	5	10	12.5	15	20
$Y(x)$	275	375	385	390	394

answers accurate to nearest apple

77.
$$f(c) = c^2 - c - 5 = 1$$
$$c^2 - c - 6 = 0$$
$$(c - 3)(c + 2) = 0$$
$$c - 3 = 0 \quad \text{or} \quad c + 2 = 0$$
$$c = 3 \qquad\qquad c = -2$$

79. 1 is not in the range of $f(x)$, since
$$1 = \frac{x - 1}{x + 1} \text{ only if } x + 1 = x - 1 \text{ or } 1 = -1.$$

81. Set the graphing utility to "dot" mode.

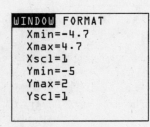

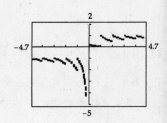

83.

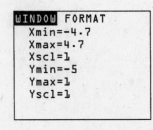

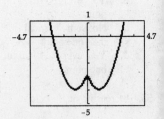

85.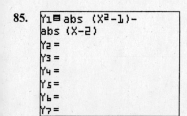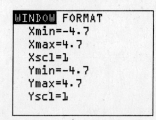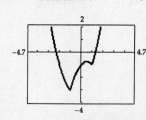

87. $f(x)\Big|_2^3 = (9-3)-(4-2)=6-2=4$

89. $f(x)\Big|_0^2 = (16-12-2)-0=2$

91. a. $f(1,7)=3(1)+5(7)-2=3+35-2=36$
    b. $f(0,3)=3(0)+5(3)-2=13$
    c. $f(-2,4)=3(-2)+5(4)-2=12$
    d. $f(4,4)=3(4)+5(4)-2=30$
    e. $f(k,2k)=3(k)+5(2k)-2=13k-2$
    f. $f(k+2,k-3)=3(k+2)+5(k-3)-2$
       $=3k+6+5k-15-2$
       $=8k-11$

93. $s=\dfrac{5+8+11}{2}=12$

    $A(5,8,11)=\sqrt{12(12-5)(12-8)(12-11)}$
    $=\sqrt{12(7)(4)(1)}=\sqrt{336}=4\sqrt{21}$

95. $a^2+3a-3=a$
    $a^2+2a-3=0$
    $(a-1)(a+3)=0$

    $a=1$ or $a=-3$

97.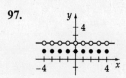

SECTION 1.4, Page 53

1.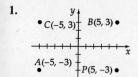

3.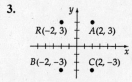

5.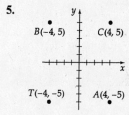

7.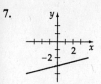

9.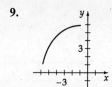

11.

13. Replacing x by $-x$ leaves the equation unaltered. Thus the graph is symmetric with respect to the y-axis.

15. Not symmetric with respect to either axis. (neither)

17. Symmetric with respect to both the x- and the y-axes.

19. Symmetric with respect to both the x- and the y-axes.

21. Symmetric with respect to both the x- and the y-axes.

23. No, since $(-y)=3(-x)-2$ simplifies to $(-y)=-3x-2$, which is not equivalent to the original equation $y=3x-2$.

25. Yes, since $(-y)=-(-x)^3$ implies $-y=x^3$ or $y=-x^3$, which is the original equation.

27. Yes, since $(-x)^2+(-y)^2=10$ simplifies to the original equation.

29. Yes, since $-y=\dfrac{-x}{|-x|}$ simplifies to the original equation.

31.

symmetric with
respect to the y-axis

33.

symmetric with
respect to the origin

35.

symmetric with
respect to the origin

37.

symmetric with
respect to the line $x = 4$

39.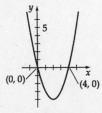

symmetric with
respect to the line $x = 2$

41.

no symmetry

43. Even, since $g(-x) = (-x)^2 - 7 = x^2 - 7 = g(x)$.

45. Odd, since $F(-x) = (-x)^5 + (-x)^3$
$= -x^5 - x^3$
$= -F(x)$.

47. Even

49. Even

51. Even

53. Even

55. Neither

57.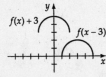

59.

61.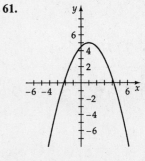

$y = -\dfrac{1}{2}m(x) + 3$

63. a.

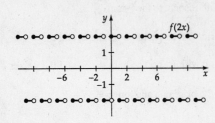

b.

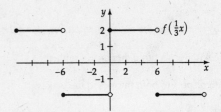

65. a.

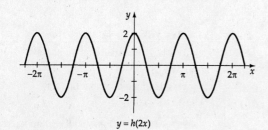

$$y = h(2x)$$

b.

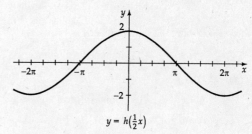

$$y = h\left(\tfrac{1}{2}x\right)$$

67.

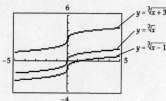

69.

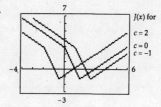

71.

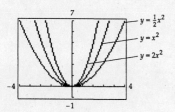

73.

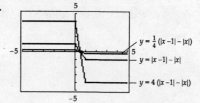

75. a.

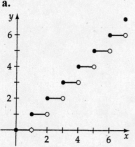

b.

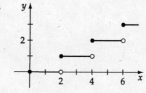

c.

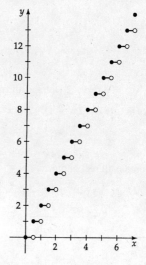

77. a. $f(x) = \dfrac{2}{(x+1)^2 + 1} + 1$

b. $f(x) = -\dfrac{2}{(x-2)^2 + 1}$

SECTION 1.5, Page 64

1.
$$f(x) + g(x) = (x^2 - 2x - 15) + (x + 3)$$
$$= x^2 - x - 12 \text{ Domain all real numbers}$$
$$f(x) - g(x) = (x^2 - 2x - 15) - (x + 3)$$
$$= x^2 - 3x - 18 \text{ Domain all real numbers}$$
$$f(x)g(x) = (x^2 - 2x - 15)(x + 3)$$
$$= x^3 + x^2 - 21x - 45 \text{ Domain all real numbers}$$
$$f(x)/g(x) = (x^2 - 2x - 15)/(x + 3)$$
$$= x - 5 \text{ Domain } \{x \mid x \neq -3\}$$

3.
$$f(x) + g(x) = (2x^2 + 8) + (x + 4)$$
$$= 3x + 12 \text{ Domain all real numbers}$$
$$f(x) - g(x) = (2x + 8) - (x + 4)$$
$$= x + 4 \text{ Domain all real numbers}$$
$$f(x)g(x) = (2x + 8)(x + 4)$$
$$= 2x^2 + 16x + 32 \text{ Domain all real numbers}$$
$$f(x)/g(x) = (2x + 8)/(x + 4)$$
$$= [2(x + 4)]/(x + 4)$$
$$= 2 \text{ Domain } \{x \mid x \neq -4\}$$

5.
$$f(x) + g(x) = (x^3 - 2x^2 + 7x) + x$$
$$= x^3 - 2x^2 + 8x \text{ Domain all real numbers}$$
$$f(x) - g(x) = (x^3 - 2x^2 + 7x) - x$$
$$= x^3 - 2x^2 + 6x \text{ Domain all real numbers}$$
$$f(x)g(x) = (x^3 - 2x^2 + 7x)x$$
$$= x^4 - 2x^3 + 7x^2 \text{ Domain all real numbers}$$
$$f(x)/g(x) = (x^3 - 2x^2 + 7x)/x$$
$$= x^2 - 2x + 7 \text{ Domain } \{x \mid x \neq 0\}$$

7.
$$f(x) + g(x) = (2x^2 + 4x - 7) + (2x^2 + 3x - 5)$$
$$= 4x^2 + 7x - 12 \text{ Domain all real numbers}$$
$$f(x) - g(x) = (2x^2 + 4x - 7) - (2x^2 + 3x - 5)$$
$$= x - 2 \text{ Domain all real numbers}$$
$$f(x)g(x) = (2x^2 + 4x - 7)(2x^2 + 3x - 5)$$
$$= 4x^4 + 6x^3 - 10x^2 + 8x^3 + 12x^2 - 20x - 14x^2 - 21x + 35$$
$$= 4x^4 + 14x^3 - 12x^2 - 41x + 35$$
Domain all real numbers
$$f(x)/g(x) = (2x^2 + 4x - 7)/(2x^2 + 3x - 5)$$
$$= 1 + \frac{x - 2}{2x^2 + 3x - 5} \text{ Domain } \left\{x \mid x \neq 1, x \neq -\frac{5}{2}\right\}$$

9.
$$f(x) + g(x) = \sqrt{x - 3} + x \qquad \text{Domain } \{x \mid x \geq 3\}$$
$$f(x) - g(x) = \sqrt{x - 3} - x \qquad \text{Domain } \{x \mid x \geq 3\}$$
$$f(x)g(x) = x\sqrt{x - 3} \qquad \text{Domain } \{x \mid x \geq 3\}$$
$$f(x)/g(x) = \frac{\sqrt{x - 3}}{x} + x \qquad \text{Domain } \{x \mid x \geq 3\}$$

11.
$$f(x) + g(x) = \sqrt{4 - x^2} + 2 + x \quad \text{Domain } \{x \mid -2 \leq x \leq 2\}$$
$$f(x) - g(x) = \sqrt{4 - x^2} - 2 - x \quad \text{Domain } \{x \mid -2 \leq x \leq 2\}$$
$$f(x)g(x) = \left(\sqrt{4 - x^2}\right)(2 + x) \quad \text{Domain } \{x \mid -2 \leq x \leq 2\}$$
$$f(x)/g(x) = \frac{\sqrt{4 - x^2}}{2 + x} \qquad \text{Domain } \{x \mid -2 \leq x \leq 2\}$$

13.
$$(f + g)(x) = x^2 - x - 2$$
$$(f + g)(5) = (5)^2 - (5) - 2$$
$$= 25 - 5 - 2$$
$$= 18$$

15.
$$(f + g)(x) = x^2 - x - 2$$
$$(f + g)\left(\frac{1}{2}\right) = \left(\frac{1}{2}\right)^2 - \left(\frac{1}{2}\right) - 2$$
$$= \frac{1}{4} - \frac{1}{2} - 2$$
$$= -\frac{9}{4}$$

17.
$$(f - g)(x) = x^2 - 5x + 6$$
$$(f - g)(-3) = (-3)^2 - 5(-3) + 6$$
$$= 9 + 15 + 6$$
$$= 30$$

19. $(f-g)(x) = x^2 - 5x + 6$

$\quad (f-g)(-1) = (-1)^2 - 5(-1) + 6$

$\qquad\qquad\quad = 1 + 5 + 6$

$\qquad\qquad\quad = 12$

21. $(fg)(x) = \left(x^2 - 3x + 2\right)(2x - 4)$

$\qquad\qquad = 2x^3 - 6x^2 + 4x - 4x^2 + 12x - 8$

$\qquad\qquad = 2x^3 - 10x^2 + 16x - 8$

$\quad (fg)(7) = 2(7)^3 - 10(7)^2 + 16(7) - 8$

$\qquad\qquad = 686 - 490 + 112 - 8$

$\qquad\qquad = 300$

23. $(fg)(x) = 2x^3 - 10x^2 + 16x - 8$

$\quad (fg)\left(\dfrac{2}{5}\right) = 2\left(\dfrac{2}{5}\right)^3 - 10\left(\dfrac{2}{5}\right)^2 + 16\left(\dfrac{2}{5}\right) - 8$

$\qquad\qquad = \dfrac{16}{125} - \dfrac{40}{25} + \dfrac{32}{5} - 8$

$\qquad\qquad = \dfrac{-384}{125}$

$\qquad\qquad = -3.072$

25. $\left(\dfrac{f}{g}\right)(x) = \dfrac{x^2 - 3x + 2}{2x - 4}$

$\quad \left(\dfrac{f}{g}\right)(x) = \dfrac{1}{2}x - \dfrac{1}{2}$

$\quad \left(\dfrac{f}{g}\right)(-4) = \dfrac{1}{2}(-4) - \dfrac{1}{2}$

$\qquad\qquad = -2 - \dfrac{1}{2}$

$\qquad\qquad = -2\dfrac{1}{2} \text{ or } -\dfrac{5}{2}$

27. $\left(\dfrac{f}{g}\right)(x) = \dfrac{1}{2}x - \dfrac{1}{2}$

$\quad \left(\dfrac{f}{g}\right)\left(\dfrac{1}{2}\right) = \dfrac{1}{2}\left(\dfrac{1}{2}\right) - \dfrac{1}{2}$

$\qquad\qquad = \dfrac{1}{4} - \dfrac{1}{2}$

$\qquad\qquad = -\dfrac{1}{4}$

29. $\dfrac{f(x+h) - f(x)}{h} = \dfrac{[2(x+h) + 4] - (2x + 4)}{h}$

$\qquad\qquad = \dfrac{2x + 2(h) + 4 - 2x - 4}{h}$

$\qquad\qquad = \dfrac{2h}{h}$

$\qquad\qquad = 2$

31. $\dfrac{f(x+h) - f(x)}{h} = \dfrac{[(x+h) - 6] - (x^2 - 6)}{h}$

$\qquad\qquad = \dfrac{x^2 + 2x(h) + (h)^2 - 6 - x^2 + 6}{h}$

$\qquad\qquad = \dfrac{2x(h) + h^2}{h}$

$\qquad\qquad = 2x + h$

33. $\dfrac{f(x+h) - f(x)}{h} = \dfrac{2(x+h)^2 + 4(x+h) - 3 - (2x^2 + 4x - 3)}{h}$

$\qquad\qquad = \dfrac{2x^2 + 4xh + 2h^2 + 4x + 4h - 3 - 2x^2 - 4x + 3}{h}$

$\qquad\qquad = \dfrac{4xh + 2h^2 + 4h}{h}$

$\qquad\qquad = 4x + 2h + 4$

35. $\dfrac{f(x+h) - f(x)}{h} = \dfrac{-4(x+h)^2 + 6 - (-4x^2 + 6)}{h}$

$\qquad\qquad = \dfrac{-4x^2 - 8xh - 4h^2 + 6 + 4x^2 - 6}{h}$

$\qquad\qquad = \dfrac{-8xh - 4h^2}{h}$

$\qquad\qquad = -8x - 4h$

37. $(g \circ f)(x) = g[f(x)]$

$\qquad\qquad = g[3x + 5]$

$\qquad\qquad = 2[3x + 5]$

$\qquad\qquad = 6x + 10 - 7$

$\qquad\qquad = 6x + 3$

$\quad (f \circ g)(x) = f[g(x)]$

$\qquad\qquad = f[2x - 7]$

$\qquad\qquad = 3[2x - 7] + 5$

$\qquad\qquad = 6x - 21 + 5$

$\qquad\qquad = 6x - 16$

39. $(g \circ f)(x) = g[x^2 + 4x - 1]$

$\qquad = [x^2 + 4x - 1] + 2$

$\qquad = x^2 + 4x + 1$

$(f \circ g)(x) = f[x + 2]$

$\qquad = [x + 2]^2 + 4[x + 2] - 1$

$\qquad = x^2 + 4x + 4 + 4x + 8 - 1$

$\qquad = x^2 + 8x + 11$

41. $(g \circ f)(x) = g[f(x)]$

$\qquad = g[x^3 + 2x]$

$\qquad = -5[x^3 + 2x]$

$\qquad = -5x^3 - 10x$

$(f \circ g)(x) = f[g(x)]$

$\qquad = f[-5x]$

$\qquad = [-5x]^3 + 2[-5x]$

$\qquad = -125x^3 - 10x$

43. $(g \circ f)(x) = g[f(x)]$

$\qquad = g\left[\dfrac{2}{x+1}\right]$

$\qquad = 3\left[\dfrac{2}{x+1}\right] - 5$

$\qquad = \dfrac{6}{x+1} - \dfrac{5(x+1)}{x+1}$

$\qquad = \dfrac{6 - 5x - 5}{x+1}$

$\qquad = \dfrac{1 - 5x}{x+1}$

$(f \circ g) = f[g(x)]$

$\qquad = f[3x - 5]$

$\qquad = \dfrac{2}{[3x-5]+1}$

$\qquad = \dfrac{2}{3x-4}$

45. $(g \circ f)(x) = g[f(x)]$

$\qquad = g\left[\dfrac{1}{x^2}\right]$

$\qquad = \sqrt{\left[\dfrac{1}{x^2}\right] - 1}$

$\qquad = \sqrt{\dfrac{1 - x^2}{x^2}}$

$\qquad = \dfrac{\sqrt{1 - x^2}}{|x|}$

$(f \circ g)(x) = f\left[\sqrt{x-1}\right]$

$\qquad = \dfrac{1}{\left[\sqrt{x-1}\right]^2}$

$\qquad = \dfrac{1}{x - 1}$

47. $(g \circ f)(x) = g\left[\dfrac{3}{|5-x|}\right]$

$\qquad = -\dfrac{2}{\left[\dfrac{3}{|5-x|}\right]}$

$\qquad = \dfrac{-2|5-x|}{3}$

$(f \circ g)(x) = f\left[-\dfrac{2}{x}\right]$

$\qquad = \dfrac{3}{\left|5 - \left[-\dfrac{2}{x}\right]\right|}$

$\qquad = \dfrac{3}{\left|5 + \dfrac{2}{x}\right|}$

$\qquad = \dfrac{3}{\dfrac{|5x + 2|}{|x|}}$

$\qquad = \dfrac{3|x|}{|5x + 2|}$

For Exercises 49 to 64, using Method 2 gives us

$(f \circ g)(x) = 2x^2 - 10x + 3$ $\qquad$ $(g \circ f)(x) = 4x^2 + 2x - 6$

$(g \circ h)(x) = 9x^4 - 9x^2 - 4$ $\qquad$ $(h \circ g)(x) = -3x^4 + 30x^3 - 75x^2 + 4$

$(f \circ f)(x) = 4x + 9$

We now use the results to work Exercises 49 to 64.

49. $(g \circ f)(x) = 4x^2 + 2x - 6$

$(g \circ f)(4) = 4(4)^2 + 2(4) - 6$

$\qquad = 64 + 8 - 6$

$\qquad = 66$

51. $(f \circ g)(x) = 2x^2 - 10x + 3$

$(f \circ g)(-3) = 2(-3)^2 - 10(-3) + 3$

$\qquad = 18 + 30 + 3$

$\qquad = 51$

53. $(g \circ h)(x) = 9x^4 - 9x^2 - 4$

$(g \circ h)(0) = 9(0)^4 - 9(0)^2 - 4$

$\qquad = -4$

55. $(f \circ f)(x) = 4x + 9$

$(f \circ f)(8) = 4(8) + 9$

$\qquad = 41$

57. $(h \circ g)(x) = -3x^4 + 30x^3 - 75x^2 + 4$

$(h \circ g)\left(\dfrac{2}{5}\right) = -3\left(\dfrac{2}{5}\right)^4 + 30\left(\dfrac{2}{5}\right)^3 - 75\left(\dfrac{2}{5}\right)^2 + 4$

$\qquad = -\dfrac{48}{625} + \dfrac{240}{125} - \dfrac{300}{25} + 4$

$\qquad = \dfrac{-48 + 1200 - 7500 + 2500}{625}$

$\qquad = -\dfrac{3848}{625}$

59. $(g \circ f)(x) = 4x^2 + 2x - 6$

$(g \circ f)(\sqrt{3}) = 4(\sqrt{3})^2 + 2(\sqrt{3}) - 6$

$\qquad = 12 + 2\sqrt{3} - 6$

$\qquad = 6 + 2\sqrt{3}$

61. $(g \circ f)(x) = 4x^2 + 2x - 6$

$(g \circ f)(2c) = 4(2c)^2 + 2(2c) - 6$

$\qquad = 16c^2 + 4c - 6$

63. $(g \circ h)(x) = 9x^4 - 9x^2 - 4$

$(g \circ h)(k+1) = 9(k+1)^4 - 9(k+1)^2 - 4$

$\qquad = 9(k^4 + 4k^3 + 6k^2 + 4k + 1) - 9k^2 - 18k - 9 - 4$

$\qquad = 9k^4 + 36k^3 + 54k^2 + 36k + 9 - 9k^2 - 18k - 13$

$\qquad = 9k^4 + 36k^3 + 45k^2 + 18k - 4$

65. a. $r = 1.5t$ and $A = \pi r^2$

so $A(t) = \pi[r(t)]^2$

$\qquad = \pi(1.5t)^2$

$\qquad = 2.25\pi(2)^2$

$\qquad = 9\pi$ square feet

$\qquad \approx 28.27$ square feet

b. $r = 1.5t$

$h = 2r = 2(1.5t) = 3t$ and

$V = \dfrac{1}{3}\pi r^2 h$ so

$V(t) = \dfrac{1}{3}\pi(1.5t)^2[3t]$

$\qquad = 2.25\pi t^3$

Note: $V = \dfrac{1}{3}\pi r^2 h = \dfrac{1}{3}(\pi r^2) = \dfrac{1}{3}hA$

$\qquad = \dfrac{1}{3}(3t)(2.25\pi t^2) = 2.25\pi t^3$

$V(3) = 2.25\pi(3)^3$

$\qquad = 60.75\pi$ cubic feet

$\qquad \approx 190.85$ cubic feet

67. a. Since $d^2 + 4^2 = s^2$,

$$d^2 = s^2 - 16$$

$$d = \sqrt{s^2 - 16}$$

$$d = \sqrt{(48-t)^2 - 16} \qquad \bullet \; s = 48 - t$$

$$= \sqrt{2304 - 96t + t^2 - 16}$$

$$= \sqrt{t^2 - 96t + 2288}$$

b. $s(35) = 48 - 35 + 13$

$$d(35) = \sqrt{35^2 - 96(35) + 2288}$$

$$= \sqrt{153} \approx 12.4$$

69. $(Y \circ F)(x) = Y(F(x))$ converts x inches to yards. F takes x inches to feet, and then Y takes feet to yards.

71. a. On $[0, 1]$, $a = 0$

$\Delta t = 1 - 0 = 1$

$C(a + \Delta t) = C(1) = 99.8$

$C(a) = C(0) = 0$

Average rate of change $= \dfrac{C(1) - C(0)}{1} = 99.8 - 0 = 99.8$

This is identical to the slope of the line through

$(0, C(0))$ and $(1, C(1))$ since $m = \dfrac{C(1) - C(0)}{1 - 0} = C(1) - C(0)$

b. On $[0, 0.5]$, $a = 0$

$\Delta t = 0.5$

Average rate of change $= \dfrac{C(0.5) - C(0)}{0.5} = \dfrac{78.1 - 0}{0.5} = 156.2$

c. On $[1, 2]$, $a = 1$

$\Delta t = 2 - 1 = 1$

Average rate of change $= \dfrac{C(2) - C(1)}{1} = \dfrac{50.1 - 99.8}{1} = -49.7$

d. On $[1, 1.5]$, $a = 1$

$\Delta t = 1.5 - 1 = 0.5$

Average rate of change $= \dfrac{C(1.5) - C(1)}{0.5} = \dfrac{84.4 - 99.8}{0.5} = \dfrac{-15.4}{0.5} = -30.8$

e. On $[1, 1.25]$, $a = 1$

$\Delta t = 1.25 - 1 = 0.25$

Average rate of change $= \dfrac{C(1.25) - C(1)}{0.25} = \dfrac{95.7 - 99.8}{0.25} = \dfrac{-4.1}{0.25} = -16.4$

f. On $[1, 1 + \Delta t]$, $\; Con(1 + \Delta t) = 25(1 + \Delta t)^3 - 150(1 + \Delta t)^2 + 225(1 + \Delta t)$

$$= 25(1 + 3(\Delta t) + 3(\Delta t)^3) - 150(1 + 2(\Delta t) + (\Delta t)^2) + 225(1 + \Delta t)$$

$$= 25 + 75(\Delta t) + 75(\Delta t)^2) + 25(\Delta t)^3 - 150 - 300(\Delta t) - 150(\Delta t)^2 + 225 + 225(\Delta t)$$

$$\doteq 100 - 75(\Delta t)^2 + 25(\Delta t)^3$$

$$Con(1) = 100$$

Average rate of change $= \dfrac{Con(1 + \Delta t) - Con(1)}{\Delta t} = \dfrac{100 - 75(\Delta t)^2 + 25(\Delta t)^3 - 100}{\Delta t}$

$$= \dfrac{-75(\Delta t)^2 + 25(\Delta t)^3}{\Delta t}$$

$$= -75(\Delta t) + 25(\Delta t)^2$$

As Δt approaches 0, the average rate of change over $[1, 1 + \Delta t]$ seems to approach 0.

73.

$$(g \circ f)(x) = g[f(x)]$$
$$= g[2x + 3]$$
$$= \frac{[2x + 3] - 3}{2}$$
$$= \frac{2x}{2}$$
$$= x$$

$$(f \circ g)(x) = f[g(x)]$$
$$= f\left[\frac{x - 3}{2}\right]$$
$$= 2\left[\frac{x - 3}{2}\right] + 3$$
$$= x - 3 + 3$$
$$= x$$

75.

$$(g \circ f)(x) = g[f(x)]$$
$$= g\left[\frac{4}{x + 1}\right]$$
$$= \frac{4 - \left[\dfrac{4}{x + 1}\right]}{\left[\dfrac{4}{x + 1}\right]}$$
$$= \frac{\dfrac{4x + 4 - 4}{x + 1}}{\dfrac{4}{x + 1}}$$
$$= \frac{4x}{x + 1} \cdot \frac{x + 1}{4}$$
$$= x$$

$$(f \circ g)(x) = f[g(x)]$$
$$= f\left[\frac{4 - x}{x}\right]$$
$$= \frac{4}{\left[\dfrac{4 - x}{x}\right] + 1}$$
$$= \frac{4}{\dfrac{4 - x + x}{x}}$$
$$= \frac{4}{\dfrac{4}{x}}$$
$$= 4 \cdot \frac{x}{4}$$
$$= x$$

77.

$$(g \circ f)(x) = g[f(x)]$$
$$= g[x^3 - 1]$$
$$= \sqrt[3]{[x^3 - 1] + 1}$$
$$= \sqrt[3]{x^3}$$
$$= x$$

$$(f \circ g)(x) = f[g(x)]$$
$$= f[\sqrt[3]{x + 1}]$$
$$= [\sqrt[3]{x + 1}]^3 - 1$$
$$= x + 1 - 1$$
$$= x$$

79. a.

$$(s \circ m)(x) = s[m(x)]$$
$$= s\left[\frac{60x + 34,000}{x}\right]$$
$$= 1.45\left[\frac{60x + 34,000}{x}\right]$$
$$= \frac{87x + 49,300}{x}$$
$$= 87 + \frac{49,300}{x}$$

b.

$$(s \circ m)(24,650) = 87 + \frac{49,300}{24,650}$$
$$= \$89$$

SECTION 1.6, Page 74

1.

$$(f \circ g)(x) = f[g(x)]$$
$$= f\left[\frac{x - 1}{2}\right]$$
$$= 2\left[\frac{x - 1}{2}\right] + 1$$
$$= x - 1 + 1$$
$$= x$$

$$(g \circ f)(x) = g[f(x)]$$
$$= g[2x + 1]$$
$$= \frac{[2x + 1] - 1}{2}$$
$$= \frac{2x}{2}$$
$$= x$$

3.

$$(f \circ g)(x) = f[g(x)]$$
$$= f\left[\frac{x + 5}{3}\right]$$
$$= 3\left[\frac{x + 5}{3}\right] - 5$$
$$= x + 5 - 5$$
$$= x$$

$$(g \circ f)(x) = g[f(x)]$$
$$= g[3x - 5]$$
$$= \frac{[3x - 5] + 5}{3}$$
$$= \frac{3x}{3}$$
$$= x$$

5. $(f \circ g)(x) = f[g(x)]$

$$= f\left[\dfrac{x-1}{4}\right]$$

$$= \dfrac{1}{\left[\dfrac{1-x}{x}\right]+1}$$

$$= \dfrac{1}{\dfrac{1-x+x}{x}}$$

$$= \dfrac{x}{1-x+x}$$

$$= x$$

$(g \circ f)(x) = g[f(x)]$

$$= g\left[\dfrac{1}{x+1}\right]$$

$$= \dfrac{1-\left[\dfrac{1}{x+1}\right]}{\left[\dfrac{1}{x+1}\right]}$$

$$= \dfrac{\dfrac{x+1-1}{x+1}}{\dfrac{1}{x+1}}$$

$$= \dfrac{x}{x+1} \cdot \dfrac{x+1}{1}$$

$$= x$$

7. $(f \circ g)(x) = f[g(x)]$

$$= f[x^3+1]$$

$$= \sqrt[3]{[x^3+1]-1}$$

$$= \sqrt[3]{x^3}$$

$$= x$$

$(g \circ f)(x) = g[f(x)]$

$$= g[\sqrt[3]{x-1}]$$

$$= [\sqrt[3]{x-1}]^3+1$$

$$= x-1+1$$

$$= x$$

9. $\{(1, -3), (2, -2), (5, 1), (-7, 4)\}$

11. $\{(1, 0), (2, 1), (4, 2), (8, 3), (16, 4)\}$

13. $f(x) = 4x+1$

$y = 4x+1$

$x = 4y+1$

$x-1 = 4y$

$\dfrac{x-1}{4} = y$

Thus, $f^{-1}(x) = \dfrac{x-1}{4}$.

15. $F(x) = -6x+1$

$y = -6x+1$

$x = -6y+1$

$x-1 = -6y$

$\dfrac{x-1}{-6} = y$

$\dfrac{1-x}{6} = y$

Thus, $F^{-1}(x) = \dfrac{1-x}{6}$.

17. $j(t) = 2t+1$

$y = 2t+1$

$t = 2y+1$

$t-1 = 2y$

$\dfrac{t-1}{2} = y$

Thus, $j^{-1}(t) = \dfrac{t-1}{2}$.

19. $f(v) = 1-v^3$

$y = 1-v^3$

$v = 1-y^3$

$v-1 = -y^3$

$\sqrt[3]{1-v} = y$

Thus, $f^{-1}(v) = \sqrt[3]{1-v}$.

21. $f(x) = \dfrac{-3x}{x+4}, \quad x \neq -4$

$y = \dfrac{-3x}{x+4}$

$x = \dfrac{-3y}{y+4}$

$xy + 4x = -3y$

$xy + 3y = -4x$

$y(x+3) = -4x$

$y = \dfrac{-4x}{x+3}$

Thus, $f^{-1}(x) = \dfrac{-4x}{x+3}, \quad x \neq -3$.

23. $M(t) = \dfrac{t-5}{t}, \quad t \neq 0$

$y = \dfrac{t-5}{t}$

$t = \dfrac{y-5}{y}$

$ty = y-5$

$ty - y = -5$

$y(t-1) = -5$

$y = \dfrac{-5}{t-1} = \dfrac{5}{1-t}$

Thus, $M^{-1}(t) = \dfrac{5}{1-t}, \quad t \neq 1$.

25. $r(t) = \dfrac{1}{t^2}$, Domain of r is $\{t | t < 0\}$ and Range of r $\{y | y > 0\}$.

$y = \dfrac{1}{t^2}$

$t = \dfrac{1}{y^2}$

$y^2 = \dfrac{1}{t}$

$y = -\sqrt{\dfrac{1}{t}}, \quad t > 0$

Thus, $r^{-1}(t) = -\sqrt{\dfrac{1}{t}}$, Domain of r^{-1} is $\{t | t > 0\}$ and Range of r^{-1} is $\{y | y < 0\}$.

27. $J(x) = x^2 + 4$, with Domain $\{x | x \geq 0\}$ and Range $\{y | y \geq 4\}$.

$$y = x^2 + 4$$
$$x = y^2 + 4$$
$$x - 4 = y^2$$
$$\sqrt{x-4} = y$$

Thus $J^{-1}(x) = \sqrt{x-4}$, with Domain $\{x | x \geq 4\}$ and Range $\{y | y \geq 0\}$.

29. $f(x) = x^2 + 3$, Domain of f is $\{x | x \geq 0\}$ and Range of f is $\{y | y \geq 3\}$.

$$y = x^2 + 3$$
$$x = y^2 + 3$$
$$x - 3 = y^2$$
$$\sqrt{x-3} = y, \ x \geq 3$$

Thus $f^{-1}(x) = \sqrt{x-3}$, Domain of f^{-1} is $\{x | x \geq 3\}$ and Range of f^{-1} is $\{y | y \geq 0\}$.

31. $f(x) = \sqrt{x}$, Domain of f is $\{x | x \geq 0\}$ and Range of f is $\{y | y \geq 0\}$.

$$y = \sqrt{x}$$
$$x = \sqrt{y}$$
$$x^2 = y$$

Thus $f^{-1}(x) = x^2$, Domain of f^{-1} is $\{x | x \geq 0\}$ and Range of f^{-1} is $\{y | y \geq 0\}$.

33. $f(x) = \sqrt{9 - x^2}$, Domain of f is $\{x | 0 \leq x \leq 3\}$ and Range of f is $\{y | 0 \leq y \leq 3\}$.

$$y = \sqrt{9 - x^2}$$
$$x = \sqrt{9 - y^2}$$
$$x^2 = 9 - y^2$$
$$x^2 - 9 = -y^2$$
$$\sqrt{9 - x^2} = y$$

Thus $f^{-1}(x) = \sqrt{9 - x^2}$, Domain of f^{-1} is $\{x | 0 \leq x \leq 3\}$ and Range of f^{-1} is $\{y | 0 \leq y \leq 3\}$.

35. $f(x) = x^2 - 4x + 1$, Domain of f is $\{x | x \geq 2\}$ and Range of f is $\{y | y \geq -3\}$.

$$y = x^2 - 4x + 1$$
$$x = y^2 - 4y + 1$$
$$x - 1 = y^2 - 4y$$
$$x + 3 = y^2 - 4y + 4 \quad \text{Complete the square.}$$
$$x + 3 = (y - 2)^2$$
$$\sqrt{x+3} = y - 2$$
$$2 + \sqrt{x+3} = y$$

Thus $f^{-1}(x) = 2 + \sqrt{x+3}$, Domain of f^{-1} is $\{x | x \geq -3\}$ and Range of f^{-1} is $\{y | y \geq 2\}$.

62   **Chapter 1/Functions and Graphs**

37. $f(x) = x^2 + 8x - 9, \ x \le -4,$ Domain of f is $\{x | x \le -4\},$ and Range of f is $\{y | y \ge -25\}.$

Thus, Domain of f^{-1} is $\{x | x \ge -25\}$ and Range of f^{-1} is $\{y | y \le -4\}.$

$$y = x^2 + 8x - 9$$
$$x = y^2 + 8y - 9$$
$$x + 9 = y^2 + 8y$$
$$x + 25 = y^2 + 8y + 16 \quad \text{Complete the square.}$$
$$x + 25 = (y + 4)^2$$
$$-\sqrt{x + 25} = y + 4 \quad \text{Choose the negative root, since the Range of } f^{-1} \text{ is } \{y | y \le -4\}.$$
$$-4 + \sqrt{x + 25} = y$$

Thus $f^{-1}(x) = -4 - \sqrt{x + 25},$ for $x \ge -25.$

39. 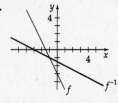   **41.** 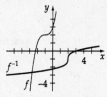   **43.**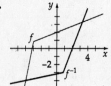

45. a.   The slope of the line containing (4, 36) and (9, 41) is $m = \dfrac{41 - 36}{9 - 4} = \dfrac{5}{5} = 1.$ Using the point-slope form of the line yields

$$y - 36 = 1(x - 4)$$
$$y - 36 = x - 4$$
$$y = x + 32$$

Thus $s(x) = x + 32.$

b. $x = y + 32$
$$x - 32 = y$$
$$s^{-1}(x) = x - 32$$

This means a woman's U.S. shoe size is her Italian shoe size less 32.

47. 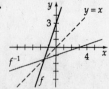   **49.** 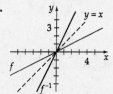   **51.**

53.   **55.** $f(x) = ax + b, \ a \ne 0$
$$y = ax + b$$
$$x = ay + b$$
$$x - b = ay$$
$$\frac{x - b}{a} = y$$

Thus $f^{-1}(x) = \frac{x-b}{a}, \ a \ne 0$

57. $f(x) = \dfrac{x-1}{x+1}, \ \{x | x \ne -1\}$
$$y = \frac{x-1}{x+1}$$
$$x = \frac{y-1}{y+1}$$
$$xy + x = y - 1$$
$$xy - y = -x - 1$$
$$y(x - 1) = -x - 1$$
$$y = \frac{-x-1}{x-1}$$
$$y = \frac{x+1}{1-x}$$

Thus $f^{-1}(x) = \frac{x+1}{1-x}, \ x \ne 1.$

59. No, $f(x)$ is not one-to-one, since $f(x) = f(-x)$ for all $x \neq 0$.

61. Yes, $p(t)$ is a one-to-one function.

63. Yes, $G(x)$ is a one-to-one function.

65. No, $F(x)$ is not one-to-one, since $F(x) = 0$ for all $x \leq 0$.

67. $f^{-1}(2) = 5$

69. $s^{-1}(60) = 4$

71. Because the graph of f is symmetric with respect to the line $y=x$, f is its own inverse.

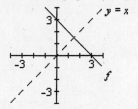

SECTION 1.7, Page 82

1. The scatter diagram suggests no relationship between x and y.

3. The scatter diagram suggests a linear relationship between x and y.

5. Figure A more approximates a graph that can be modeled by an equation than does Figure B. Thus Figure A has a coefficient of determination closer to 1.

7. Enter the data on your calculator. The technique for a TI-83 calculator is illustrated here.

```
EDIT CALC TESTS
1:Edit
2:SortA(
3:SortD(
4:ClrList
5:SetUpEditor
```

L1	L2	L3 2
2	6	------
3	6	
4	8	
6	11	
8	18	
L2(6)=		

```
EDIT CALC TEST
1:1-Var Stats
2:2-Var Stats
3:Med-Med
4:LinReg (ax+b)
5:QuadReg
6:CubicReg
7↓QuartReg
```

```
LinReg
y=ax+b
a=2.00862069
b=.5603448276
r²=.9285885331
r=.9636329867
```

$y = 2.00862069x + 0.5603448276$

9. Enter the data on your calculator. The technique for a TI-83 calculator is illustrated here.

```
EDIT CALC TESTS
1:Edit
2:SortA(
3:SortD(
4:ClrList
5:SetUpEditor
```

L1	L2	L3 2
-3	11.8	------
-1	9.5	
0	8.6	
2	8.7	
5	5.4	
L2(6)=		

```
EDIT CALC TEST
1:1-Var Stats
2:2-Var Stats
3:Med-Med
4:LinReg (ax+b)
5:QuadReg
6:CubicReg
7↓QuartReg
```

```
LinReg
y=ax+b
a=-.7231182796
b=9.233870968
r²=.9218901289
r=-.9601510969
```

$y = -0.7231182796x + 9.233870968$

11. Enter the data on your calculator. The technique for a TI-83 calculator is illustrated here.

```
EDIT CALC TESTS
1:Edit
2:SortA(
3:SortD(
4:ClrList
5:SetUpEditor
```

L1	L2	L3 2
1.3	-4.1	------
2.6	-.9	
5.4	1.2	
6.2	7.6	
7.5	10.5	
L2(6)=		

```
EDIT CALC TEST
1:1-Var Stats
2:2-Var Stats
3:Med-Med
4:LinReg (ax+b)
5:QuadReg
6:CubicReg
7↓QuartReg
```

```
LinReg
y=ax+b
a=2.222641509
b=-7.364150943
r²=.8956132837
r=-.9463684714
```

$y = 2.222641509x - 7.364150943$

13. Enter the data on your calculator. The technique for a TI-83 calculator is illustrated here.

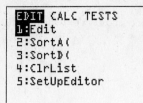

```
EDIT CALC TESTS
1:Edit
2:SortA(
3:SortD(
4:ClrList
5:SetUpEditor
```

```
 L1    L2    L3   2
 1     -1    ------
 2      1
 4      8
 5     14
 6     25
       -----

L2(6)=
```

```
EDIT CALC TEST
1:1-Var Stats
2:2=Var Stats
3:Med-Med
4:LinReg (ax+b)
5:QuadReg
6:CubicReg
7↓QuartReg
```

```
QuadReg
y=ax²+bx+c
a=1.095779221
b=-2.696428571
c=1.136363636
R²=.9943480823
```

$y = 1.095779221x^2 - 2.69642857x + 1.136363636$

15. Enter the data on your calculator. The technique for a TI-83 calculator is illustrated here.

```
EDIT CALC TESTS
1:Edit
2:SortA(
3:SortD(
4:ClrList
5:SetUpEditor
```

```
 L1    L2     L3   2
 1.5   -2.2   ------
 2.2   -4.8
 3.4   -11.2
 5.1   -20.6
 6.3   -28.7
       -----

L2(6)=
```

```
EDIT CALC TEST
1:1-Var Stats
2:2=Var Stats
3:Med-Med
4:LinReg (ax+b)
5:QuadReg
6:CubicReg
7↓QuartReg
```

```
QuadReg
y=ax²+bx+c
a=-.2987274717
b=-3.20998141
c=3.416463667
R²=.9996315893
```

$y = -0.2987274717x^2 - 3.20998141x + 3.416463667$

17. Enter the data on your calculator. The technique for a TI-83 calculator is illustrated here.

```
EDIT CALC TESTS
1:Edit
2:SortA(
3:SortD(
4:ClrList
5:SetUpEditor
```

```
 L1    L2    L3   3
 24    600   -----
 32    750
 22    430
 17    370
 13    270
 4.4   68
 3.2   53
L3(1)=
```

```
EDIT CALC TEST
1:1-Var Stats
2:2=Var Stats
3:Med-Med
4:LinReg (ax+b)
5:QuadReg
6:CubicReg
7↓QuartReg
```

```
LinReg
y=ax+b
a=23.55706665
b=-24.4271215
r²=.9763673085
r=.9881130039
```

a. $y = 23.55706665x - 24.4271215$

b. $y = 23.55706665(54) - 24.4271215 \approx 1247.7$ cm

19. Enter the data on your calculator. The technique for a TI-83 calculator is illustrated here.

```
EDIT CALC TESTS
1:Edit
2:SortA(
3:SortD(
4:ClrList
5:SetUpEditor
```

```
 L1    L2    L3   3
 2     14    -----
 23    51
 20    52
 13    45
 17    45
 16    31
 14    47
L3(1)=
```

```
EDIT CALC TEST
1:1-Var Stats
2:2=Var Stats
3:Med-Med
4:LinReg (ax+b)
5:QuadReg
6:CubicReg
7↓QuartReg
```

```
LinReg
y=ax+b
a=1.671510024
b=16.32830605
r²=.6719162729
r=.8197049914
```

a. $y = 1.671510024x + 16.32830605$

b. $y = 1.671510024(18) + 16.32830605 \approx 46.4$ cm

21. Enter the data on your calculator. The technique for a TI-83 calculator is illustrated here.

```
EDIT CALC TESTS
1:Edit
2:SortA(
3:SortD(
4:ClrList
5:SetUpEditor
```

```
 L1    L2    L3   3
 110   17    -----
 120   19
 125   20
 135   21
 140   22
 145   23
 150   24
L3(1)=
```

```
EDIT CALC TEST
1:1-Var Stats
2:2=Var Stats
3:Med-Med
4:LinReg (ax+b)
5:QuadReg
6:CubicReg
7↓QuartReg
```

```
LinReg
y=ax+b
a=.1628623408
b=-.6875682232
r²=.9976804345
r=.9988395439
```

a. $y = 0.1628623408x - 0.6875682232$

b. $y = 0.1628623408(158) - 0.6875682232 \approx 25$

23. Enter the data on your calculator. The technique for a TI-83 calculator is illustrated here.

EDIT CALC TESTS
1:Edit
2:SortA(
3:SortD(
4:ClrList
5:SetUpEditor

L1	L2	L3	3
7.3	9.4	-----	
11.9	2.8		
5.6	5.6		
14.2	4.9		
7.9	7		
8.5	8.6		
7.9	4.3		
L3(1)=			

EDIT **CALC** TEST
1:1-Var Stats
2:2=Var Stats
3:Med-Med
4:LinReg (ax+b)
5:QuadReg
6:CubicReg
7↓QuartReg

LinReg
y=ax+b
a=7.9539822E-4
b=6.501958524
r^2=1.08788E-6
r=.001043015

The value of r is close to 0. Therefore, no, there is not a strong linear relationship between the current and the torque.

25. Enter the data on your calculator. The technique for a TI-83 calculator is illustrated here.

EDIT CALC TESTS
1:Edit
2:SortA(
3:SortD(
4:ClrList
5:SetUpEditor

L1	L2	L3	2
0	79.4	------	
15	65.1		
35	45.7		
65	19.2		
75	12.1		

L2(6)=			

EDIT **CALC** TEST
1:1-Var Stats
2:2=Var Stats
3:Med-Med
4:LinReg (ax+b)
5:QuadReg
6:CubicReg
7↓QuartReg

LinReg
y=ax+b
a=-.9033088235x
b=78.62573529
r^2=.9987054499
r=-.9993525153

The value of r is close to −1, so, yes, there is a strong linear correlation.

   a. $y = -0.9033088235x + 78.62573529$

   b. $y = -0.9033088235(25) + 78.62573529 \approx 56$ years

27. Enter the data on your calculator. The technique for a TI-83 calculator is illustrated here.

EDIT CALC TESTS
1:Edit
2:SortA(
3:SortD(
4:ClrList
5:SetUpEditor

L1	L2	L3	3
18	24	-----	
17	24		
21	29		
17	27		
18	23		
17	24		
18	25		
L3(1)=			

EDIT **CALC** TEST
1:1-Var Stats
2:2=Var Stats
3:Med-Med
4:LinReg (ax+b)
5:QuadReg
6:CubicReg
7↓QuartReg

LinReg
y=ax+b
a=.9565217391
b=7.869565217
r^2=.4661529994
r=.6827539816

The value of r is not very close to 1, so no, there is not a strong linear correlation between city mpg and highway mpg for these cars.

29. Enter the data on your calculator. The technique for a TI-83 calculator is illustrated here.

EDIT CALC TESTS
1:Edit
2:SortA(
3:SortD(
4:ClrList
5:SetUpEditor

L1	L2	L3	3
20	40	-----	
21	47		
22	52		
23	61		
24	64		
25	64		
26	68		
L3(1)=			

EDIT **CALC** TEST
1:1-Var Stats
2:2=Var Stats
3:Med-Med
4:LinReg (ax+b)
5:QuadReg
6:CubicReg
7↓QuartReg

QuadReg
y=ax^2+bx+c
a=-.6328671329
b=33.6160839
c=-379.4405594
R^2=.9757488948

$y = -0.6328671329x^2 + 33.6160839x - 379.4405594$

31. Enter the data on your calculator. The technique for a TI-83 calculator is illustrated here.

EDIT CALC TESTS
1:Edit
2:SortA(
3:SortD(
4:ClrList
5:SetUpEditor

L1	L2	L3	3
25	29	-----	
30	32		
35	33		
40	35		
45	34		
50	33		
55	31		
L3(1)=			

EDIT **CALC** TEST
1:1-Var Stats
2:2=Var Stats
3:Med-Med
4:LinReg (ax+b)
5:QuadReg
6:CubicReg
7↓QuartReg

QuadReg
y=ax^2+bx+c
a=-.0165034965
b= 1.366713287
c=5.685314685
R^2=.9806348021

   a. $y = -0.0165034965x^2 + 1.366713287x + 5.685314685$

   b. $y = -0.0165034965(50)^2 + 1.366713287(50) + 5.685314685 \approx 32.8$ mpg

33. a. Enter the data for each ball on your calculator. The technique for a TI-83 calculator is illustrated here.

5-lb ball

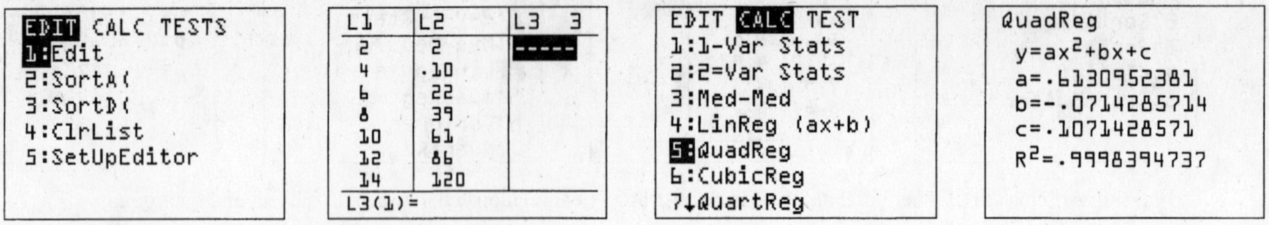

```
EDIT CALC TESTS
1:Edit
2:SortA(
3:SortD(
4:ClrList
5:SetUpEditor
```

```
L1    L2    L3    3
2     2     -----
4    .10
6     22
8     39
10    61
12    86
14   120
L3(1)=
```

```
EDIT CALC TEST
1:1-Var Stats
2:2=Var Stats
3:Med-Med
4:LinReg (ax+b)
5:QuadReg
6:CubicReg
7↓QuartReg
```

```
QuadReg
y=ax²+bx+c
a=.6130952381
b=-.0714285714
c=.1071428571
R²=.9998394737
```

$$y = 0.6130952381t^2 - 0.0714285714t + 0.1071428571$$

10-lb ball

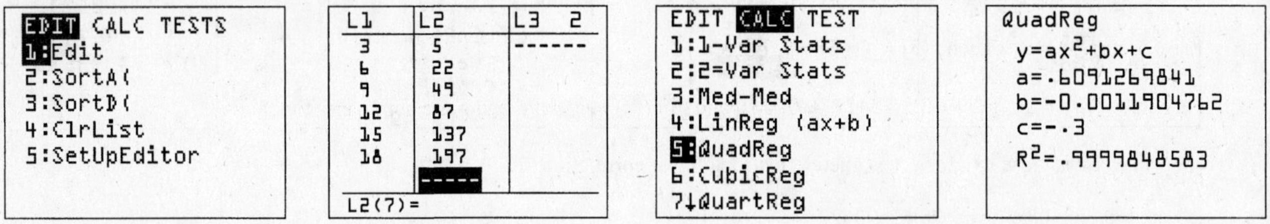

```
EDIT CALC TESTS
1:Edit
2:SortA(
3:SortD(
4:ClrList
5:SetUpEditor
```

```
L1    L2    L3    2
3     5     ------
6     22
9     49
12    87
15   137
18   197
L2(7)=
```

```
EDIT CALC TEST
1:1-Var Stats
2:2=Var Stats
3:Med-Med
4:LinReg (ax+b)
5:QuadReg
6:CubicReg
7↓QuartReg
```

```
QuadReg
y=ax²+bx+c
a=.6091269841
b=-0.0011904762
c=-.3
R²=.9999848583
```

$$y = 0.6091269841t^2 - 0.0011904762t - 0.3$$

15-lb ball

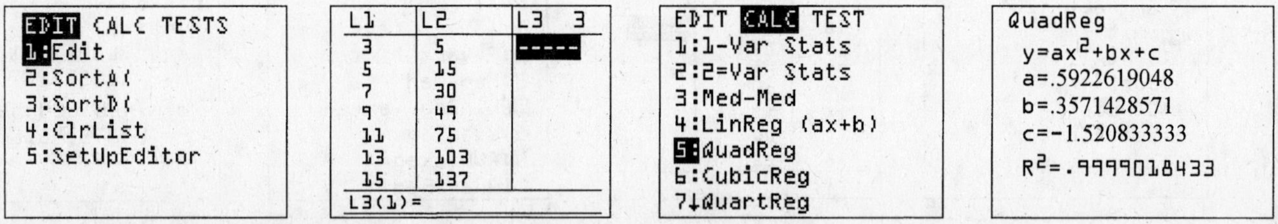

```
EDIT CALC TESTS
1:Edit
2:SortA(
3:SortD(
4:ClrList
5:SetUpEditor
```

```
L1    L2    L3    3
3     5     -----
5     15
7     30
9     49
11    75
13   103
15   137
L3(1)=
```

```
EDIT CALC TEST
1:1-Var Stats
2:2=Var Stats
3:Med-Med
4:LinReg (ax+b)
5:QuadReg
6:CubicReg
7↓QuartReg
```

```
QuadReg
y=ax²+bx+c
a=.5922619048
b=.3571428571
c=-1.520833333
R²=.9999018433
```

$$y = 0.5922619048t^2 + 0.3571428571t - 1.520833333$$

b. All the regression equations are approximately the same. Therefore, there is one equation of motion.

35. Enter the data on your calculator. The technique for a TI-83 calculator is illustrated here.
Compute r for the linear regression model and R^2 for the quadratic model.

```
EDIT CALC TESTS
1:Edit
2:SortA(
3:SortD(
4:ClrList
5:SetUpEditor
```

```
L1    L2    L3    2
20    70    -----
40   135
60   178
80   210
100  260
120  280
140  301
L2(6)=
```

```
EDIT CALC TEST
1:1-Var Stats
2:2=Var Stats
3:Med-Med
4:LinReg (ax+b)
5:QuadReg
6:CubicReg
7↓QuartReg
```

```
LinReg
y=ax+b
a=.6575174825
b=144.2727273
r²=.4193509866
r=.6475731515
```

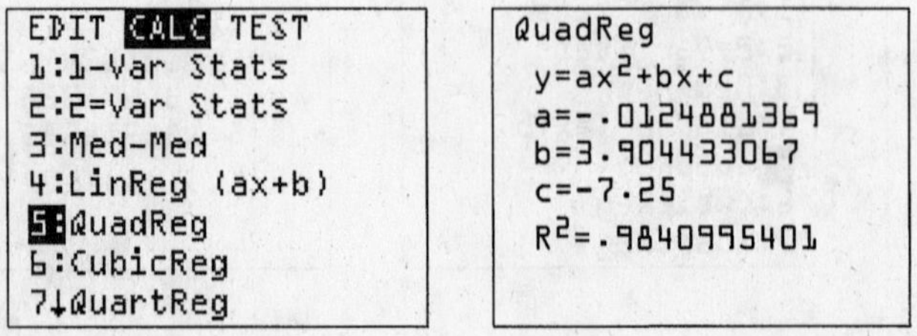

```
EDIT CALC TEST
1:1-Var Stats
2:2=Var Stats
3:Med-Med
4:LinReg (ax+b)
5:QuadReg
6:CubicReg
7↓QuartReg
```

```
QuadReg
y=ax²+bx+c
a=-.0124881369
b=3.904433067
c=-7.25
R²=.9840995401
```

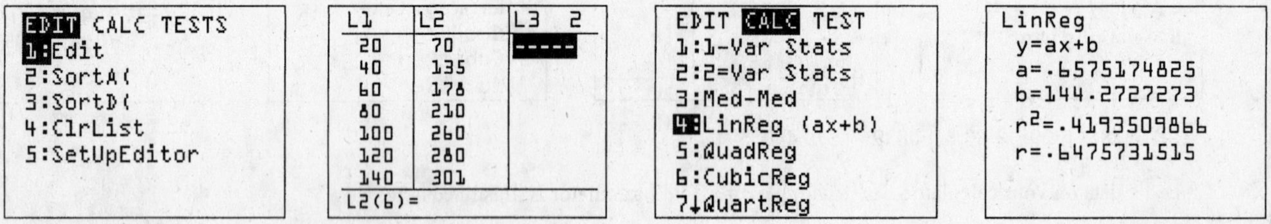

$$r = 0.6475731515$$

$$R^2 = 0.9840995401$$

The quadratic regression model is the better fit for this data. R^2 is closer to 1 for the quadratic model than r is for the linear model.

EXPLORING CONCEPTS WITH TECHNOLOGY, Page 89

1. Use Dot mode. Enter the function as
   $Y_1 = X^2 * (X<2) - X * (X \geq 2)$
   and graph this in the standard viewing window.

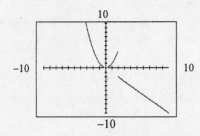

3. Use Dot mode. Enter the function as
   $Y_1 = (-X^2 + 1) * (X<0) + (X^2 - 1) * (X \geq 0)$
   and graph this in the standard viewing window.

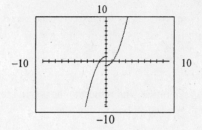

CHAPTER 1 TRUE/FALSE EXERCISES, Page 91

1. False. Let $f(x) = x^2$. Then $f(3) = f(-3) = 9$, but $3 \neq -3$.

3. False. Let $f(x) = 2x$, $g(x) = 3x$. Then $f(g(0)) = 0$ and $g(f(0)) = 0$, but f and g are not inverse functions.

5. False. Let $f(x) = 3x$. $[f(x)]^2 = 9x^2$, whereas $f[f(x)] = f(3x) = 3(3x) = 9x$.

7. True   9. True   11. True   13. True

CHAPTER 1 REVIEW EXERCISES, Page 91

1. $3 - 4z = 12$
   $-4z = 9$
   $z = -\dfrac{9}{4}$

3. $2x - 3(2 - 3x) = 14x$
   $2x - 6 + 9x = 14x$
   $-6 = 3x$
   $-2 = x$

5. $y^2 - 3y - 18 = 0$
   $(y - 6)(y + 3) = 0$
   $y - 6 = 0$ or $y + 3 = 0$
   $y = 6$ or $y = -3$

7. $3v^2 + v = 1$
   $3v^2 + v - 1 = 0$
   $v = \dfrac{-1 \pm \sqrt{1^2 - 4(3)(-1)}}{2(3)}$
   $v = \dfrac{-1 \pm \sqrt{13}}{6}$

9. $3c - 5 \leq 5c + 7$
   $-2c \leq 12$
   $c \geq -6$

11. $x^2 - x - 12 \geq 0$
    $(x - 4)(x + 3) \geq 0$
    Critical values are 4 and -3.
    $(-\infty, -3] \cup [4, \infty)$

13. $|2x - 5| > 3$
    $2x - 5 > 3$ or $2x - 5 < -3$
    $2x > 8$ $2x < 2$
    $x > 4$ $x < 1$
    $(-\infty, 1) \cup (4, \infty)$

15. $c^2 = a^2 + b^2$
    $c^2 = 6^2 + 8^2$
    $c = \sqrt{36 + 64}$
    $c = \sqrt{100} = 10$

17. $c^2 = a^2 + b^2$
    $13^2 = a^2 + 12^2$
    $a = \sqrt{169 - 144}$
    $a = \sqrt{25}$
    $a = 5$

19. $d = \sqrt{(7 - (-3))^2 + (11 - 2)^2}$
    $= \sqrt{100 + 81}$
    $= \sqrt{181}$

21. $\left(\dfrac{2 + (-3)}{2}, \dfrac{8 + 12}{2} \right) = \left(-\dfrac{1}{2}, 10 \right)$

23.

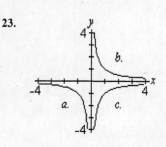

25. $y = x^2 - 7$

Replace y with $-y$.

$$-y = x^2 - 7$$
$$y = -x^2 + 7$$

Thus, y is not symmetric with respect to the x-axis.

Replace x with $-x$.

$$y = (-x)^2 - 7$$
$$y = x^2 - 7$$

Thus, y is symmetric with respect to the y-axis.

Replace x with $-x$ and replace y with $-y$.

$$-y = (-x)^2 - 7$$
$$y = -x^2 + 7$$

Thus, y is not symmetric with respect to the origin.

Therefore, the graph of $y = x^2 - 7$ is symmetric with respect to the y-axis.

27. $y = x^3 - 4x$

Replace y with $-y$.

$$-y = x^3 - 4x$$
$$y = -x^3 + 4x$$

Thus, y is not symmetric with respect to the x-axis.

Replace x with $-x$.

$$y = (-x)^3 - 4(-x)$$
$$y = -x^3 + 4x$$

Thus, y is not symmetric with respect to the y-axis.

Replace x with $-x$ and replace y with $-y$.

$$-y = (-x)^3 - 4(-x)$$
$$-y = -x^3 + 4x$$
$$y = x^3 - 4x$$

Thus, y is symmetric with respect to the origin.

Therefore, the graph of $y = x^3 - 4x$ is symmetric with respect to the origin.

29. $\dfrac{x^2}{3^2} + \dfrac{y^2}{4^2} = 1$

Replace y with $-y$.

$$\frac{x^2}{3^2} + \frac{(-y)^2}{4^2} = 1 \Rightarrow \frac{x^2}{3^2} + \frac{y^2}{4^2} = 1$$

Thus, y is symmetric with respect to the x-axis.

Replace x with $-x$.

$$\frac{(-x)^2}{3^2} + \frac{y^2}{4^2} = 1 \Rightarrow \frac{x^2}{3^2} + \frac{y^2}{4^2} = 1$$

Thus, y is symmetric with respect to the y-axis.

Replace x with $-x$ and replace y with $-y$.

$$\frac{(-x)^2}{3^2} + \frac{(-y)^2}{4^2} = 1 \Rightarrow \frac{x^2}{3^2} + \frac{y^2}{4^2} = 1$$

Thus, y is symmetric with respect to the origin.

Therefore, the graph of $\dfrac{x^2}{3^2} + \dfrac{y^2}{4^2} = 1$ is symmetric with respect to the x-axis, the y-axis, and the origin.

31. $|y| = |x|$

Replace y with $-y$.

$$|-y| = |x| \Rightarrow |y| = |x|$$

Thus, y is symmetric with respect to the x-axis.

Replace x with $-x$.

$$|y| = |-x| \Rightarrow |y| = |x|$$

Thus, y is symmetric with respect to the y-axis.

Replace x with $-x$ and replace y with $-y$.

$$|-y| = |-x| \Rightarrow |y| = |x|$$

Thus, y is symmetric with respect to the origin.

Therefore, the graph of $|y| = |x|$ is symmetric with respect to the x-axis, the y-axis, and the origin.

33. center $(3, -4)$, radius 9

35. $(x - 2)^2 + (y + 3)^2 = 5^2$

37. a. $f(1) = 3(1)^2 + 4(1) - 5$
$$= 2$$

    b. $f(-3) = 27 - 12 - 5$
$$= 10$$

    c. $f(t) = 3t^2 + 4t - 5$

    d. $f(x+h) = 3(x+h)^2 + 4(x+h) - 5$
$$= 3x^2 + 6xh + 3h^2 + 4x + 4h - 5$$

    e. $3f(t) = 9t^2 + 12t - 15$

    f. $f(3t) = 3(3t)^2 + 4(3t) - 5$
$$= 27t^2 + 12t - 5$$

39. a. $(f \circ g)(3) = f[g(3)]$
$$= f[3 - 8]$$
$$= f[-5]$$
$$= (-5)^2 + 4(-5)$$
$$= 5$$

    b. $(g \circ f)(-3) = g[f(-3)]$
$$= g[(-3)^2 + 4(-3)]$$
$$= g[-3]$$
$$= [-3 - 8]$$
$$= -11$$

    c. $(f \circ g)(x) = f[g(x)]$
$$= f[x - 8]$$
$$= (x - 8)^2 + 4(x - 8)$$
$$= x^2 - 16x + 64 + 4x - 32$$
$$= x^2 - 12x + 32$$

    d. $(g \circ f)(x) = g[f(x)]$
$$= g[x^2 + 4x]$$
$$= [x^2 + 4x] - 8$$
$$= x^2 + 4x - 8$$

41. $\dfrac{f(x+h) - f(x)}{h} = \dfrac{4(x+h)^2 - 3(x+h) - 1 - (4x^2 - 3x - 1)}{h}$

$$= \frac{4x^2 + 8xh + 4h^2 - 3x - 3h - 1 - 4x^2 + 3x + 1}{h}$$

$$= \frac{8xh + 4h^2 - 3h}{h}$$

$$= 8x + 4h - 3$$

43. $f(x) = -2x^2 + 3$

Domain: All real numbers

45. $f(x) = \sqrt{25 - x^2}$
$$25 - x^2 \geq 0$$
$$(5 - x)(5 + x) \geq 0$$

Critical values -5 and 5.

Domain: $\{x | -5 \leq x \leq 5\}$

47.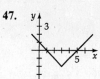

f is increasing on $[3, \infty)$

f is decreasing on $(-\infty, 3]$

49.

(graph)

f is increasing on $[-2, 2]$

f is constant on $(-\infty, -2] \cup [2, \infty)$

51.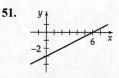

f is increasing on $(-\infty, \infty)$

53.

  a. Domain $\{x|x \text{ is a real number}\}$

  Range $\{y|y \le 4\}$

  b. g is an even function

55.

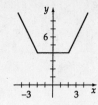

  a. Domain all real numbers

  Range $\{y|y \ge 4\}$

  b. g is an even function

57.

  a. Domain $\{x|x \text{ is a real number}\}$

  Range $\{y|y \text{ is a real number}\}$

  b. g is an odd function

59. $(f+g)(x) = x^2 - 9 + x + 3$

$$= x^2 + x - 6$$

Domain of $(f+g)(x)$ is $\{x|x \text{ is a real number}\}$.

$(f-g)(x) = x^2 - 9 - (x+3)$

$$= x^2 - x - 12$$

Domain of $(f-g)(x)$ is $\{x|x \text{ is a real number}\}$.

$(fg)(x) = (x^2 - 9)(x+3)$

$$= x^3 + 3x^2 - 9x - 27$$

Domain of $(fg)(x)$ is $\{x|x \text{ is a real number}\}$.

$$\left(\frac{f}{g}\right)(x) = \frac{x^2 - 9}{x+3}$$

$$= x - 3$$

Domain of $\left(\dfrac{f}{g}\right)(x)$ is $\{x|x \ne -3\}$.

61. $F[G(x)] = 2\left(\dfrac{x+5}{2}\right) - 5$ and $G[F(x)] = \dfrac{(2x-5)+5}{2}$

$= x + 5 - 5$ $= \dfrac{2x}{2}$

$= x$ $= x$

Because $F[G(x)] = x$ and $G[F(x)] = x$ for all real numbers x, F and G are inverses.

63. $l[m(x)] = \dfrac{\dfrac{3}{x-1} + 3}{\dfrac{3}{x-1}}$ $m[l(x)] = m\left[\dfrac{x+3}{x}\right]$

$= \dfrac{3+3x-3}{x-1} , \dfrac{x-1}{3}$ $= \dfrac{3}{\dfrac{x+3}{x} - 1}$

$= \dfrac{3x}{3}$ $= \dfrac{3}{\dfrac{x+3-x}{x}}$

$= x$ $= 3 \cdot \dfrac{x}{3} = x$

Thus, l and m are inverse functions.

65. $f(x) = 3x - 4$

$y = 3x - 4$

$x = 3y - 4$

$x + 4 = 3y$

$\dfrac{x+4}{3} = y$

Thus $f^{-1}(x) = \dfrac{x+4}{3} = \dfrac{1}{3}x + \dfrac{4}{3}$.

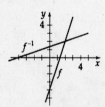

67. $h(x) = -\dfrac{1}{2}x - 2$

$$y = -\dfrac{1}{2}x - 2$$

$$x = -\dfrac{1}{2}y - 2$$

$$x + 2 = -\dfrac{1}{2}y$$

$$-2x - 4 = y$$

Thus, $h^{-1}(x) = -2x - 4$.

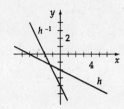

69.   Let $x =$ one of the numbers and $50 - x =$ the other number. Their product y is given by

$$y = x(50 - x) = 50x - x^2 = x^2 + 50x$$

Now y takes on its maximum value when

$$x = \dfrac{-b}{2a} = \dfrac{-50}{2(-1)} = 25$$

Thus, the two numbers are 25 and $(50 - 25) = 25$. That is, both numbers are 25.

71. $h(t) = -16t^2 + 220$ and $h(t) = 0$ when

$$-16t^2 + 220 = 0$$

$$220 = 16t^2$$

$$\dfrac{220}{16} = t^2$$

$$\sqrt{\dfrac{220}{16}} = t$$

$$\dfrac{2\sqrt{55}}{4} = t$$

Thus, $t \approx 3.7$ seconds.

73.   a.   Enter the data on your calculator. The technique for a TI-83 is illustrated here.

L1	L2	L3 3
10.5	.2	▮▮▮▮▮
12.9	.24	
15	.27	
20	.36	
60	1.09	
75	1.42	
110	2.01	
L3(1)=		

```
EDIT CALC TESTS
1:1-Var Stats
2:2-Var Stats
3:Med-Med
4:LinReg(ax+b)
5:QuadReg
6:CubicReg
7↓QuartReg
```

```
LinReg
y=ax+b
a=.018024687
b=5.0045744E-4
r²=.9982908274
r=.9991450482
```

$y = 0.018024687x + 0.00050045744$

b.   Yes. $r \approx 0.999$, which is very close to 1.

c. $y = 0.018024687(100) + 0.00050045744$

$= 1.8024687 + 0.00050045744$

≈ 1.8 seconds

CHAPTER 1 TEST, Page 93

1. $4x - 2(2 - x) = 5 - 3(2x + 1)$

$4x - 4 + 2x = 5 - 6x - 3$

$6x - 4 = 2 - 6x$

$12x = 6$

$x = \dfrac{1}{2}$

2. $6 - 3x \geq 3 - 4(2 - 2x)$

$6 - 3x \geq 3 - 8 + 8x$

$-11x \geq -11$

$x \leq 1$

3. $2x^2 - 3x = 2$

$2x^2 - 3x - 2 = 0$

$(2x + 1)(x - 2) = 0$

$x = -\dfrac{1}{2}$ or $x = 2$

4. $3x^2 - x = 2$

$3x^2 - x - 2 = 0$

$(3x + 2)(x - 1) = 0$

$x = -\frac{2}{3}$ or $x = 1$

5. $|4 - 5x| > 6$

$4 - 5x < -6$ or $4 - 5x > 6$

$-5x < -10$ $-5x > 2$

$x > 2$ $x < -\frac{2}{5}$

$\left(-\infty, -\frac{2}{5}\right) \cup (2, \infty)$

6. $d = \sqrt{(x_2 - x_1)^2 + (y_2 - y_1)^2}$

$d = \sqrt{[4 - (-2)]^2 + (-2 - 5)^2}$

$d = \sqrt{36 + 49}$

$d = \sqrt{85}$

7. $x_m = \dfrac{4 - 2}{2} = 1$ length $= \sqrt{(x_2 - x_1)^2 + (y_2 - y_1)^2}$

$y_m = \dfrac{-1 + 3}{2} = 1$ $= \sqrt{[4 - (-2)]^2 + (-1 - 3)^2}$

midpoint $= (1, 1)$ $= \sqrt{6^2 + (-4)^2}$

$= \sqrt{36 + 16} = \sqrt{52}$

$= 2\sqrt{13}$

8. $x = 2y^2 - 4$

$x = 0$ $2y^2 - 4 = 0$

$2y^2 = 4$

$y^2 = 2$

$y = \pm\sqrt{2}$

If $y = 0$, $x = -4$

intercepts $(0, -\sqrt{2})$, $(0, \sqrt{2})$, $(-4, 0)$

9.

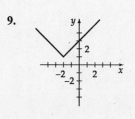

10. $x^2 - 4x + y^2 + 2y - 4 = 0$

$(x^2 - 4x + 4) + (y^2 + 2y + 1) = 4 + 4 + 1$

$(x - 2)^2 + (y + 1)^2 = 9$

center: $(2, -1)$

radius: 3

11. $f(x) = -\sqrt{25 - x^2}$

$f(-3) = -\sqrt{25 - (-3)^2}$

$f(-3) = -\sqrt{16}$

$f(-3) = -4$

12. $x^2 - 16 \geq 0$

$(x - 4)(x + 4) \geq 0$

Critical values 4 and -4.

Domain $\{x \mid x \geq 4 \text{ or } x \leq -4\}$

13.

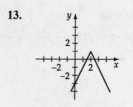

f is increasing on $(-\infty, 2]$

f is decreasing on $[2, \infty)$

14. First shift the graph of $f(x)$ horizontally 2 units to the left. Next, reflect the graph across the x-axis. Finally, shift the graph vertically down 1 unit.

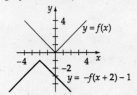

15. a. $f(-x) = (-x)^4 - (-x)^2$

$$= x^4 - x^2$$

$$= f(x)$$

The function of $f(x) = x^4 - x^2$ is an even function.

b. $f(-x) = (-x)^3 - (-x)$

$$= -x^3 + x$$

$$= -(x^3 - x)$$

$$= -f(x)$$

The function $f(x) = x^3 - x$ is an odd function.

c. $f(-x) = -x - 1$

The function $f(x) = x - 1$ is neither an even nor an odd function.

Thus, only **b** defines an odd function.

16. $(f + g)(x) = (x^2 - 1) + (x - 2)$

$$= x^2 + x - 3$$

$$\left(\frac{f}{g}\right)(x) = \frac{x^2 - 1}{x - 2}, \quad x \neq 2$$

17. $\dfrac{f(x+h) - f(x)}{h} = \dfrac{\left[(x+h)^2 + 1\right] - (x^2 + 1)}{h}$

$$= \frac{x^2 + 2xh + h^2 + 1 - x^2 - 1}{h}$$

$$= \frac{2xh + h^2}{h}$$

$$= 2x + h$$

18. $(f \circ g)(x) = (\sqrt{x-2})^2 - 2\sqrt{x-2} + 1$

$$= x - 2 - 2\sqrt{x-2} + 1$$

$$= x - 2\sqrt{x-2} - 1$$

19. $y = \dfrac{x}{x+1}$

Interchange x and y. Then solve for y.

$$x = \frac{y}{y+1}$$

$$x(y+1) = y$$

$$xy + x = y$$

$$xy - y = -x$$

$$y(x-1) = -x$$

$$y = \frac{-x}{x-1} = \frac{x}{1-x}$$

$$f^{-1}(x) = \frac{x}{1-x}$$

20. a. Enter the data on your calculator. The technique for a TI-83 is illustrated here.

L1	L2	L3	3
93.2	28		
92.3	26		
91.9	39		
89.5	56		
89.6	56		
90.5	36		
91.9	32		
L3(1)=			

```
EDIT CALC TESTS
1:1-Var Stats
2:2-Var Stats
3:Med-Med
4:LinReg(ax+b)
5:QuadReg
6:CubicReg
7↓QuartReg
```

```
LinReg
y=ax+b
a=-7.98245614
b=767.122807
r²=.805969575
r=-.8977580826
```

$$y = -7.98245614x + 767.122807$$

b. $y = -7.98245614(89) + 767.122807$

$$\approx 57 \text{ Calories}$$

SECTION 2.1, page 107

1. $90^0 - 15^0 = 75^0$

$180^0 - 15^0 = 165^0$

3.

$$\begin{array}{cc} 90^\circ & 89^\circ 60' \\ \underline{-70^\circ 15'} & = \underline{-70^\circ 15'} \\ & 19^\circ 45' \end{array} \qquad \begin{array}{cc} 180^\circ & 179^\circ 60' \\ \underline{-70^\circ 15'} & = \underline{-70^\circ 15'} \\ & 109^\circ 45' \end{array}$$

5.

$$\begin{array}{cc} 90^\circ & 89^\circ 59' 60'' \\ \underline{-56^\circ 33' 15''} & = \underline{-56^\circ 33' 15''} \\ & 33^\circ 26' 45'' \end{array} \qquad \begin{array}{cc} 180^\circ & 179^\circ 59' 60'' \\ \underline{-56^\circ 33' 15''} & = \underline{-56^\circ 33' 15''} \\ & 123^\circ 26' 45'' \end{array}$$

7. $\dfrac{\pi}{2} - 1$

$\pi - 1$

9. $\dfrac{\pi}{2} - \dfrac{\pi}{4} = \dfrac{\pi}{4}$

$\pi - \dfrac{\pi}{4} = \dfrac{3\pi}{4}$

11. $\dfrac{\pi}{2} - \dfrac{\pi}{2} = 0$

$\pi - \dfrac{\pi}{2} = \dfrac{\pi}{2}$

13. $610^\circ = 250^\circ + 360^\circ$

α is a quadrant III angle coterminal with an angle of

measure 250°.

15. $-975^\circ = 105^\circ - 3 \cdot 360^\circ$

α is a quadrant II angle coterminal with an angle of

measure 105°.

17. $2456^\circ = 296^\circ + 6 \cdot 360^\circ$

α is a quadrant IV angle coterminal with an angle of

measure 296°.

19. On a TI-83 graphing calculator, the degree symbol, $^\circ$, and the DMS function are located in the ANGLE menu.

$24.56^\circ = 24^\circ 33' 36''$

21. On a TI-83 graphing calculator, the degree symbol, $^\circ$, and the DMS function are located in the ANGLE menu.

$64.158^\circ = 64^\circ 9' 28.8''$

23. On a TI-83 graphing calculator, the degree symbol, $^\circ$, and the DMS function are located in the ANGLE menu.

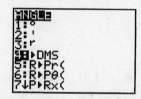 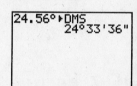

$3.402^\circ = 3^\circ 24' 7.2''$

25. A TI-83 calculator needs to be in degree mode to convert a DMS measure to its equivalent degree measure. On a TI-83 both the degree symbol, $^\circ$, and the minute symbol, ', are located in the ANGLE menu. The second symbol, '', is entered by pressing ALPHA followed by ["] which is located on the plus sign, [+], key.

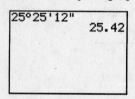

$25^\circ 25' 12'' = 25.42^\circ$

27. A TI-83 calculator needs to be in degree mode to convert a DMS measure to its equivalent degree measure. On a TI-83 both the degree symbol, $^\circ$, and the minute symbol, ', are located in the ANGLE menu. The second symbol, '', is entered by pressing ALPHA followed by ["] which is located on the plus sign, [+], key.

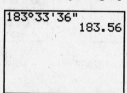

$183^\circ 33' 36'' = 183.56^\circ$

29. A TI-83 calculator needs to be in degree mode to convert a DMS measure to its equivalent degree measure. On a TI-83 both the de-gree symbol, °, and the minute symbol, ', are located in the ANGLE menu. The second sym-bol, ", is entered by pressing ALPHA followed by ["] which is located on the plus sign, [+], key.

```
211°46'48"
        211.78
```

$211°46'48'' = 211.78°$

31. $30° = 30°\left(\dfrac{\pi}{180°}\right) = \dfrac{\pi}{6}$

33. $90° = 90°\left(\dfrac{\pi}{180°}\right) = \dfrac{\pi}{2}$

35. $165° = 165°\left(\dfrac{\pi}{180°}\right) = \dfrac{11\pi}{12}$

37. $420° = 420°\left(\dfrac{\pi}{180°}\right) = \dfrac{7\pi}{3}$

39. $585° = 585°\left(\dfrac{\pi}{180°}\right) = \dfrac{13\pi}{4}$

41. $\dfrac{\pi}{5} = \dfrac{\pi}{5}\left(\dfrac{180°}{\pi}\right) = 36°$

43. $\dfrac{\pi}{6} = \dfrac{\pi}{6}\left(\dfrac{180°}{\pi}\right) = 30°$

45. $\dfrac{3\pi}{8} = \dfrac{3\pi}{8}\left(\dfrac{180°}{\pi}\right) = 67.5°$

47. $\dfrac{11\pi}{3} = \dfrac{11\pi}{3}\left(\dfrac{180°}{\pi}\right) = 660°$

49. $1.5 = 1.5\left(\dfrac{180°}{\pi}\right) \approx 85.94°$

51. $133° = 133°\left(\dfrac{\pi}{180°}\right) \approx 2.32$

53. $8.25 = 8.25\left(\dfrac{180°}{\pi}\right) \approx 472.69°$

55. $\theta = \dfrac{s}{r}$

$= \dfrac{8}{2} = 4$

$= 4\left(\dfrac{180°}{\pi}\right) \approx 229.18°$

57. $\theta = \dfrac{s}{r}$

$= \dfrac{12.4}{5.2} \approx 2.38$

$= 2.38\left(\dfrac{180°}{\pi}\right) \approx 136.63°$

59. $s = r\theta$

$= (8)\dfrac{\pi}{4}$

≈ 6.28 in.

61. $s = r\theta$

$= 25\,(42°)\left(\dfrac{\pi}{180°}\right)$

≈ 18.33 cm

63. $\theta = \dfrac{3}{2}(2\pi)$

$= 3\pi$

65. $\theta_2 = \dfrac{r_1}{r_2}\theta_1$

$= \dfrac{14}{28}\left(150°\right)\left(\dfrac{\pi}{180°}\right)$

$= \dfrac{5\pi}{12}$ radians or $75°$

67. $\omega = \dfrac{\theta}{t}$

$= \dfrac{2\pi}{60}$

$= \dfrac{\pi}{30}$ radians/sec

69. $\omega = \dfrac{\theta}{t}$

$= \dfrac{50(2\pi)}{60}$

$= \dfrac{5\pi}{3}$ radians/sec

71. $\omega = \dfrac{\theta}{t}$

$= \dfrac{2\pi\left(33\frac{1}{3}\right)}{60}$

$= \dfrac{10\pi}{9}$ radians/sec

≈ 3.49 radians per second

73. $v = \omega r$

$= \dfrac{450 \cdot 2\pi \cdot 60 \cdot 15}{12 \cdot 5280}$

≈ 40 mph

75. $s = r\theta$

$$= 93{,}000{,}000(31') \left(\frac{1°}{60'}\right)\left(\frac{\pi}{180°}\right)$$

$$\approx 840{,}000 \text{ mi}$$

77. a. $\omega = \dfrac{\theta}{t}$

$$= \frac{2\pi}{1.61 \text{ hours}}$$

$$\approx 3.9 \text{ radians per hour}$$

b. $v = \dfrac{s}{t}$

$$= \frac{2\pi r}{1.61 \text{ hours}}$$

$$= \frac{2\pi(625 \text{ km} + 6370 \text{ km})}{1.61 \text{ hours}}$$

$$= \frac{2\pi(6995 \text{ km})}{1.61 \text{ hours}}$$

$$\approx 27{,}300 \text{ km per hour}$$

79. a. When the rear tire makes one revolution, the bicycle travels $s = r\theta = 2\pi r = 2\pi(30 \text{ inches}) = 60\pi$ inches.

The angular velocity of point A is

$$\omega = \frac{\theta}{t} = \frac{2\pi}{t} = 2\left(\frac{\pi}{t}\right).$$

When the bicycle travels 60π inches, point B on the front tire travels though an angle of

$$\theta = \frac{s}{r} = \frac{60\pi \text{ inches}}{20 \text{ inches}} = 3\pi.$$

The angular velocity of point B is

$$\omega = \frac{\theta}{t} = \frac{3\pi}{t} = 3\left(\frac{\pi}{t}\right).$$

Thus, point B has the greater angular velocity.

b. Point A and point B travel a linear distance of 60π inches in the same amount of time. Therefore, both points have the same linear velocity.

81. a. $1' = 1' \cdot \left(\dfrac{1°}{60'}\right) \cdot \left(\dfrac{\pi}{180°}\right) = \dfrac{\pi}{10{,}800}$ radians

1 nautical mile $= s = r\theta = (3960 \text{ statute miles})\left(\dfrac{\pi}{10{,}800}\right) \approx 1.15$ statute miles

b. Earth's circumference $= 2\pi r \approx 2\pi(3960 \text{ statute miles})\left(\dfrac{1 \text{ nautical mile}}{1.15 \text{ statute miles}}\right) \approx \dfrac{7920\pi}{1.15}$ nautical miles

The question, then, is what percent of $\dfrac{7920\pi}{1.15}$ is 2217.

$$\frac{2217}{7920\pi/1.15} \approx 0.10 = 10\%$$

83. $A = \frac{1}{2}r^2\theta$

$$= \frac{1}{2}(5^2)\left(\frac{\pi}{3}\right)$$

$$\approx 13 \text{ in}^2$$

85. $A = \frac{1}{2}r^2\theta$

$$= \frac{1}{2}(120)^2 0.65$$

$$= 4680 \text{ cm}^2$$

87. $A = \frac{1}{2}r^2\theta$

$$= \frac{1}{2}(20)^2 125°\left(\frac{\pi}{180°}\right)$$

$$\approx 436 \text{ m}^2$$

89. a. $\quad s = r\theta$
$\qquad = 11.6(14.25)2\pi$
$\qquad \approx 1039 \text{ m}$

b. $\quad v = \omega r$
$\qquad = \dfrac{14.25(2\pi)}{5 \cdot 60} \cdot 11.6$
$\qquad \approx 3.5 \text{ m/sec}$

91. $\quad v = \omega r$
$\qquad v = \dfrac{18 \cdot 2\pi}{60} \cdot 10$
$\qquad \approx 18.8 \text{ ft/sec}$

93. $\quad A = \dfrac{1}{2}r^2\theta - \dfrac{1}{2}bh$
$\qquad A = \dfrac{1}{2} \cdot 9^2 \cdot \dfrac{\pi}{2} - \dfrac{1}{2}(9)(9)$
$\qquad \approx 63.6 - 40.5$
$\qquad = 23.1 \text{ in}^2$

95. $\quad 25°47' = 25° + 47'\left(\dfrac{1°}{60'}\right)$

$\qquad\qquad = 25\dfrac{47}{60}°$

Convert to radians.

$25\dfrac{47}{60}° \cdot \dfrac{\pi}{180°} = \dfrac{1547\pi}{10,800}$ radians

$\qquad s = r\theta$

$s = 3960\left(\dfrac{1547\pi}{10,800}\right)$

$\qquad \approx 1780$

To the nearest 10 miles, Miami is 1780 miles north of the equator.

SECTION 2.2, Page 118

1. $\quad r = \sqrt{5^2 + 12^2}$
$\qquad r = \sqrt{25 + 144} + \sqrt{169}$
$\qquad r = 13$

$\sin\theta = \dfrac{y}{r} = \dfrac{12}{13}$ $\qquad$ $\csc\theta = \dfrac{r}{y} = \dfrac{13}{12}$

$\cos\theta = \dfrac{x}{r} = \dfrac{5}{13}$ $\qquad$ $\sec\theta = \dfrac{r}{x} = \dfrac{13}{5}$

$\tan\theta = \dfrac{y}{x} = \dfrac{12}{5}$ $\qquad$ $\cot\theta = \dfrac{x}{y} = \dfrac{5}{12}$

3. $\quad x = \sqrt{7^2 - 4^2}$
$\qquad x = \sqrt{49 - 16} = \sqrt{33}$

$\sin\theta = \dfrac{y}{r} = \dfrac{4}{7}$ $\qquad$ $\csc\theta = \dfrac{r}{y} = \dfrac{7}{4}$

$\cos\theta = \dfrac{x}{r} = \dfrac{\sqrt{33}}{7}$ $\qquad$ $\sec\theta = \dfrac{r}{x} = \dfrac{7}{\sqrt{33}} = \dfrac{7\sqrt{33}}{33}$

$\tan\theta = \dfrac{y}{x} = \dfrac{4}{\sqrt{33}} = \dfrac{4\sqrt{33}}{33}$ $\qquad$ $\cot\theta = \dfrac{x}{y} = \dfrac{\sqrt{33}}{4}$

5. $\quad r = \sqrt{2^2 + 5^2}$
$\qquad r = \sqrt{4 + 25} = \sqrt{29}$

$\sin\theta = \dfrac{y}{r} = \dfrac{5}{\sqrt{29}} = \dfrac{5\sqrt{29}}{29}$ $\qquad$ $\csc\theta = \dfrac{r}{y} = \dfrac{\sqrt{29}}{5}$

$\cos\theta = \dfrac{x}{r} = \dfrac{2}{\sqrt{29}} = \dfrac{2\sqrt{29}}{29}$ $\qquad$ $\sec\theta = \dfrac{r}{x} = \dfrac{\sqrt{29}}{2}$

$\tan\theta = \dfrac{y}{x} = \dfrac{5}{2}$ $\qquad$ $\cot\theta = \dfrac{x}{y} = \dfrac{2}{5}$

7. $\quad x = \sqrt{2 + \left(\sqrt{3}\right)^2}$
$\qquad x = \sqrt{4 + 3} = \sqrt{7}$

$\sin\theta = \dfrac{y}{r} = \dfrac{\sqrt{3}}{\sqrt{7}} = \dfrac{\sqrt{21}}{7}$ $\qquad$ $\csc\theta = \dfrac{r}{y} = \dfrac{\sqrt{7}}{\sqrt{3}} = \dfrac{\sqrt{21}}{3}$

$\cos\theta = \dfrac{x}{r} = \dfrac{2}{\sqrt{7}} = \dfrac{2\sqrt{7}}{7}$ $\qquad$ $\sec\theta = \dfrac{r}{x} = \dfrac{\sqrt{7}}{2}$

$\tan\theta = \dfrac{y}{x} = \dfrac{\sqrt{3}}{2}$ $\qquad$ $\cot\theta = \dfrac{x}{y} = \dfrac{2}{\sqrt{3}} = \dfrac{2\sqrt{3}}{3}$

9. $y = \sqrt{\left(\sqrt{15}\right)^2 - \left(\sqrt{7}\right)^2}$

$y = \sqrt{15 - 7} = \sqrt{8}$

$y = 2\sqrt{2}$

$\sin\theta = \dfrac{y}{r} = \dfrac{2\sqrt{2}}{\sqrt{15}} = \dfrac{2\sqrt{30}}{15}$

$\cos\theta = \dfrac{x}{r} = \dfrac{\sqrt{7}}{\sqrt{15}} = \dfrac{\sqrt{105}}{15}$

$\tan\theta = \dfrac{y}{x} = \dfrac{2\sqrt{2}}{\sqrt{7}} = \dfrac{2\sqrt{14}}{7}$

$\csc\theta = \dfrac{r}{y} = \dfrac{\sqrt{15}}{2\sqrt{2}} = \dfrac{\sqrt{30}}{4}$

$\sec\theta = \dfrac{r}{x} = \dfrac{\sqrt{15}}{\sqrt{7}} = \dfrac{\sqrt{105}}{7}$

$\cot\theta = \dfrac{x}{y} = \dfrac{\sqrt{7}}{2\sqrt{2}} = \dfrac{\sqrt{14}}{4}$

11. opposite side $= \sqrt{6^2 - 3^2}$

$\quad = \sqrt{36 - 9}$

$\quad = \sqrt{27} = 3\sqrt{3}$

$\sin\theta = \dfrac{\text{opp}}{\text{hyp}} = \dfrac{3\sqrt{3}}{6} = \dfrac{\sqrt{3}}{2}$

$\cos\theta = \dfrac{\text{adj}}{\text{hyp}} = \dfrac{3}{6} = \dfrac{1}{2}$

$\tan\theta = \dfrac{\text{opp}}{\text{adj}} = \dfrac{3\sqrt{3}}{3} = \sqrt{3}$

$\csc\theta = \dfrac{\text{hyp}}{\text{opp}} = \dfrac{6}{3\sqrt{3}} = \dfrac{2}{\sqrt{3}} = \dfrac{2\sqrt{3}}{3}$

$\sec\theta = \dfrac{\text{hyp}}{\text{adj}} = \dfrac{6}{3} = 2$

$\cot\theta = \dfrac{\text{adj}}{\text{opp}} = \dfrac{3}{3\sqrt{3}} = \dfrac{1}{\sqrt{3}} = \dfrac{\sqrt{3}}{3}$

13. hypotenuse $= \sqrt{5^2 + 6^2}$

$\quad = \sqrt{25 + 36}$

$\quad = \sqrt{61}$

$\sin\theta = \dfrac{\text{opp}}{\text{hyp}} = \dfrac{6}{\sqrt{61}} = \dfrac{6\sqrt{61}}{61}$

$\cos\theta = \dfrac{\text{adj}}{\text{hyp}} = \dfrac{5}{\sqrt{61}} = \dfrac{5\sqrt{61}}{61}$

$\tan\theta = \dfrac{\text{opp}}{\text{adj}} = \dfrac{6}{5}$

$\csc\theta = \dfrac{\text{hyp}}{\text{opp}} = \dfrac{\sqrt{61}}{6}$

$\sec\theta = \dfrac{\text{hyp}}{\text{adj}} = \dfrac{\sqrt{61}}{5}$

$\cot\theta = \dfrac{\text{adj}}{\text{opp}} = \dfrac{5}{6}$

For exercises 15 to 17, since $\sin\theta = \dfrac{y}{r} = \dfrac{3}{5}$, $y = 3$, $r = 5$, and $x = \sqrt{5^2 - 3^2} = 4$.

15. $\tan\theta = \dfrac{y}{x} = \dfrac{3}{4}$

17. $\cos\theta = \dfrac{x}{r} = \dfrac{4}{5}$

For exercises 18 to 20, since $\tan\theta = \dfrac{y}{x} = \dfrac{4}{3}$, $y = 4$, $x = 3$, and $r = \sqrt{3^2 + 4^2} = 5$.

19. $\cot\theta = \dfrac{x}{y} = \dfrac{3}{4}$

For exercises 21 to 23, since $\sec\beta = \dfrac{r}{x} = \dfrac{13}{12}$, $r = 13$, $x = 12$, and $y = \sqrt{13^2 - 12^2} = \sqrt{25} = 5$.

21. $\cos\beta = \dfrac{x}{r} = \dfrac{12}{13}$

23. $\csc\beta = \dfrac{r}{y} = \dfrac{13}{5}$

For exercises 24 to 26, since $\cos\theta\,\dfrac{x}{r} = \dfrac{2}{3}$, $x = 2$, $r = 3$, and $y = \sqrt{3^2 - 2^2} = \sqrt{9 - 4} = \sqrt{5}$.

25. $\sec\theta = \dfrac{r}{x} = \dfrac{3}{2}$

27. $\sin 45^\circ + \cos 45^\circ = \dfrac{\sqrt{2}}{2} + \dfrac{\sqrt{2}}{2} = \sqrt{2}$

29. $\sin 30^\circ \cos 60^\circ - \tan 45^\circ = \dfrac{1}{2}\cdot\dfrac{1}{2} - 1$

$\quad = \dfrac{1}{4} - 1$

$\quad = -\dfrac{3}{4}$

31. $\sin 30^\circ \cos 60^\circ + \tan 45^\circ = \dfrac{1}{2}\cdot\dfrac{1}{2} + 1 = \dfrac{1}{4} + 1 = \dfrac{5}{4}$

33. $2\sin 60^\circ - \sec 45^\circ \tan 60^\circ = 2\left(\dfrac{\sqrt{3}}{2}\right) - \sqrt{2}\cdot\sqrt{3} = \sqrt{3} - \sqrt{6}$

35. $\sin\dfrac{\pi}{3} + \cos\dfrac{\pi}{6} = \dfrac{\sqrt{3}}{2} + \dfrac{\sqrt{3}}{2} = 2\cdot\dfrac{\sqrt{3}}{2} = \sqrt{3}$

37. $\sin\dfrac{\pi}{4} + \tan\dfrac{\pi}{6} = \dfrac{\sqrt{2}}{2} + \dfrac{\sqrt{3}}{3} = \dfrac{3\sqrt{2} + 2\sqrt{3}}{6}$

39. $\sec\dfrac{\pi}{3}\cos\dfrac{\pi}{3} - \tan\dfrac{\pi}{6} = 2\cdot\dfrac{1}{2} - \dfrac{\sqrt{3}}{3} = 1 - \dfrac{\sqrt{3}}{3} = \dfrac{3 - \sqrt{3}}{3}$

41. $2\csc\dfrac{\pi}{4} - \sec\dfrac{\pi}{3}\cos\dfrac{\pi}{6} = 2\cdot\sqrt{2} - 2\cdot\dfrac{\sqrt{3}}{2} = 2\sqrt{2} - \sqrt{3}$

43. $\tan 32° \approx 0.6249$

45. $\cos 63°20' \approx 0.4488$

47. $\cos 34.7° \approx 0.8221$

49. $\sec 5.9° \approx 1.0053$

51. $\tan \frac{\pi}{7} \approx 0.4816$

53. $\csc 1.2 \approx 1.0729$

55. $\cos 1.25 \approx 0.3153$

57. $\sec \frac{5}{8} \approx 1.2331$

59.

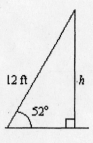

$$\sin 52° = \frac{h}{12}$$
$$h = 12 \sin 52°$$
$$h \approx 9.5 \text{ ft}$$

61.

$$m = \frac{y_2 - y_1}{x_2 - x_1} = \frac{a}{b}$$
$$\tan \theta = \frac{a}{b}$$

Therefore, $\tan \theta = m$.

63.

$$d = rt$$
$$t = \frac{d}{r} \quad \cos 30° = \frac{d}{40}, \quad d = 40 \cos 30°$$
$$t = \frac{40 \cos 30°}{10}$$
$$t \approx 3.46 \text{ h}$$
$$t \approx 3 \text{ hr } 28 \text{ min}$$

Time of closest approach: 1:00 P.M.+3 hr 28 min is 4:28 P.M.

65.

$$\cos 38° = \frac{d + 0.33}{6}$$
$$d = 6 \cos 38° + 0.33$$
$$d \approx 5.1 \text{ ft}$$

67.

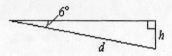

$$d = \frac{240 \text{ mi}}{\text{hr}} \left(\frac{1 \text{ hr}}{60 \text{ min}} \right) 4 \text{ min}$$
$$d = 16 \text{ mi}$$
$$\sin 6° = \frac{h}{d}$$
$$h = 16 \sin 6°$$
$$h \approx 1.7 \text{ mi}$$

69.

$$\tan 78° = \frac{h}{300}$$
$$h = 300 \tan 78°$$
$$h \approx 1411.4 \text{ ft}$$
$$h \approx 1400 \text{ ft} \text{ (to two significant digits)}$$

71.

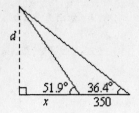

$$\tan 36.4 = \frac{d}{350+x} \qquad \tan 51.9° = \frac{d}{x}$$

$$x = \frac{d}{\tan 51.9°}$$

$$\tan 36.4° = \frac{d}{350 + \dfrac{d}{\tan 51.9°}}$$

$$d = \frac{350 \tan 36.4°}{1 - \dfrac{\tan 36.4°}{\tan 51.9°}}$$

$$d \approx 612 \text{ ft}$$

73.

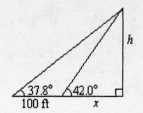

$$\tan 42° = \frac{h}{x} \qquad \tan 37.8° = \frac{h}{100+x}$$

$$x = \frac{h}{\tan 42°} \qquad \tan 37.8° = \frac{h}{100 + \dfrac{h}{\tan 42°}}$$

$$h = \frac{100 \tan 37.8°}{1 - \dfrac{\tan 37.8°}{\tan 42.0°}}$$

$$h \approx 5.60 \times 10^2 \text{ ft}$$

$$h \approx 560 \text{ ft}$$

75. a.

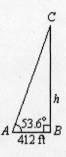

$$\tan 53.6° = \frac{h}{412}$$

$$h = 412 \tan 53.6°$$

$$h \approx 559 \text{ feet}$$

b.

$$(AC)^2 = 412^2 + 559^2$$

$$AC = \sqrt{412^2 + 559^2}$$

$$AC = \sqrt{482,225}$$

$$AC \approx 694.4 \text{ feet}$$

$$\tan 15.5° = \frac{x}{694.4}$$

$$x = 694.4 \tan 15.5°$$

$$x \approx 193 \text{ feet}$$

77.

Consider the right triangle formed by A, B and the midpoint of AC.

$$r = \sqrt{6^2 - 3^2}$$

$$r = \sqrt{27}$$

$$r = 3\sqrt{3}$$

$$r \approx 5.2 \text{ m}$$

79.

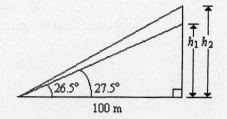

$$\tan 26.5° = \frac{h_1}{100} \qquad \tan 27.5° = \frac{h_2}{100}$$
$$h_1 = 100\tan 26.5° \qquad h_2 = 100\tan 27.5°$$
$$h_1 = 49.9 \text{ m} \qquad h_2 = 52.1 \text{ m}$$

$$49.9 \text{ m} \le h < 52.1 \text{ m}$$

81.

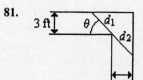

if $\theta = 45°$ $d_1 = d_2$

$$\sin\theta = \frac{3}{d_1}$$

$$d_1 = \frac{3}{\sin 45°}$$

$$d = 2d_1 = \frac{6}{\sin 45°}$$

$$d \approx 8.5 \text{ ft}$$

83.

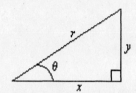

a. $\sin\theta = \dfrac{y}{r} \qquad \cos\theta = \dfrac{x}{r}$

$$\sin^0\theta + \cos^2\theta = \frac{y^2}{r^2} + \frac{x^2}{r^2} = \frac{y^2 + x^2}{r^2}$$
$$= \frac{r^2}{r^2} \qquad \bullet\, y^2 + x^2 = r^2$$
$$= 1$$

b. $\dfrac{\sin\theta}{\cos\theta} = \dfrac{\dfrac{y}{r}}{\dfrac{x}{r}} = \dfrac{y}{r}\cdot\dfrac{r}{x} = \dfrac{y}{x}$

$$\frac{\sin\theta}{\cos\theta} = \tan\theta \qquad \bullet\, \frac{y}{x} = \tan\theta$$

SECTION 2.3, Page 129

1. $x = 2, y = 3, r = \sqrt{2^2 + 3^2} = \sqrt{13}$

$$\sin\theta = \frac{y}{r} = \frac{3}{\sqrt{13}} = \frac{3\sqrt{13}}{13} \qquad \csc\theta = \frac{\sqrt{13}}{3}$$

$$\cos\theta = \frac{x}{r} = \frac{2}{\sqrt{13}} = \frac{2\sqrt{13}}{13} \qquad \sec\theta = \frac{\sqrt{13}}{2}$$

$$\tan\theta = \frac{y}{x} = \frac{3}{2} \qquad \cot\theta = \frac{2}{3}$$

3. $x = -2, y = 3, r = \sqrt{(-2)^2 + (3)^2} = \sqrt{13}$

$$\sin\theta = \frac{y}{r} = \frac{3}{\sqrt{13}} = \frac{3\sqrt{13}}{13} \qquad \csc\theta = \frac{\sqrt{13}}{3}$$

$$\cos\theta = \frac{x}{r} = \frac{-2}{\sqrt{13}} = -\frac{2\sqrt{13}}{13} \qquad \sec\theta = -\frac{\sqrt{13}}{2}$$

$$\tan\theta = \frac{y}{x} = \frac{3}{-2} = -\frac{3}{2} \qquad \cot\theta = -\frac{2}{3}$$

5. $x = -8, y = -5, c = \sqrt{(-8)^2 + (-5)^2} = \sqrt{89}$

$$\sin\theta = \frac{y}{r} = \frac{-5}{\sqrt{89}} = -\frac{5\sqrt{89}}{89} \qquad \csc\theta = -\frac{\sqrt{89}}{5}$$

$$\cos\theta = \frac{x}{r} = \frac{-8}{\sqrt{89}} = -\frac{8\sqrt{89}}{89} \qquad \sec\theta = -\frac{\sqrt{89}}{8}$$

$$\tan\theta = \frac{y}{x} = \frac{-5}{-8} = \frac{5}{8} \qquad \cot\theta = \frac{8}{5}$$

7. $x = -5, y = 0, r = \sqrt{(-5)^2 + (0)^2} = 5$

$$\sin\theta = \frac{y}{r} = \frac{0}{5} = 0 \qquad \csc\theta \text{ is undefined}$$

$$\cos\theta = \frac{x}{r} = \frac{-5}{5} = -1 \qquad \sec\theta = -1$$

$$\tan\theta = \frac{y}{x} = \frac{0}{-5} = 0 \qquad \cot\theta \text{ is undefined}$$

9. $\sin\theta > 0$ in quadrants I and II.
$\cos\theta > 0$ in quadrants I and IV.
quadrant I

11. $\cos\theta > 0$ in quadrants I and IV.
$\tan\theta < 0$ in quadrants II and IV.
quadrant IV

13. $\sin\theta < 0$ in quadrants III and IV.
$\cos\theta < 0$ in quadrants II and III.
quadrant III

15. $\sin\theta = -\dfrac{1}{2} = \dfrac{y}{r}, y = -1, r = 2, x = \pm\sqrt{2^2 - (-1)^2} = \pm\sqrt{3}, x = -\sqrt{3}$ in quadrant III

$\tan\theta = \dfrac{y}{x} = \dfrac{-y}{-\sqrt{3}} = \dfrac{\sqrt{3}}{3}$

17. $\csc\theta = \sqrt{2} = \dfrac{r}{y}, r = \sqrt{2}, y = 1, x = \pm\sqrt{(\sqrt{2})^2 - 1^2} = \pm 1, x = -1$ in quadrant II

$\cot\theta = \dfrac{-1}{1} = -1$

19. θ is in quadrant IV, $\sin\theta = -\dfrac{1}{2} = \dfrac{y}{r}, y = -1, r = 2, x = \sqrt{2^2 - 1^2} = \sqrt{3}$

$\tan\theta = \dfrac{-1}{\sqrt{3}} = -\dfrac{\sqrt{3}}{3}$

21. $\cos\theta = \dfrac{1}{2}, \theta$ is in quadrant I or IV.
$\tan\theta = \sqrt{3}, \theta$ is in quadrant I or III.
θ is in quadrant I, $x = 1, y = \sqrt{3}, r = 2$
$\csc\theta = \dfrac{r}{y} = \dfrac{2}{\sqrt{3}} = \dfrac{2\sqrt{3}}{3}$

23. $\cos\theta = -\dfrac{1}{2}, \theta$ is in quadrant II or III.
$\sin\theta = \dfrac{\sqrt{3}}{2}, \theta$ is in quadrant I or II.
θ is in quadrant II, $x = -1, y = \sqrt{3}, r = 2$
$\cot\theta = \dfrac{x}{y} = \dfrac{-1}{\sqrt{3}} = -\dfrac{\sqrt{3}}{3}$

25. $\theta = 160°$
Since $90° < \theta < 180°$,
$\theta + \theta' = 180°$
$\theta' = 20°$

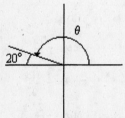

27. $\theta = 351°$
Since $270° < \theta < 360°$,
$\theta = \theta' = 360°$
$\theta' = 9°$

29. $\theta = \dfrac{11\pi}{5}$
$\theta > 2\pi = \dfrac{10\pi}{5}$,
θ is coterminal with $\alpha = \dfrac{11\pi}{5} - \dfrac{10\pi}{5} = \dfrac{\pi}{5}$.
Since $0 < \alpha < \dfrac{\pi}{2}$,
$\alpha' = \alpha = \theta'$
$\theta' = \dfrac{\pi}{5}$

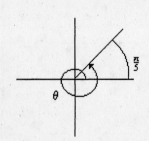

31. $\theta = \dfrac{8}{3}$

Since $\dfrac{\pi}{2} < \theta < \pi$,

$\theta + \theta' = \pi$

$\theta' = \pi - \dfrac{8}{3}$

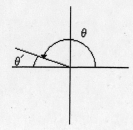

33. $\theta = 1406° = 326° + 3 \cdot 360°$

θ is coterminal with $\alpha = 326°$.

Since $270° < \alpha < 360°$,

$\alpha + \alpha' = 360°$

$\alpha' = 34°$

$\theta' = 34°$

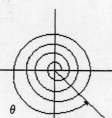

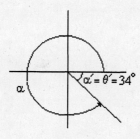

35. $\theta = -475° = 245° - 2 \cdot 360°$

θ is coterminal with $\alpha = 245°$

Since $180° < \alpha < 270°$,

$\alpha' + 180° = \alpha$

$\alpha' = 245° - 180°$

$\alpha' = 65°$

$\theta' = 65°$

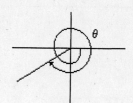

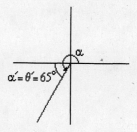

37. $\theta = 225°$ is in quadrant III.

$225° - 180° = 45°$ so $\theta' = 45°$.

Thus, $\sin 225° = -\sin 45° = -\dfrac{\sqrt{2}}{2}$.

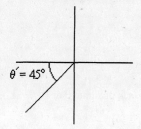

39. $\theta = 405°$ is in quadrant I.

$405° - 360° = 45°$ so $\theta' = 45°$.

Thus, $\tan 405° = \tan 45° = 1$.

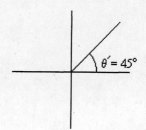

41. $\theta = \dfrac{4}{3}\pi$ is in quadrant III.

$\dfrac{4}{3}\pi - \pi = \dfrac{\pi}{3}$ so $\theta' = \dfrac{\pi}{3}$.

Thus, $\csc \dfrac{4\pi}{3} = \dfrac{1}{\sin 4\pi/3} = \dfrac{1}{-\sin \pi/3}$

$= \dfrac{1}{-\sqrt{3}/2} = -\dfrac{2}{\sqrt{3}} = -\dfrac{2\sqrt{3}}{3}$.

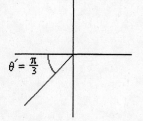

43. $\theta = \dfrac{17\pi}{4} = \dfrac{16\pi}{4} + \dfrac{\pi}{4}$ is coterminal

with $\dfrac{\pi}{4}$ in quadrant I and $\theta' = \dfrac{\pi}{4}$,

so $\cos \dfrac{17\pi}{4} = \cos \dfrac{\pi}{4} = \dfrac{\sqrt{2}}{2}$.

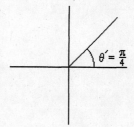

45. $\theta = 765° = 720° + 45°$ is coterminal

with $45°$ in quadrant I and $\theta' = 45°$,

so $\sec 765° = \sec 45° = \dfrac{1}{\cos 45°} = \dfrac{1}{\sqrt{2}/2} = \sqrt{2}.$

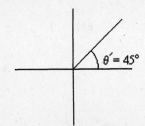

47. $\theta = 540° = 360° + 180°$ is coterminal

with $180°$, so $\cot 540° = \cot 180° = \dfrac{\cos 180°}{\sin 180°} = \dfrac{-1}{0},$

which is undefined.

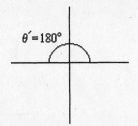

49. $\sin 127° \approx 0.798636$

51. $\cos(-116°) \approx -0.438371$

53. $\sec 578° \approx -1.26902$

55. $\sin\left(-\dfrac{\pi}{5}\right) \approx -0.587785$

57. $\csc \dfrac{9\pi}{5} \approx -1.70130$

59. $\sec(-4.45) \approx -3.85522$

61. $\sin 210° - \cos 330° \tan 330° = -\dfrac{1}{2} - \dfrac{\sqrt{3}}{2}\left(-\dfrac{\sqrt{3}}{3}\right)$

$\qquad\qquad = -\dfrac{1}{2} + \dfrac{1}{2} = 0$

63. $\sin^2 30° + \cos^2 30° = \left(\dfrac{1}{2}\right)^° + \left(\dfrac{\sqrt{3}}{2}\right)^2 = \dfrac{1}{4} + \dfrac{3}{4} = 1$

65. $\sin \dfrac{3\pi}{2} \tan \dfrac{\pi}{4} - \cos \dfrac{\pi}{3} = (-1)(1) - \dfrac{1}{2} = -1 - \dfrac{1}{2} = -\dfrac{3}{2}$

67. $\sin^2 \dfrac{5\pi}{4} + \cos^2 \dfrac{5\pi}{4} = \left(-\dfrac{\sqrt{2}}{2}\right)^2 + \left(-\dfrac{\sqrt{2}}{2}\right)^2 = \dfrac{1}{2} + \dfrac{1}{2} = 1$

69. $\sin\theta = \dfrac{1}{2}$, θ is in quadrant I or quadrant II

$\theta = 30°, 150°$

71. $\cos\theta = \dfrac{-\sqrt{3}}{2}$, θ is in quadrant II or quadrant III

$\theta = 150°, 120°$

73. $\csc\theta = -\sqrt{2}$

θ is in quadrant III or IV

$\theta = 225°, 315°$

75. $\tan\theta = -1$

θ is in quadrant II or IV

$\theta = \dfrac{3\pi}{4}, \dfrac{7\pi}{4}$

77. $\tan\theta = \dfrac{-\sqrt{3}}{3}$

θ is in quadrant II or IV

$\theta = \dfrac{5\pi}{6}, \dfrac{11\pi}{6}$

79. $\sin\theta = \dfrac{\sqrt{3}}{2}$

θ is in quadrant I or II

$\theta = \dfrac{\pi}{3}, \dfrac{2\pi}{3}$

81. $1 + \tan^2\theta = \sec^2\theta$

$1 + \dfrac{y^2}{x^2} = \dfrac{x^2 + y^2}{x^2}$

$\qquad = \dfrac{r^2}{x^2}$

$\qquad = \sec^2\theta$

83. $\tan\theta = \dfrac{\sin\theta}{\cos\theta}$

$\qquad = \dfrac{\dfrac{y}{r}}{\dfrac{x}{r}}$

$\qquad = \dfrac{y}{r} \cdot \dfrac{r}{x}$

$\tan\theta = \dfrac{y}{x}$

85.

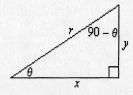

$$\cos(90° - \theta) = \sin\theta$$
$$\frac{y}{r} = \frac{y}{r}$$

87.

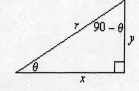

$$\tan(90° - \theta) = \cot\theta$$
$$\frac{x}{y} = \frac{x}{y}$$

89.

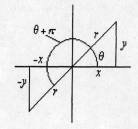

$$\sin(\theta + \pi) = -\frac{y}{r}$$
$$= -\sin\theta$$

91. $(\cos 78°, \sin 78°)$
     $(0.2079, 0.9781)$

93. $(\cos 3, \sin 3)$
     $(-0.9900, 0.1411)$

95. $(\cos(-68°), \sin(-68°))$
     $(0.3746, -0.9272)$

97. $\sin 0° = 0$
       $\sin 15° \approx 0.2588$
       $\sin 30° = 0.5$
       $\sin 45° \approx 0.7071$
       $\sin 60° \approx 0.8660$
       $\sin 75° \approx 0.9659$
       $\sin 90° = 1$
       $\sin 105° \approx 0.9659$
       $\sin 120° \approx 0.8660$
       $\sin 135° \approx 0.7071$
       $\sin 150° = 0.5$
       $\sin 165° \approx 0.2588$
       $\sin 180° = 0$

   a. 1

   b. increasing from 0 to 1

   c. decreasing from 1 to 0

SECTION 2.4, Page 139

1. $t = \dfrac{\pi}{6}$

   $y = \sin t$ $x = \cos t$

   $= \sin\dfrac{\pi}{6}$ $= \cos\dfrac{\pi}{6}$

   $= \dfrac{1}{2}$ $= \dfrac{\sqrt{3}}{2}$

   The point on the unit circle corresponding to $t = \frac{\pi}{6}$ is

   $\left(\dfrac{\sqrt{3}}{2}, \dfrac{1}{2}\right)$.

3. $t = \dfrac{7\pi}{6}$

   $y = \sin t$ $x = \cos t$

   $= \sin\dfrac{7\pi}{6}$ $= \cos\dfrac{7\pi}{6}$

   $= -\dfrac{1}{2}$ $= -\dfrac{\sqrt{3}}{2}$

   The point on the unit circle corresponding to $t = \frac{7\pi}{6}$ is

   $\left(-\dfrac{\sqrt{3}}{2}, -\dfrac{1}{2}\right)$.

5. $t = \dfrac{5\pi}{3}$

$$y = \sin t \qquad\qquad x = \cos t$$

$$= \sin\dfrac{5\pi}{3} \qquad\qquad = \cos\dfrac{5\pi}{3}$$

$$= -\dfrac{\sqrt{3}}{2} \qquad\qquad = \dfrac{1}{2}$$

The point on the unit circle corresponding to $t = \frac{5\pi}{3}$ is $\left(\dfrac{1}{2}, -\dfrac{\sqrt{3}}{2}\right)$.

7. $t = \dfrac{11\pi}{6}$

$$y = \sin t \qquad\qquad x = \cos t$$

$$= \sin\dfrac{11\pi}{6} \qquad\qquad = \cos\dfrac{11\pi}{6}$$

$$= -\dfrac{1}{2} \qquad\qquad = \dfrac{\sqrt{3}}{2}$$

The point on the unit circle corresponding to $t = \frac{11\pi}{6}$ is $\left(\dfrac{\sqrt{3}}{2}, -\dfrac{1}{2}\right)$.

9. $t = \pi$

$$y = \sin t \qquad x = \cos t$$

$$= \sin\pi \qquad = \cos\pi$$

$$= 0 \qquad = -1$$

The point on the unit circle corresponding to $t = \pi$ is $(-1, 0)$.

11. $t = -\dfrac{2\pi}{3}$

$$y = \sin t \qquad\qquad x = \cos t$$

$$= \sin\left(-\dfrac{2\pi}{3}\right) \qquad\qquad = \cos\left(-\dfrac{2\pi}{3}\right)$$

$$= -\sin\dfrac{2\pi}{3} \qquad\qquad = \cos\dfrac{2\pi}{3}$$

$$= -\dfrac{\sqrt{3}}{2} \qquad\qquad = -\dfrac{1}{2}$$

The point on the unit circle corresponding to $t = -\frac{2\pi}{3}$ is $\left(-\dfrac{1}{2}, -\dfrac{\sqrt{3}}{2}\right)$.

13. $\tan\dfrac{11\pi}{6} = -\tan\dfrac{\pi}{6} = -\dfrac{\sqrt{3}}{3}$

15. $\cos\left(-\dfrac{2\pi}{3}\right) = -\cos\dfrac{\pi}{3} = -\dfrac{1}{2}$

17. $\csc\left(-\dfrac{\pi}{3}\right) = -\csc\dfrac{\pi}{3} = -\dfrac{2\sqrt{3}}{3}$

19. $\sin\dfrac{3\pi}{2} = -\sin\dfrac{\pi}{2} = -1$

21. $\sec\left(-\dfrac{7\pi}{6}\right) = -\sec\dfrac{\pi}{6} = -\dfrac{2\sqrt{3}}{3}$

23. $\sin 1.22 \approx 0.9391$

25. $\csc(-1.05) \approx -1.1528$

27. $\tan\dfrac{11\pi}{12} \approx -0.2679$

29. $\cos\left(-\dfrac{\pi}{5}\right) \approx 0.8090$

31. $\sec 1.55 \approx 48.0889$

33. $f(-x) = -4\sin(-x)$

$$= 4\sin x$$

$$= -f(x)$$

The function defined by $f(x) = -4\sin x$ is an odd function.

35. $G(-x) = \sin(-x) + \cos(-x)$

$$= -\sin x + \cos x$$

The function defined by $G(x) = \sin x + \cos x$ is neither an even nor an odd function.

37. $S(-x) = \dfrac{\sin(-x)}{-x}$

$$= -\dfrac{\sin x}{-x} = \dfrac{\sin x}{x}$$

$$= S(x)$$

The function defined by $S(x) = \dfrac{\sin(x)}{x}$ is an even function.

39. $V(-x) = 2\sin(-x)\cos(-x)$

$$= -2\sin x\cos x$$

$$= -V(x)$$

The function defined by $V(x) = 2\sin x\cos x$ is an odd function.

41.

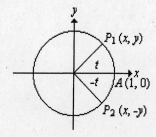

$$\cos t = x$$
$$\cos(-t) = x$$
$$\cos(-t) = \cos t$$

43.

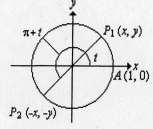

$$\cos t = x$$
$$\cos(\pi + t) = -x$$
$$\cos t = -\cos(\pi + t)$$

45.

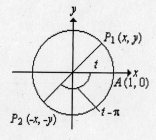

$$\sin(t - \pi) = -y$$
$$\sin t = y$$
$$\sin(t - \pi) = -\sin t$$

47.

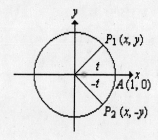

$$\csc t = \frac{1}{y}$$
$$\csc(-t) = -\frac{1}{y}$$
$$\csc(-t) = -\csc t$$

49. $\tan t \cos t = \dfrac{\sin t}{\cos t} \cdot \cos t$

$\qquad\qquad = \sin t$

51. $\dfrac{\csc t}{\cot t} = \dfrac{\dfrac{1}{\sin t}}{\dfrac{\cos t}{\sin t}}$

$\qquad = \dfrac{1}{\sin t} \cdot \dfrac{\sin t}{\cos t}$

$\qquad = \dfrac{1}{\cos t} = \sec t$

53. $1 - \sec^2 t = 1 - \dfrac{1}{\cos^2 t}$

$\qquad = \dfrac{\cos^2 t - 1}{\cos^2 t}$

$\qquad = \dfrac{-\sin^2 t}{\cos^2 t} = -\tan^2 t$

55. $\tan t - \dfrac{\sec^2 t}{\tan t} = \tan t - \dfrac{1 + \tan^2 t}{\tan t}$

$\qquad\qquad = \tan t - \dfrac{1}{\tan t} - \dfrac{\tan^2 t}{\tan t}$

$\qquad\qquad = \tan t - \cot t - \tan t$

$\qquad\qquad = -\cot t$

57. $\dfrac{1 - \cos^2 t}{\tan^2 t} = \dfrac{\sin^2 t}{\dfrac{\sin^2 t}{\cos^2 t}}$

$\qquad\qquad = \sin^2 t \cdot \dfrac{\cos^2 t}{\sin^2 t}$

$\qquad\qquad = \cos^2 t$

59. $\dfrac{1}{1 - \cos t} + \dfrac{1}{1 + \cos t} = \dfrac{1 + \cos t + 1 - \cos t}{(1 - \cos t)(1 + \cos t)}$

$\qquad\qquad = \dfrac{2}{1 - \cos^2 t}$

$\qquad\qquad = \dfrac{2}{\sin^2 t} = 2\csc^2 t$

61. $\dfrac{\tan t + \cot t}{\tan t} = \dfrac{\dfrac{\sin t}{\cos t} + \dfrac{\cos t}{\sin t}}{\dfrac{\sin t}{\cos t}}$

$= \left(\dfrac{\sin^2 t + \cos^2 t}{\sin t \cdot \cos t} \right) \dfrac{\cos t}{\sin t}$

$= \dfrac{\sin^2 t + \cos^2 t}{\sin^2 t}$

$= \dfrac{1}{\sin^2 t}$

$= \csc^2 t$

63. $\sin^2 t \left(1 + \cot^2 t \right) = \sin^2 t \left(\csc^2 t \right)$

$= \sin^2 t \cdot \dfrac{1}{\sin^2 t}$

$= 1$

65. $\sin^2 t + \cos^2 t = 1$

$\sin^2 t = 1 - \cos^2 t$

$\sin t = \pm\sqrt{1 - \cos^2 t}$

Because $0 < t < \dfrac{\pi}{2}$, $\sin t$ is positive.

Thus, $\sin t = \sqrt{1 - \cos^2 t}$.

67. $\csc^2 t = 1 + \cot^2 t$

$\csc t = \pm\sqrt{1 + \cot^2 t}$

Because $\dfrac{\pi}{2} < t < \pi$, $\csc t$ is positive.

Thus, $\csc t = \sqrt{1 + \cot^2 t}$.

69. $d(t) = 1970\cos\left(\dfrac{\pi}{64} t \right)$

$d(24) = 1970\cos\left(\dfrac{\pi}{64} \cdot 24 \right)$

$= 1970\cos\left(\dfrac{3\pi}{8} \right)$

≈ 750 miles

71. $\cos t - \dfrac{1}{\cos t} = \dfrac{\cos^2 t - 1}{\cos t} = -\dfrac{\sin^2 t}{\cos t}$

73. $\cot t + \dfrac{1}{\cot t} = \dfrac{\cot^2 t + 1}{\cot t} = \dfrac{\csc^2 t}{\cot t} = \dfrac{\dfrac{1}{\sin^2 t}}{\dfrac{\cos t}{\sin t}} = \dfrac{1}{\sin^2 t} \cdot \dfrac{\sin t}{\cos t} = \dfrac{1}{\sin t \cos t} = \dfrac{1}{\sin t} \cdot \dfrac{1}{\cos t} = \csc t \sec t$

75. $(1 - \sin t)^2 = 1 - 2\sin t + \sin^2 t$

77. $(\sin t - \cos t)^2 = \sin^2 t - 2\sin t \cos t + \cos^2 t$

$= 1 - 2\sin t \cos t$

79. $(1 - \sin t)(1 + \sin t) = 1 - \sin^2 t$

$= \cos^2 t$

81. $\dfrac{\sin t}{1 + \cos t} + \dfrac{1 + \cos t}{\sin t} = \dfrac{(\sin t)(\sin t) + (1 + \cos t)(1 + \cos t)}{\sin t (1 + \cos t)}$

$= \dfrac{\sin^2 t + 1 + 2\cos t + \cos^2 t}{\sin t (1 + \cos t)}$

$= \dfrac{2 + 2\cos t}{\sin t (1 + \cos t)} = \dfrac{2(1 + \cos t)}{\sin t (1 + \cos t)}$

$= \dfrac{2}{\sin t} = 2\csc t$

83. $\cos^2 t - \sin^2 t = (\cos t - \sin t)(\cos t + \sin t)$

85. $\tan^2 t - \tan t - 6 = (\tan t + 2)(\tan t - 3)$

87. $2\sin^2 t - \sin t - 1 = (2\sin t + 1)(\sin t - 1)$

89. $\cos^4 t - \sin^4 t = \left(\cos^2 t - \sin^2 t \right)\left(\cos^2 t + \sin^2 t \right)$

$= (\cos t - \sin t)(\cos t + \sin t)\left(\cos^2 t + \sin^2 t \right)$

$= (\cos t - \sin t)(\cos t + \sin t)$

91. $\csc t = \sqrt{2}, \quad 0 < t < \dfrac{\pi}{2}$

$\sin t = \dfrac{1}{\csc t} = \dfrac{\sqrt{2}}{2}$

$\cos^2 t + \sin^2 t = 1$

$\cos t = \pm\sqrt{1-\sin^2 t}$

$\cos t$ is positive in quadrant I.

$\cos t = \sqrt{1-\left(\dfrac{\sqrt{2}}{2}\right)^2} = \sqrt{\dfrac{1}{2}}$

$\cos t = \dfrac{\sqrt{2}}{2}$

93. $\sin t = \dfrac{1}{2}, \dfrac{\pi}{2} < t < \pi$

$\tan t = \dfrac{\sin t}{\cos t}$

$\tan t = \dfrac{\sin t}{\pm\sqrt{1-\sin^2 t}}$

Because $\dfrac{\pi}{2} < t < \pi, \tan t$ is negative.

$\tan t = -\dfrac{\dfrac{1}{2}}{\sqrt{1-\left(\dfrac{1}{2}\right)^2}}$

$\tan t = -\dfrac{\dfrac{1}{2}}{\dfrac{\sqrt{3}}{2}}$

$\tan t = -\dfrac{\sqrt{3}}{3}$

95. $\dfrac{\sin^2 t + \cos^2 t}{\sin^2 t} = \dfrac{1}{\sin^2 t}$

$\qquad = \csc^2 t$

97. $(\cos t - 1)(\cos t + 1) = \cos^2 t - 1$

$\qquad = -\left(1-\cos^2 t\right)$

$\qquad = -\sin^2 t$

SECTION 2.5, Page 148

1. $y = 2\sin x$

$a = 2, p = 2\pi$

3. $y = \sin 2x$

$a = 1, p = \dfrac{2\pi}{2} = \pi$

5. $y = \dfrac{1}{2}\sin 2\pi x$

$a = \dfrac{1}{2}, p = \dfrac{2\pi}{2\pi} = 1$

7. $y = -2\sin\dfrac{x}{2}$

$a = |-2| = 2, p = \dfrac{2\pi}{1/2} = 4\pi$

9. $y = \dfrac{1}{2}\cos x$

$a = \dfrac{1}{2}, p = 2\pi$

11. $y = \cos\dfrac{x}{4}$

$a = 1, p = \dfrac{2\pi}{1/4} = 8\pi$

13. $y = 2\cos\dfrac{\pi x}{3}$

$a = 2, p = \dfrac{2\pi}{\pi/3} = 6$

15. $y = -3\cos\dfrac{2x}{3}$

$a = |-3| = 3, p = \dfrac{2\pi}{2/3} = 3\pi$

17. $y = \dfrac{1}{2}\sin x, a = \dfrac{1}{2}, p = 2\pi$

19. $y = 3\cos x, a = 3, p = 2\pi$

21. $f(x) = -\dfrac{7}{2}\cos x, \quad a = \left|-\dfrac{7}{2}\right| = \dfrac{7}{2}, \quad p = 2\pi$

23. $f(x) = -4\sin x,$ $a = |-4| = 4,$ $p = 2\pi$

25. $f(x) = \cos 3x,$ $a = 1,$ $p = \dfrac{2\pi}{3}$

27. $f(x) = \sin\dfrac{3x}{2},$ $a = 1,$ $p = \dfrac{4\pi}{3}$

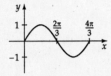

29. $f(x) = \cos\dfrac{\pi}{2}x,$ $a = 1,$ $p = 4$

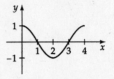

31. $f(x) = \sin 2\pi x,$ $a = 1,$ $p = 1$

33. $f(x) = 4\cos\dfrac{x}{2},$ $a = 4,$ $p = \dfrac{2\pi}{1/2} = 4\pi$

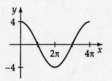

35. $f(x) = -2\cos\dfrac{x}{3},$ $a = |-2| = 2,$ $p = \dfrac{2\pi}{1/3} = 6\pi$

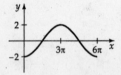

37. $f(x) = 2\sin\pi x,$ $a = 2,$ $p - \dfrac{2\pi}{\pi} - 2$

39. $f(x) = \dfrac{3}{2}\cos\dfrac{\pi x}{2},$ $a = \dfrac{3}{2},$ $p = \dfrac{2\pi}{\pi/2} = 4$

41. $f(x) = 4\sin\dfrac{2\pi}{3}x, a = |4| = 4, p = \dfrac{2\pi}{2\pi/3} = 3$

43. $f(x) = 2\cos 2x, a = 2, p = \dfrac{2\pi}{2} = \pi$

45. $f(x) = -2\sin 1.5x, a = |-2| = 2, p = \dfrac{2\pi}{1.5} = \dfrac{4\pi}{3}$

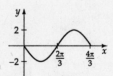

47. $f(x) = \left| 2\sin\dfrac{x}{2} \right|$

49. $f(x) = \left| -2\cos 3x \right|$

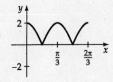

51. $f(x) = -\left| 2\sin\dfrac{x}{3} \right|$

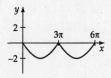

53. $f(x) = -\left| 3\cos\pi x \right|$

55. $\dfrac{2\pi}{b} = \pi, b = 2, a = 1$

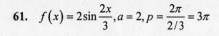

$y = \cos 2x$

57. $\dfrac{2x}{b} = 3\pi, b = \dfrac{2}{3}, a = 2$

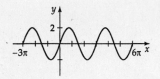

$y = 2\sin\dfrac{2}{3}x$

59. $\dfrac{2\pi}{b} = 2, b = \pi, a = 2$

$y = -2\cos\pi x$

61. $f(x) = 2\sin\dfrac{2x}{3}, a = 2, p = \dfrac{2\pi}{2/3} = 3\pi$

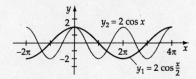

63. $f(x) = 2\cos\dfrac{x}{2}, a = 2, p = \dfrac{2\pi}{1/2} = 4\pi, g(x) = 2\cos x, a = 2, p = 2\pi$

65. $y = \cos^2 x$

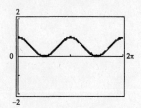

67. $y = \cos|x|$

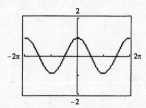

69. $y = \dfrac{1}{2}x\sin x$

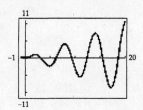

71. $y = -x\cos x$

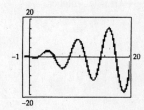

73.

x	−0.1	−0.05	−0.01	−0.001	0.001	0.01	0.05	0.1
$\dfrac{\sin x}{x}$	0.99833417	0.99958339	0.99998333	0.99999983	0.99999983	0.99998333	0.99958339	0.99833417

The value of $\dfrac{\sin x}{x} \to 1$ as $x \to 0$. The graph of f does not have a vertical asymptote at $x = 0$.

75. $y = e^{\sin x}$

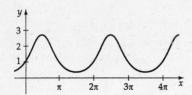

The maximum value of $e^{\sin x}$ is e.

The minimum value of $e^{\sin x}$ is $\dfrac{1}{e} \approx 0.3679$.

The function defined by $y = e^{\sin x}$ is periodic with a period of 2π.

77. $a = 2$

$$p = \frac{2\pi}{b} = 3\pi$$

$$b = \frac{2}{3}$$

$$y = 2\sin\frac{2}{3}x$$

79. $a = 4$

$$p = \frac{2\pi}{b} = 2$$
$$b = \pi$$

$$y = 4\sin\pi x$$

81. $a = 3$

$$p = \frac{2\pi}{b} = \frac{\pi}{2}$$
$$b = 4$$

$$y = 3\cos 4x$$

83. $a = 3$

$$p = \frac{2\pi}{b} = 2.5$$

$$b = \frac{4\pi}{5}$$

$$y = 3\cos\frac{4\pi}{5}x$$

85.

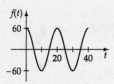

$a = 60, p = \dfrac{2\pi}{B} = 20, B = \dfrac{\pi}{10}$

$f(t) = A\cos Bt$

$f(t) = 60\cos\dfrac{\pi}{10}t$

87. $\dfrac{2\pi}{b} = \dfrac{2\pi}{3}$

$b = 3$

a. $y = 4\sin 3t$

b. $y = 4\sin 3(0.75)$

$y = 3.1$ ft

SECTION 2.6, Page 156

1. $y = \tan x$ is undefined for $\dfrac{\pi}{2} + k\pi$, k an integer.

3. $y = \sec x$ is undefined for $\dfrac{\pi}{2} + k\pi$, k an integer.

5. $p = 2\pi$

7. $p = \pi$

9. $p = \dfrac{\pi}{1/2} = 2\pi$

11. $p = \dfrac{2\pi}{3}$

13. $p = \dfrac{\pi}{3}$

15. $p = \dfrac{2\pi}{1/4} = 8\pi$

17. $p = \dfrac{\pi}{\pi} = 1$

19. $p = \dfrac{2\pi}{\pi/2} = 4$

21. $y = 3\tan x, p = \pi$

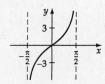

23. $y = \dfrac{3}{2}\cot x, p = \pi$

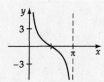

25. $y(x) = 2\sec x, p = 2\pi$

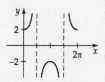

27. $y = \dfrac{1}{2}\csc x, p = 2\pi$

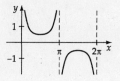

29. $y = 2\tan\dfrac{x}{2}, p = \dfrac{\pi}{1/2} = 2\pi$

31. $y = -3\cot\dfrac{x}{2}, p = \dfrac{\pi}{1/2} = 2\pi$

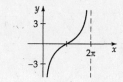

33. $y = -2\csc\dfrac{x}{3}, p = \dfrac{2\pi}{1/3} = 6\pi$

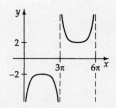

35. $y = \dfrac{1}{2}\sec 2x, p = \dfrac{2\pi}{2} = \pi$

37. $y = -2\sec\pi x, p = \dfrac{2\pi}{\pi} = 2$

39. $y = 3\tan 2\pi x, p = \dfrac{\pi}{2\pi} = \dfrac{1}{2}$

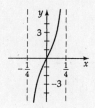

41. $y = 2\csc 3x, p = \dfrac{2\pi}{3}$

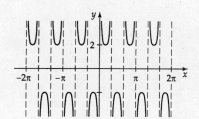

43. $y = 3\sec\pi x, p = \dfrac{2\pi}{\pi} = 2$

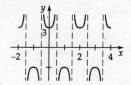

45. $y = 2\cot 2x, p = \dfrac{\pi}{2}$

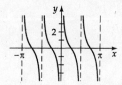

47. $y = 3\tan\pi x, p = \dfrac{\pi}{\pi} = 1$

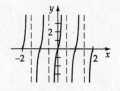

49. $\dfrac{\pi}{b} = \dfrac{2\pi}{3}, b = \dfrac{3}{2}$

$y = \cot\dfrac{3}{2}x$

51. $\dfrac{2\pi}{b} = 3\pi, b = \dfrac{2}{3}$

$y = \csc\dfrac{2}{3}x$

53. $\dfrac{2\pi}{b} = \dfrac{8\pi}{3}, b = \dfrac{3}{4}$

$y = \sec\dfrac{3}{4}x$

55. $y = \tan|x|$

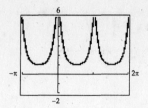

57. $y = |\csc x|$

59.

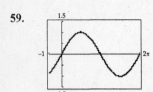 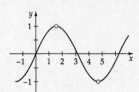

61. $y = \tan x, x = \tan y$

63. $\dfrac{\pi}{b} = \dfrac{\pi}{3}, b = 3$

$y = \tan 3x$

65. $\dfrac{2\pi}{b} = \dfrac{3\pi}{4}, b = \dfrac{8}{3}$

$y = \sec\dfrac{8}{3}x$

67. $\dfrac{\pi}{b} = 2, b = \dfrac{\pi}{2}$

$y = \cot\dfrac{\pi}{2}x$

69. $\dfrac{2\pi}{b} = 1.5, b = \dfrac{4}{3}\pi$

$y = \csc\dfrac{4\pi}{3}x$

SECTION 2.7, Page 164

1. $a = 2, p = 2\pi,$ phase shift $= \dfrac{\pi}{2}$

3. $a = 1, p = \dfrac{2\pi}{2} = \pi,$ phase shift $= \dfrac{\pi/4}{2} = \dfrac{\pi}{8}$

5. $a = |-4| = 4, p = \dfrac{2\pi}{2/3} = 3\pi,$ phase shift $= \dfrac{-\pi/6}{2/3} = -\dfrac{\pi}{4}$

7. $a = \dfrac{5}{4}, p = \dfrac{2\pi}{3},$ phase shift $= \dfrac{2\pi}{3}$

9. $p = \dfrac{\pi}{2},$ phase shift $= \dfrac{\pi/4}{2} = \dfrac{\pi}{8}$

11. $p = \dfrac{2\pi}{1/3} = 6\pi,$ phase shift $= \dfrac{-\pi}{1/3} = -3\pi$

13. $p = \dfrac{2\pi}{2} = \pi,$ phase shift $= \dfrac{\pi/8}{2} = \dfrac{\pi}{16}$

15. $p = \dfrac{\pi}{1/4} = 4\pi,$ phase shift $= \dfrac{-3\pi}{1/4} = -12\pi$

17. $y = \sin\left(x - \dfrac{\pi}{2}\right)$

$0 \le x - \dfrac{\pi}{2} \le 2\pi$

$\dfrac{\pi}{2} \le x \le \dfrac{5\pi}{2}$

period $= 2\pi$, phase shift $\dfrac{\pi}{2}$

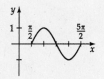

19. $y = \cos\left(\dfrac{x}{2} + \dfrac{\pi}{3}\right)$

$0 \le \dfrac{x}{2} + \dfrac{\pi}{3} \le 2\pi$

$-\dfrac{\pi}{3} \le \dfrac{x}{2} \le \dfrac{5\pi}{3}$

$-\dfrac{2\pi}{3} \le x \le \dfrac{10\pi}{3}$

period $= 4\pi$, phase shift $= -\dfrac{2\pi}{3}$

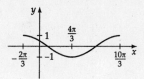

21. $y = \tan\left(x + \dfrac{\pi}{4}\right)$

$-\dfrac{\pi}{2} < x + \dfrac{\pi}{4} < \dfrac{\pi}{2}$

$-\dfrac{3\pi}{4} < x < \dfrac{\pi}{4}$

period $= \pi$, phase shift $= -\dfrac{\pi}{4}$

23. $y = 2\cot\left(\dfrac{x}{2} - \dfrac{\pi}{8}\right)$

$0 < \dfrac{x}{2} - \dfrac{\pi}{8} < \pi$

$\dfrac{\pi}{8} < \dfrac{x}{2} < \dfrac{9\pi}{8}$

$\dfrac{\pi}{4} < x < \dfrac{9\pi}{4}$

period $= 2\pi$, phase shift $= \dfrac{\pi}{4}$

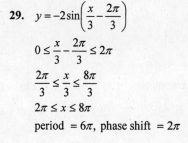

25. $y = \sec\left(x + \dfrac{\pi}{4}\right)$

$0 \le x + \dfrac{\pi}{4} \le 2\pi$

$-\dfrac{\pi}{4} \le x \le \dfrac{7\pi}{4}$

period $= 2\pi$, phase shift $= -\dfrac{\pi}{4}$

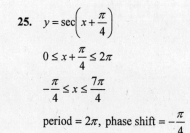

27. $y = \csc\left(\dfrac{x}{3} - \dfrac{\pi}{2}\right)$

$0 \le \dfrac{x}{3} - \dfrac{\pi}{2} \le 2\pi$

$\dfrac{\pi}{2} \le \dfrac{x}{3} \le \dfrac{5\pi}{2}$

$\dfrac{3\pi}{2} \le x \le \dfrac{15\pi}{2}$

period $= 6\pi$, phase shift $= \dfrac{3\pi}{2}$

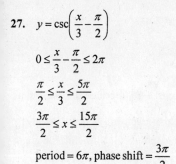

29. $y = -2\sin\left(\dfrac{x}{3} - \dfrac{2\pi}{3}\right)$

$0 \le \dfrac{x}{3} - \dfrac{2\pi}{3} \le 2\pi$

$\dfrac{2\pi}{3} \le \dfrac{x}{3} \le \dfrac{8\pi}{3}$

$2\pi \le x \le 8\pi$

period $= 6\pi$, phase shift $= 2\pi$

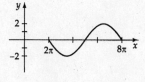

31. $y = -3\cos\left(3x + \dfrac{\pi}{4}\right)$

$0 \le 3x + \dfrac{\pi}{4} \le 2\pi$

$-\dfrac{\pi}{4} \le 3x \le \dfrac{7\pi}{4}$

$-\dfrac{\pi}{12} \le x \le \dfrac{7\pi}{12}$

period $= \dfrac{2\pi}{3}$, phase shift $= -\dfrac{\pi}{12}$

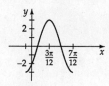

33. $y = \sin x + 1, p = 2\pi$

35. $y = -\cos x - 2, p = 2\pi$

37. $y = \sin 2x - 2, p = \pi$

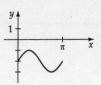

39. $y = 4\cos(\pi x - 2) + 1, p = 2$

phase shift $= \dfrac{2}{\pi}$

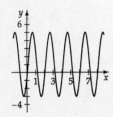

41. $y = -\sin(\pi x + 1) - 2, p = 2$

phase shift $= -\dfrac{1}{\pi}$

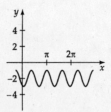

43. $y = \sin\left(x - \dfrac{\pi}{2}\right) - \dfrac{1}{2}, p = 2\pi$, phase shift $= \dfrac{\pi}{2}$

45. $y = \tan\dfrac{x}{2} - 4, p = 2\pi$

47. $y = \sec 2x - 2, p = \pi$

49. $y = \csc\dfrac{x}{2} - 1, p = 4\pi$

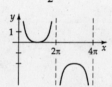

51. a.   phase shift: $-\dfrac{c}{b} = -\dfrac{-1.25\pi}{\pi/6} = 1.25(6) = 7.5$ months

period: $\dfrac{2\pi}{b} = \dfrac{2\pi}{\pi/6} = 2(6) = 12$ months

b.   First graph $y_1 = 4.1\cos\left(\dfrac{\pi}{6}t\right)$. Because the phase shift is 7.5 months, shift the graph of y_1 7.5 units to the right to produce the graph of y_2. Now shift the graph of y_2 upward 7 units to product the graph of S.

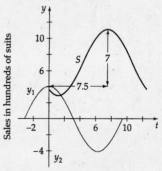

c.   7.5 months after January 1 is the middle of August.

53. $y = x - \sin x$

55. $y = x + \sin 2x$

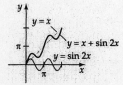

57. $y = \sin x + \cos x$

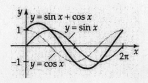

59. sine curve, $a = 1$, $\dfrac{2\pi}{b} = \pi$, $b = 2$

phase shift $= -\dfrac{c}{b} = \dfrac{\pi}{6}$, $c = -\dfrac{\pi}{3}$

$y = \sin\left(2x - \dfrac{\pi}{3}\right)$

61. cosecant curve, $\dfrac{2\pi}{b} = 4\pi$, $b = \dfrac{1}{2}$

phase shift $= -\dfrac{c}{b} = 2\pi$, $c = -\pi$

$y = \csc\left(\dfrac{x}{2} - \pi\right)$

63. secant curve, $\dfrac{2\pi}{b} = 2\pi$, $b = 1$

phase shift $= -\dfrac{c}{b} = \dfrac{\pi}{2}$, $c = -\dfrac{\pi}{2}$

$y = \sec\left(x - \dfrac{\pi}{2}\right)$

65. $y = 2.3 \sin 2\pi t + 1.25t + 315$

$t = 20$ between 1972 and 1992

$y = 2.3 \sin 2\pi(20) + 1.25(20) + 315$

$y \approx 340$ ppm

difference $\approx 340 - 315$

≈ 25 ppm

67. 5 rpm $a = 7$

$p = 0.2$ min $\qquad p = \dfrac{2\pi}{b}$

$0.2 = \dfrac{2\pi}{b}$

$b = 10\pi$

$s = 7 \cos 10\pi t + 5$

69. Change 6 rpm to radians/second.

$6 \text{ rpm} = \dfrac{6 \cdot 2\pi \text{ radians}}{1 \text{ minute}} \cdot \dfrac{1 \text{ minute}}{60 \text{ sec}}$

$= \dfrac{\pi}{5} \text{ radians/sec}$

in t seconds, θ increased by $\dfrac{\pi}{5}t$ radians.

$\tan \dfrac{\pi}{5}t = \dfrac{s}{400}$

$400 \tan \dfrac{\pi}{5}t = s$ for $0 \le t < 2.5$ or $7.5 < t \le 10$

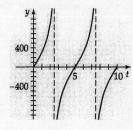

71. amplitude $A = 3$

$k = 9$

$24 = 2$ cycles

$12 = 1$ cycle

$\dfrac{2\pi}{B} = 12 \Rightarrow B = \dfrac{\pi}{6}$

$y = 3 \cos \dfrac{\pi}{6}t + 9$

At 6:00 P.M., $t = 12$.

$y = 3 \cos \dfrac{\pi}{6} \cdot 12 + 9$

$y = 3 \cos 2\pi + 9$

$y = 3 + 9$

$y = 12$ ft

73. $y = \sin x - \cos \dfrac{x}{2}$

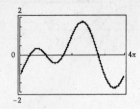

75. $y = 2\cos x + \sin \dfrac{x}{2}$

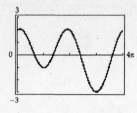

77. $y = \dfrac{x}{2}\sin x$

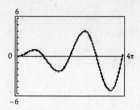

79. $y = x\sin \dfrac{x}{2}$

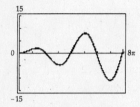

81. $y = x\sin\left(x + \dfrac{\pi}{2}\right)$

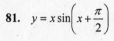

83.

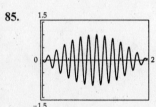

85.

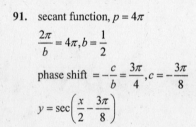

87. sine function, $a = 2, p = \pi$

$\dfrac{2\pi}{b} = \pi, b = 2$

phase shift $= -\dfrac{c}{b} = \dfrac{\pi}{3}, c = -\dfrac{2\pi}{3}$

$y = 2\sin\left(2x - \dfrac{2\pi}{3}\right)$

89. tangent function, $p = 2\pi$

$\dfrac{\pi}{b} = 2\pi \Rightarrow b = \dfrac{1}{2}$

phase shift $= -\dfrac{c}{b} = \dfrac{\pi}{2} \Rightarrow c = -\dfrac{\pi}{4}$

$y = \tan\left(\dfrac{x}{2} - \dfrac{\pi}{4}\right)$

91. secant function, $p = 4\pi$

$\dfrac{2\pi}{b} = 4\pi, b = \dfrac{1}{2}$

phase shift $= -\dfrac{c}{b} = \dfrac{3\pi}{4}, c = -\dfrac{3\pi}{8}$

$y = \sec\left(\dfrac{x}{2} - \dfrac{3\pi}{8}\right)$

93. $g(x) + h(x) = \sin^2 x + \cos^2 x = 1$

95. $g\big[h(x)\big] = (\cos x)^2 + 2$

$\quad\quad = \cos^2 x + 2$

97. $\dfrac{\sin x}{x} \to 1$ as $x \to 0$

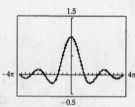

99. $y = |x|\sin x$

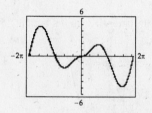

SECTION 2.8, Page 172

1. $y = 2\sin 2t$

amplitude $= 2$

$p = \dfrac{2\pi}{2} = \pi$

frequency $= \dfrac{1}{p} = \dfrac{1}{\pi}$

3. $y = 3\cos \dfrac{2t}{3}$

amplitude $= 3$

$p = \dfrac{2\pi}{2/3} = 3\pi$

frequency $= \dfrac{1}{p} = \dfrac{1}{3\pi}$

5. $y = 4\cos \pi t$

amplitude $= 4$

$p = \dfrac{2\pi}{\pi} = 2$

frequency $= \dfrac{1}{p} = \dfrac{1}{2}$

7. $y = \dfrac{3}{4} \sin \dfrac{\pi t}{2}$

amplitude $= \dfrac{3}{4}$

$p = \dfrac{2\pi}{\pi/2} = 4$

frequency $= \dfrac{1}{p} = \dfrac{1}{4}$

9. $a = 4$

$p = \dfrac{1}{\text{frequency}} = \dfrac{1}{1.5} = \dfrac{2}{3}$

$\dfrac{2\pi}{b} = \dfrac{2}{3} \Rightarrow b = 3\pi$

$y = 4\cos 3\pi t$

11. $p = 1.5$

amplitude $= \dfrac{3}{2}$

$\dfrac{2\pi}{b} = 1.5 \Rightarrow b = \dfrac{4\pi}{3}$

$y = \dfrac{3}{2} \cos \dfrac{4\pi}{3} t$

13. amplitude $= 2$

$p = \pi$

$\dfrac{2\pi}{b} = \pi \Rightarrow b = 2$

$y = 2\sin 2t$

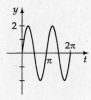

15. amplitude $= 1$

$p = 2$

$\dfrac{2\pi}{b} = 2 \Rightarrow b = \pi$

$y = \sin \pi t$

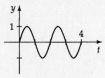

17. amplitude $= 2$

frequency $= 1$

$p = \dfrac{1}{\text{frequency}} = \dfrac{1}{1} = 1$

$\dfrac{2\pi}{b} = 1 \Rightarrow b = 2\pi$

$y = 2\sin 2\pi t$

19. amplitude $= \dfrac{1}{2}$

frequency $= \dfrac{2}{\pi} \Rightarrow p = \dfrac{\pi}{2}$

$\dfrac{2\pi}{b} = \dfrac{\pi}{2} \Rightarrow b = 4$

$y = \dfrac{1}{2}\cos 4t$

21. amplitude $= 2.5$

frequency $= 0.5 \Rightarrow p = 2$

$\dfrac{2\pi}{b} = 2 \Rightarrow b = \pi$

$y = 2.5\cos \pi t$

23. amplitude $= \dfrac{1}{2}$

$p = 3$

$\dfrac{2\pi}{b} = 3 \Rightarrow b = \dfrac{2}{3}\pi$

$y = \dfrac{1}{2}\cos \dfrac{2\pi}{3} t$

25. amplitude $= 4$

$p = \dfrac{\pi}{2}$

$\dfrac{2\pi}{b} = \dfrac{\pi}{2} \Rightarrow b = 4$

$y = 4\cos 4t$

27. amplitude $= 2$ feet, $a = -2$

frequency $= f = \dfrac{1}{2\pi}\sqrt{\dfrac{k}{m}} = \dfrac{1}{2\pi}\sqrt{\dfrac{8}{32}} = \dfrac{1}{2\pi} \cdot \dfrac{1}{2} = \dfrac{1}{4\pi}$

period $= p = \dfrac{1}{f} = 4\pi$

$y = a\cos 2\pi f t = -2\cos 2\pi \left(\dfrac{1}{4\pi}\right) t$

$= -2\cos \dfrac{1}{2} t$

29. a. The pseudoperiod is $\frac{2\pi}{2} = \pi$. There are $\frac{10}{\pi} \approx 3$ complete oscillations of length π in $0 \le t \le 10$.

b. $|f(t)| < 0.01$ for all $t > 59.8$ (nearest tenth).

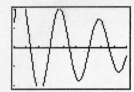

Xmin $= 56$, Xmax $= 65$, Xscl $= 1$,
Ymin $= -0.01$, Ymax $= 0.01$, Yscl $= 0.005$

31. a. The pseudoperiod is $\frac{2\pi}{2\pi} = 1$. There are 10 complete oscillations of length 1 in $0 \le t \le 10$.

b. $|f(t)| < 0.01$ for all $t > 71.0$ (nearest tenth).

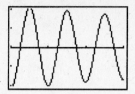

Xmin $= 70$, Xmax $= 73$, Xscl $= 1$
Ymin $= -0.01$, Ymax $= 0.01$, Yscl $= 0.005$

33. a. The pseudoperiod is $\frac{2\pi}{2\pi} = 1$. There are 10 complete oscillations of length 1 in $0 \le t \le 10$.

b. $|f(t)| < 0.01$ for all $t > 9.1$ (nearest tenth).

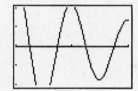

Xmin $= 8$, Xmax $= 10$, Xscl $= 1$,
Ymin $= -0.01$, Ymax $= 0.01$, Yscl $= 0.005$

35. a. The pseudoperiod is $\frac{2\pi}{2\pi} = 1$. There are 10 complete oscillations of length 1 in $0 \le t \le 10$.

b. $|f(t)| < 0.01$ for all $t > 6.1$ (nearest tenth).

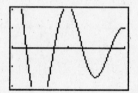

Xmin $= 5$, Xmax $= 7$, Xscl $= 1$,
Ymin $= -0.01$, Ymax $= .01$, Yscl $= 0.005$

37. $p_1 = 2\pi\sqrt{\dfrac{m}{k}}$, $p_2 = 2\pi\sqrt{\dfrac{9m}{k}} = 3p_1$

Increasing the main mass to $9m$ will triple the period.

39. yes

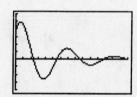

Xmin $= 0$, Xmax $= 15$, Xscl $= 1$,
Ymin $= -1$, Ymax $= 1.5$, Yscl $= 0.25$

41. yes

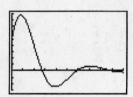

Xmin $= 0$, Xmax $= 10$, Xscl $= 1$,
Ymin $= -3$, Ymax $= 9$, Yscl $= 1$

EXPLORING CONCEPTS WITH TECHNOLOGY, Page 173

1.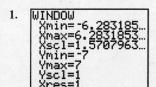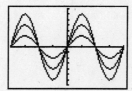

All three sine graphs have, a period of 2π, x-intercepts at $n\pi$, and no phase shift, but their amplitudes are 2, 4, and 6 respectively.

3.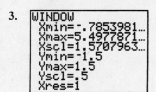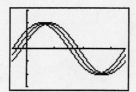

All three sine graphs have a period of 2π and an amplitude of 1, but their phase shifts are $-\pi/4$, $-\pi/6$, and $-\pi/12$, respectively.

CHAPTER 2 TRUE/FALSE EXERCISES, Page 175

1. False; the initial side must be along the positive x-axis.

3. True

5. False; $\sec^2\theta - \tan^2\theta = 1$ is an identity.

7. False; the period is 2π.

9. False; $\sin 45° + \cos(90° - 45°) = \sin 45° + \cos 45°$
$$= \frac{\sqrt{2}}{2} + \frac{\sqrt{2}}{2} = \sqrt{2}.$$

11. False; $\sin^2\frac{\pi}{6} = \left(\sin\frac{\pi}{6}\right)^2$ $\sin\left(\frac{\pi}{6}\right)^2 = \sin\frac{\pi^2}{36}$
$$= \left(\frac{1}{2}\right)^2 = \frac{1}{4}$$ $\approx 0.2707.$

13. True

15. False, the graph lies on or between the graphs of $y = 2^{-x}$ and $y = -2^{-x}$.

CHAPTER 2 REVIEW EXERCISES, Page 175

1. complement: $90° - 65° = 25°$
supplement: $180° - 65° = 115°$

3. $2 = 2\left(\frac{180°}{\pi}\right)$
$= 114.59°$

5. $s = r\theta = 3(75°)\left(\frac{\pi}{180°}\right)$
$= 3.93$ meters

7. $w = \frac{V}{r} = \frac{50}{16}\cdot\frac{63360}{3600}$
≈ 55 rad/sec

For exercises 8 to 11, $\csc\theta = \frac{3}{2} = \frac{r}{y}$, $r = 3$, $y = 2$, and $x = \sqrt{3^2 - 2^2} = \sqrt{5}$.

9. $\cot\theta = \frac{x}{y} = \frac{\sqrt{5}}{2}$

11. $\sec\theta = \frac{r}{x} = \frac{3}{\sqrt{5}} = \frac{3\sqrt{5}}{5}$

13. a. $\sec 150° = \frac{2}{-\sqrt{3}} = -\frac{2\sqrt{3}}{3}$

b. $\tan\left(-\frac{3\pi}{4}\right) = 1$

c. $\cot(-225°) = -1$

d. $\cos\frac{2\pi}{3} = -\frac{1}{2}$

102   **Chapter 2/Trigonometric Functions**

15. $\cos\phi = -\dfrac{\sqrt{3}}{2} = \dfrac{x}{r}$, $x = -\sqrt{3}$, $r = 2$, $y = -\sqrt{2^2 - \left(-\sqrt{3}\right)^2} = -1$

    a. $\sin\phi = \dfrac{y}{r} = -\dfrac{1}{2}$

    b. $\tan\phi = \dfrac{y}{x} = \dfrac{-1}{-\sqrt{3}} = \dfrac{\sqrt{3}}{3}$

17. $\sin\phi = -\dfrac{\sqrt{2}}{2}$, $y = -\sqrt{2}$, $r = 2$, $x = -\sqrt{2^2 - \left(-\sqrt{2}\right)^2} = \sqrt{2}$

    a. $\cos\phi = \dfrac{x}{r} = \dfrac{\sqrt{2}}{2}$

    b. $\cot\phi = \dfrac{x}{y} = \dfrac{\sqrt{2}}{-\sqrt{2}} = -1$

19. $f(x) = \sin(x)\tan(x)$

$f(-x) = \sin(-x)\tan(-x) = (-\sin x)(-\tan x)$

         $= \sin x \tan x$

         $= f(x)$

The function defined by $f(x) = \sin(x)\tan(x)$ is an even function.

21.

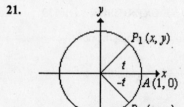

$\tan(-t) = -\dfrac{y}{x}$

$\tan t = \dfrac{y}{x}$

$\tan(-t) = -\tan t$

23. $\dfrac{\tan\phi + 1}{\cot\phi + 1} = \dfrac{\dfrac{\sin\phi}{\cos\phi} + 1}{\dfrac{\cos\phi}{\sin\phi} + 1}$

$= \dfrac{\dfrac{\sin\phi + \cos\phi}{\cos\phi}}{\dfrac{\cos\phi + \sin\phi}{\sin\phi}}$

$= \dfrac{\sin\phi(\sin\phi + \cos\phi)}{\cos\phi(\cos\phi + \sin\phi)}$

$= \tan\phi$

25. $\sin^2\phi(\tan^2\phi + 1) = \sin^2\phi\sec^2\phi$

$= \dfrac{\sin^2\phi}{\cos^2\phi}$

$= \tan^2\phi$

27. $\dfrac{\cos^2\phi}{1-\sin^2\phi} - 1 = \dfrac{1-\sin^2\phi}{1-\sin^2\phi} - 1$

         $= 1 - 1$

         $= 0$

29. $y = 2\tan 3x$

no amplitude; period $= \dfrac{\pi}{b} = \dfrac{\pi}{3}$

phase shift $= 0$

31. $y = \cos\left(2x - \dfrac{2\pi}{3}\right) + 2$

$a = |1| = 1$; period $= \dfrac{2\pi}{b} = \dfrac{2\pi}{2} = \pi$

phase shift $= -\dfrac{c}{b} = -\dfrac{-2\pi/3}{2} = \dfrac{\pi}{3}$

33. $y = 2\csc\left(x - \dfrac{\pi}{4}\right) - 3$

no amplitude; period $= \dfrac{2\pi}{b} = \dfrac{2\pi}{1} = 2\pi$

phase shift $= -\dfrac{c}{b} = -\dfrac{-\pi/4}{1} = \dfrac{\pi}{4}$

35. $y = -\sin\dfrac{2x}{3}$, $p = \dfrac{2\pi}{2/3} = 3\pi$

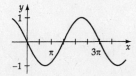

37. $y = \cos\left(x - \dfrac{\pi}{2}\right)$, $p = 2\pi$, phase shift $= \dfrac{\pi}{2}$

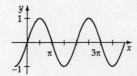

39. $y = 3\cos 3(x - \pi)$, $p = \dfrac{2\pi}{3}$, phase shift $= \pi$

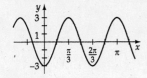

41. $y = 2\cot 2x$, $p = \dfrac{\pi}{2}$

43. $y = -\cot\left(2x + \dfrac{\pi}{4}\right)$, $p = \dfrac{\pi}{2}$, phase shift $= -\dfrac{\pi}{8}$

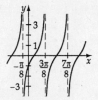

45. $y = 3\sec\left(x + \dfrac{\pi}{4}\right)$, $p = 2\pi$, phase shift $= -\dfrac{\pi}{4}$

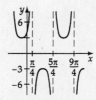

47. $y = 2\cos 3x + 3$

49. $y = 3\sin\left(4x - \dfrac{2\pi}{3}\right) - 3$

51. $y = \sin x - \sqrt{3}\cos x$

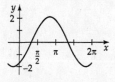

53.

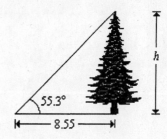

$\tan 55.3° = \dfrac{h}{8.55}$

$h = 8.55\tan 55.3 \approx 12.3$ feet

55.

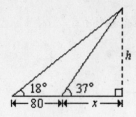

(1) $\cot 18° = \dfrac{80 + x}{h} = \dfrac{80}{h} + \dfrac{x}{h}$

(2) $\cot 37° = \dfrac{x}{h}$

Substitute for $\dfrac{x}{h}$ in equation (1).

$$\cot 18° = \dfrac{80}{h} + \cot 37°$$

Solve for h. $\dfrac{80}{h} = \cot 18° - \cot 37°$

$$\dfrac{h}{80} = \dfrac{1}{\cot 18° - \cot 37°}$$

$$h = \dfrac{80}{\cot 18° - \cot 37°} \approx 46 \text{ ft}$$

57. amplitude $= 0.5$

$$f = \dfrac{1}{2\pi}\sqrt{\dfrac{k}{m}} = \dfrac{1}{2\pi}\sqrt{\dfrac{20}{5}} = \dfrac{1}{\pi}$$

$$p = \pi$$

$$y = -0.5\cos 2\pi f t = -0.5\cos 2\pi\left(\dfrac{1}{\pi}\right)t$$

$$y = -0.5\cos 2t$$

CHAPTER 2 TEST, Page 177

1. $150° = 150°\left(\dfrac{\pi}{180°}\right)$

$\quad = \dfrac{5\pi}{6}$

2. $\pi - \dfrac{11}{12}\pi = \dfrac{\pi}{12}$

3. $s = r\theta$

$s = 10(75°)\left(\dfrac{\pi}{180°}\right)$

$s \approx 13.1 \text{ cm}$

4. $w = 6\dfrac{rev}{sec}$

$w = 6\dfrac{rev}{sec}\left(\dfrac{2\pi\ rad}{rev}\right)$

$w = 12\pi\ \text{rad/sec}$

5. $v = rw$

$\quad = 8 \cdot 10$

$\quad = 80 \text{ cm/sec}$

6.

$r = \sqrt{7^2 + 3^2}$

$r = \sqrt{58}$

$\sec\theta = \dfrac{\sqrt{58}}{7}$

7. $\csc 67° \approx 1.0864$

8. $\tan\dfrac{\pi}{6}\cos\dfrac{\pi}{3} - \sin\dfrac{\pi}{2} = \dfrac{1}{\sqrt{3}}\cdot\dfrac{1}{2} - 1$

$\quad = \dfrac{1}{2\sqrt{3}} - 1$

$\quad = \dfrac{\sqrt{3}}{6} - 1$

$\quad = \dfrac{\sqrt{3} - 6}{6}$

9. $t = \dfrac{11\pi}{6}$

$x = \cos t \qquad y = \sin t$

$= \dfrac{\sqrt{3}}{2} \qquad = -\dfrac{1}{2}$

$P(x,y) = P\left(\dfrac{\sqrt{3}}{2}, -\dfrac{1}{2}\right)$

10. $\dfrac{\sec^2 t - 1}{\sec^2 t} = \dfrac{\dfrac{1}{\cos^2 t} - 1}{\dfrac{1}{\cos^2 t}}$

$= \dfrac{\dfrac{1 - \cos^2 t}{\cos^2 t}}{\dfrac{1}{\cos^2 t}}$

$= 1 - \cos^2 t$

$= \sin^2 t$

11. $\text{period} = \dfrac{\pi}{b} = \dfrac{\pi}{3}$

12. $a = |-3| = 3; \ \ \text{period} = \dfrac{2\pi}{b} = \dfrac{2\pi}{2} = \pi$

$\text{phase shift} = -\dfrac{\pi}{4}$

13. $\text{period} = \dfrac{\pi}{\pi/3} = 3$

$\text{phase shift} = -\dfrac{c}{b} = -\dfrac{\pi/6}{\pi/3} = -\dfrac{1}{2}$

14. $y = 3\cos\dfrac{1}{2}x, \ p = 4\pi$

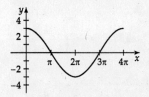

15. $y = -2\sec\dfrac{1}{2}x, \qquad p = 4\pi$

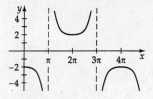

16. Shift the graph [of $y = 2\sin(2x)$] $\dfrac{\pi}{4}$ units to the right and down 1 unit.

17. $y = 2 - \sin\dfrac{x}{2}$

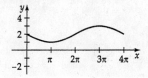

18. $y = \sin x - \cos 2x$

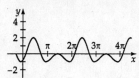

19.

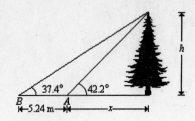

$$\tan 42.2° = \frac{h}{x}$$

$$x = \frac{h}{\tan 42.2°}$$

$$= h \cot 422.2°$$

$$\tan 37.4° = \frac{h}{5.24x}$$

$$= \frac{h}{5.24 + h \cot 42.2°}$$

Solve for h.

$$\tan 37.4° = \frac{h}{5.24 + h \cot 42.2°}$$

$$\tan 37.4°(5.24 + h \cot 42.2°) = h$$

$$5.24 \tan 37.4° + h \tan 37.4° \cot 42.2° = h$$

$$h - h \tan 37.4° \cot 42.2° = 5.24 \tan 37.4°$$

$$h(1 - \tan 37.4° \cot 42.2°) = 5.24 \tan 37.4°$$

$$h = \frac{5.24 \tan 37.4°}{1 - \tan 37.4° \cot 42.2°}$$

$$h \approx 25.5 \text{ meters}$$

The height of the tree is approximately 25.5 meters.

20. $p = 5$ $5 = \frac{2\pi}{b}, b = \frac{2\pi}{5}$

$a = 13$

$$y = 13 \cos \frac{2\pi}{5} t \text{ or } y = 13 \sin \frac{2\pi}{5} t$$

SECTION 3.1, Page 184

1. Not an identity. If $x = \pi/4$, the left side is 2 and the right side is 1.

3. Not an identity. If $x = 0°$, the left side is $\sqrt{3}/2$ and the right side is $(2+\sqrt{3})/2$.

5. Not an identity. If $x = 0$, the left side is -1 and the right side is 1.

7. Not an identity. If $x = 0$, the left side is -1 and the right side is 1.

9. Not an identity. The left side is 2 and the right side is $\sqrt{3}/2$.

11. Not an identity. If $x = \pi/2$, the left side is 1 and the right side is 2.

13. $\tan x \csc x \cos x = \dfrac{\sin x}{\cos x} \cdot \dfrac{1}{\sin x} \cdot \cos x = 1$

15. $\dfrac{4\sin^2 x - 1}{2\sin x + 1} = \dfrac{(2\sin x - 1)(2\sin x + 1)}{2\sin x + 1} = 2\sin x - 1$

17. $(\sin x - \cos x)(\sin x + \cos x) = \sin^2 x - \cos^2 x$
$$= 1 - \cos^2 x - \cos^2 x$$
$$= 1 - 2\cos^2 x$$

19. $\dfrac{1}{\sin x} - \dfrac{1}{\cos x} = \dfrac{\cos x}{\sin x \cos x} - \dfrac{\sin x}{\sin x \cos x}$
$$= \dfrac{\cos x - \sin x}{\sin x \cos x}$$

21. $\dfrac{\cos x}{1 - \sin x} = \dfrac{\cos x(1 + \sin x)}{(1 - \sin x)(1 + \sin x)}$
$$= \dfrac{\cos x(1 + \sin x)}{1 - \sin^2 x}$$
$$= \dfrac{\cos x(1 + \sin x)}{\cos^2 x}$$
$$= \dfrac{(1 + \sin x)}{\cos x} = \dfrac{1}{\cos x} + \dfrac{\sin x}{\cos x}$$
$$= \sec x + \tan x$$

23. $\dfrac{1 - \tan^4 x}{\sec^2 x} = \dfrac{(1 + \tan^2 x)(1 - \tan^2 x)}{\sec^2 x}$
$$= \dfrac{\sec^2 x(1 - \tan^2 x)}{\sec^2 x}$$
$$= 1 - \tan^2 x$$

25. $\dfrac{1 + \tan^3 x}{1 + \tan x} = \dfrac{(1 + \tan x)(1 - \tan x + \tan^2 x)}{1 + \tan x}$
$$= 1 - \tan x + \tan^2 x$$

27. $\dfrac{\sin x - 2 + \frac{1}{\sin x}}{\sin x - \frac{1}{\sin x}} = \dfrac{\sin x - 2 + \frac{1}{\sin x}}{\sin x - \frac{1}{\sin x}} \cdot \dfrac{\sin x}{\sin x}$
$$= \dfrac{\sin^2 x - 2\sin x + 1}{\sin^2 x - 1}$$
$$= \dfrac{(\sin x - 1)(\sin x - 1)}{(\sin x - 1)(\sin x + 1)}$$
$$= \dfrac{\sin x - 1}{\sin x + 1}$$

29. $(\sin x + \cos x)^2 = \sin^2 x + 2\sin x \cos x + \cos^2 x$
$$= \sin^2 x + \cos^2 x + 2\sin x \cos x$$
$$= 1 + 2\sin x \cos x$$

31. $\dfrac{\cos x}{1 + \sin x} = \dfrac{\cos x(1 - \sin x)}{(1 + \sin x)(1 - \sin x)}$
$$= \dfrac{\cos x(1 - \sin x)}{1 - \sin^2 x}$$
$$= \dfrac{\cos x(1 - \sin x)}{\cos^2 x}$$
$$= \dfrac{1 - \sin x}{\cos x}$$
$$= \dfrac{1}{\cos x} - \dfrac{\sin x}{\cos x}$$
$$= \sec x - \tan x$$

33. $\dfrac{\cot x + \tan x}{\sec x} = \dfrac{\dfrac{\cos x}{\sin x} + \dfrac{\sin x}{\cos x}}{\dfrac{1}{\cos x}}$

$\qquad = \dfrac{\dfrac{\cos x}{\sin x} + \dfrac{\sin x}{\cos x}}{\dfrac{1}{\cos x}} \cdot \dfrac{\sin x \cos x}{\sin x \cos x}$

$\qquad = \dfrac{\cos^2 x + \sin^2 x}{\sin x}$

$\qquad = \dfrac{1}{\sin x}$

$\qquad = \csc x$

35. $\dfrac{\cos x \tan x + 2\cos x - \tan x - 2}{\tan x + 2} = \dfrac{\cos x(\tan x + 2) - (\tan x + 2)}{\tan x + 2}$

$\qquad = \dfrac{(\tan x + 2)(\cos x - 1)}{\tan x + 2}$

$\qquad = \cos x - 1$

37. $\dfrac{1 - \sin x}{\cos x} = \dfrac{1}{\cos x} - \dfrac{\sin x}{\cos x} = \sec x - \tan x$

39. $\sin^2 x - \cos^2 x = \sin^2 x - (1 - \sin^2 x)$

$\qquad = \sin^2 x - 1 + \sin^2 x$

$\qquad = 2\sin^2 x - 1$

41. $\dfrac{1}{\sin^2 x} + \dfrac{1}{\cos^2 x} = \dfrac{\cos^2 x + \sin^2 x}{\sin^2 x \cos^2 x}$

$\qquad = \dfrac{1}{\sin^2 x \cos^2 x}$

$\qquad = \csc^2 x \sec^2 x$

43. $\sec x - \cos x = \dfrac{1}{\cos x} - \cos x$

$\qquad = \dfrac{1 - \cos^2 x}{\cos x}$

$\qquad = \dfrac{\sin^2 x}{\cos x}$

$\qquad = \sin x \tan x$

45. $\dfrac{\dfrac{1}{\sin x} + 1}{\dfrac{1}{\sin x} - 1} = \dfrac{\dfrac{1}{\sin x} + 1}{\dfrac{1}{\sin x} - 1} \cdot \dfrac{\sin x}{\sin x}$

$\qquad = \dfrac{1 + \sin x}{1 - \sin x}$

$\qquad = \dfrac{(1 + \sin x)}{1 - \sin x} \cdot \dfrac{(1 + \sin x)}{1 + \sin x}$

$\qquad = \dfrac{1 + 2\sin x + \sin^2 x}{1 - \sin^2 x}$

$\qquad = \dfrac{1 + 2\sin x + \sin^2 x}{\cos^2 x}$

$\qquad = \dfrac{1}{\cos^2 x} + \dfrac{2\sin x}{\cos^2 x} + \dfrac{\sin^2 x}{\cos^2 x}$

$\qquad = \sec^2 x + 2\tan x \sec x + \tan^2 x$

47. $\sin^4 x - \cos^4 x = (\sin^2 x + \cos^2 x)(\sin^2 x - \cos^2 x)$

$\qquad = 1(\sin^2 x - \cos^2 x)$

$\qquad = \sin^2 x - (1 - \sin^2 x)$

$\qquad = \sin^2 x - 1 + \sin^2 x$

$\qquad = 2\sin^2 x - 1$

49. $\dfrac{1}{1 - \cos x} = \dfrac{1}{1 - \cos x} \cdot \dfrac{1 + \cos x}{1 + \cos x}$

$\qquad = \dfrac{1 + \cos x}{1 - \cos^2 x}$

$\qquad = \dfrac{1 + \cos x}{\sin^2 x}$

51. $\dfrac{\sin x}{1 - \sin x} - \dfrac{\cos x}{1 - \sin x} = \dfrac{\sin x - \cos x}{1 - \sin x}$

$\qquad = \dfrac{\dfrac{\sin x}{\sin x} - \dfrac{\cos x}{\sin x}}{\dfrac{1}{\sin x} - \dfrac{\sin x}{\sin x}}$

$\qquad = \dfrac{1 - \cot x}{\csc x - 1}$

53. $\dfrac{1}{1+\cos x} - \dfrac{1}{1-\cos x} = \dfrac{(1-\cos x)-(1+\cos x)}{(1+\cos x)(1-\cos x)}$

$\qquad = \dfrac{1-\cos x - 1 - \cos x}{1-\cos^2 x}$

$\qquad = \dfrac{-2\cos x}{\sin^2 x}$

$\qquad = -2\cot x \csc x$

55. $\dfrac{\dfrac{1}{\sin x} + \csc x}{\dfrac{1}{\sin x} - \sin x} = \dfrac{\dfrac{1}{\sin x} + \csc x}{\dfrac{1}{\sin x} - \sin x} \cdot \dfrac{\sin x}{\sin x}$

$\qquad = \dfrac{1+1}{1-\sin^2 x}$

$\qquad = \dfrac{2}{\cos^2 x}$

57. $\sqrt{\dfrac{1+\sin x}{1-\sin x}} = \sqrt{\dfrac{1+\sin x}{1-\sin x} \cdot \dfrac{1+\sin x}{1+\sin x}}$

$\qquad = \sqrt{\dfrac{(1+\sin x)^2}{1-\sin^2 x}}$

$\qquad = \sqrt{\dfrac{(1+\sin x)^2}{\cos^2 x}}$

$\qquad = \dfrac{1+\sin x}{\cos x}, \;\; \cos x > 0$

59. $\dfrac{\sin^3 x + \cos^3 x}{\sin x + \cos x} = \dfrac{(\sin x + \cos x)(\sin^2 x - \sin x \cos x + \cos^2 x)}{\sin x + \cos x}$

$\qquad = \sin^2 x - \sin x \cos x + \cos^2 x$

$\qquad = 1 - \sin x \cos x$

61. $\dfrac{\sec x - 1}{\sec x + 1} - \dfrac{\sec x + 1}{\sec x - 1} = \dfrac{(\sec x - 1)^2 - (\sec x + 1)^2}{(\sec x - 1)(\sec x + 1)}$

$\qquad = \dfrac{\sec^2 x - 2\sec x + 1 - \sec^2 x - 2\sec x - 1}{\sec^2 x - 1}$

$\qquad = \dfrac{-4\sec x}{\tan^2 x}$

$\qquad = \dfrac{-4}{\cos x} \cdot \dfrac{\cos^2 x}{\sin^2 x}$

$\qquad = -4\csc x \cot x$

63. $\dfrac{1+\sin x}{\cos x} - \dfrac{\cos x}{1-\sin x} = \dfrac{(1+\sin x)(1-\sin x) - \cos x(\cos x)}{\cos x(1-\sin x)}$

$\qquad = \dfrac{1-\sin^2 x - \cos^2 x}{\cos x(1-\sin x)}$

$\qquad = \dfrac{\cos^2 x - \cos^2 x}{\cos x(1-\sin x)}$

$\qquad = 0$

65. $\dfrac{\sec x + \tan x}{\sec x - \tan x} = \dfrac{\dfrac{1}{\cos x} + \dfrac{\sin x}{\cos x}}{\dfrac{1}{\cos x} - \dfrac{\sin x}{\cos x}} \cdot \dfrac{\cos x}{\cos x}$

$\qquad = \dfrac{1+\sin x}{1-\sin x}$

$\qquad = \dfrac{1+\sin x}{1-\sin x} \cdot \dfrac{1+\sin x}{1+\sin x}$

$\qquad = \dfrac{(1+\sin x)^2}{1-\sin^2 x}$

$\qquad = \dfrac{(1+\sin x)^2}{\cos^2 x}$

67. $\sin^2 x + \cos^2 x = 1$

$\qquad \cos x^2 = 1 - \sin^2 x$

$\qquad \cos x = \pm\sqrt{1-\sin^2 x}$

69. $\sec x = \dfrac{1}{\cos x}$

$\qquad = \dfrac{1}{\pm\sqrt{\cos^2 x}}$

$\qquad = \dfrac{1}{\pm\sqrt{1-\sin^2 x}}$

$\qquad = \dfrac{\pm\sqrt{1-\sin^2 x}}{1-\sin^2 x}$

71.

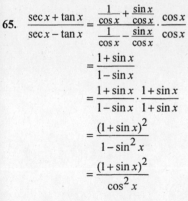

73.

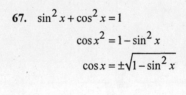

75.

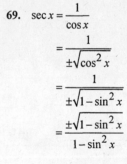

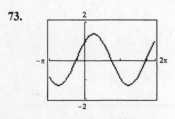

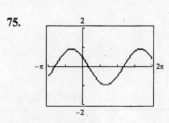

77.

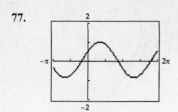

79. $\dfrac{1-\sin x+\cos x}{1+\sin x+\cos x} = \dfrac{1-\sin x+\cos x}{1+\sin x+\cos x}\cdot\dfrac{1+\sin x+\cos x}{1+\sin x-\cos x}$

$\qquad = \dfrac{1-\sin^2 x+2\sin x\cos x-\cos^2 x}{1+2\sin x+\sin^2 x-\cos^2 x}$

$\qquad = \dfrac{1-(\sin^2 x+\cos^2 x)+2\sin x\cos x}{1+2\sin x+\sin^2 x-(1-\sin^2 x)}$

$\qquad = \dfrac{1-1+2\sin x\cos x}{1+2\sin x+\sin^2 x-1+\sin^2 x}$

$\qquad = \dfrac{2\sin x\cos x}{2\sin x+2\sin^2 x} = \dfrac{2\sin x\cos x}{2\sin x(1+\sin x)}$

$\qquad = \dfrac{\cos x}{1+\sin x}$

81. $\dfrac{2\sin^4 x+2\sin^2 x\cos^2 x-3\sin^2 x-3\cos^2 x}{2\sin^2 x} = \dfrac{2\sin^2 x(\sin^2 x+\cos^2 x)-3(\sin^2 x+\cos^2 x)}{2\sin^2 x}$

$\qquad = \dfrac{(2\sin^2 x-3)(\sin^2 x+\cos^2 x)}{2\sin^2 x}$

$\qquad = \dfrac{2\sin^2 x-3}{2\sin^2 x}$

$\qquad = \dfrac{2\sin^2 x}{2\sin^2 x}-\dfrac{3}{2\sin^2 x}$

$\qquad = 1-\dfrac{3}{2}\csc^2 x$

83. $\dfrac{\sin x(\tan x+1)-2\tan x\cos x}{\sin x-\cos x} = \dfrac{\sin x\tan x+\sin x-2\dfrac{\sin x}{\cos x}\cos x}{\sin x-\cos x}$

$\qquad = \dfrac{\sin x\tan x+\sin x-2\sin x}{\sin x-\cos x}$

$\qquad = \dfrac{\sin x(\tan x-1)}{\sin x-\cos x}$

$\qquad = \dfrac{\dfrac{\sin x(\tan x-1)}{\cos x}}{\dfrac{\sin x}{\cos x}-\dfrac{\cos x}{\cos x}}$

$\qquad = \dfrac{\tan x(\tan x-1)}{\tan x-1}$

$\qquad = \tan x$

85. $\sin^4 x+\cos^4 x = \sin^4 x+2\sin^2 x\cos^2 x+\cos^4 x-2\sin^2 x\cos^2 x$

$\qquad = (\sin^2 x+\cos^2 x)^2-2\sin^2 x\cos^2 x$

$\qquad = 1-2\sin^2 x\cos^2 x$

SECTION 3.2, Page 192

1. $\sin(45°+30°) = \sin 45°\cos 30°+\cos 45°\sin 30°$

$\qquad = \dfrac{\sqrt{2}}{2}\cdot\dfrac{\sqrt{3}}{2}+\dfrac{\sqrt{2}}{2}\cdot\dfrac{1}{2}$

$\qquad = \dfrac{\sqrt{6}}{4}+\dfrac{\sqrt{2}}{4} = \dfrac{\sqrt{6}+\sqrt{2}}{4}$

3. $\cos(45°-30°) = \cos 45°\cos 30°+\sin 45°\sin 30°$

$\qquad = \dfrac{\sqrt{2}}{2}\cdot\dfrac{\sqrt{3}}{2}+\dfrac{\sqrt{2}}{2}\cdot\dfrac{1}{2}$

$\qquad = \dfrac{\sqrt{6}}{4}+\dfrac{\sqrt{2}}{4} = \dfrac{\sqrt{6}+\sqrt{2}}{4}$

5. $\tan(45° - 30°) = \dfrac{\tan 45° - \tan 30°}{1 + \tan 45° \tan 30°}$

$= \dfrac{1 - \dfrac{\sqrt{3}}{3}}{1 + 1\left(\dfrac{\sqrt{3}}{3}\right)} = \dfrac{\dfrac{3 - \sqrt{3}}{3}}{\dfrac{3 + \sqrt{3}}{3}} = \dfrac{3 - \sqrt{3}}{3 + \sqrt{3}} = 2 - \sqrt{3}$

7. $\sin\left(\dfrac{5\pi}{4} - \dfrac{\pi}{6}\right) = \sin\dfrac{5\pi}{4}\cos\dfrac{\pi}{6} - \cos\dfrac{5\pi}{4}\sin\dfrac{\pi}{6}$

$= -\dfrac{\sqrt{2}}{2} \cdot \dfrac{\sqrt{3}}{2} - \left(-\dfrac{\sqrt{2}}{2}\right) \cdot \dfrac{1}{2}$

$= -\dfrac{\sqrt{6}}{4} + \dfrac{\sqrt{2}}{4} = \dfrac{-\sqrt{6} + \sqrt{2}}{4}$

9. $\cos\left(\dfrac{3\pi}{4} + \dfrac{\pi}{6}\right) = \cos\dfrac{3\pi}{4}\cos\dfrac{\pi}{6} - \sin\dfrac{3\pi}{4}\sin\dfrac{\pi}{6}$

$= -\dfrac{\sqrt{2}}{2} \cdot \dfrac{\sqrt{3}}{2} - \dfrac{\sqrt{2}}{2} \cdot \dfrac{1}{2}$

$= -\dfrac{\sqrt{6}}{4} - \dfrac{\sqrt{2}}{4} = -\dfrac{\sqrt{6} + \sqrt{2}}{4}$

11. $\tan\left(\dfrac{\pi}{6} + \dfrac{\pi}{4}\right) = \dfrac{\tan\dfrac{\pi}{6} + \tan\dfrac{\pi}{4}}{1 - \tan\dfrac{\pi}{6}\tan\dfrac{\pi}{4}}$

$= \dfrac{\dfrac{\sqrt{3}}{3} + 1}{1 - \dfrac{\sqrt{3}}{3} \cdot 1} = \dfrac{\dfrac{\sqrt{3}+3}{3}}{\dfrac{3-\sqrt{3}}{3}}$

$= \dfrac{\sqrt{3}+3}{3-\sqrt{3}} \cdot \dfrac{3+\sqrt{3}}{3+\sqrt{3}} = \dfrac{9 + 6\sqrt{3} + 3}{9 - 3}$

$= \dfrac{12 + 6\sqrt{3}}{6} = 2 + \sqrt{3}$

13. $\cos 212° \cos 122° + \sin 212° \sin 122° = \cos(212° - 122°) = \cos 90° = 0$

15. $\sin\dfrac{5\pi}{12}\cos\dfrac{\pi}{4} - \cos\dfrac{5\pi}{12}\sin\dfrac{\pi}{4} = \sin\left(\dfrac{5\pi}{12} - \dfrac{\pi}{4}\right) = \sin\dfrac{\pi}{6} = \dfrac{1}{2}$

17. $\dfrac{\tan\dfrac{7\pi}{12} - \tan\dfrac{\pi}{4}}{1 + \tan\dfrac{7\pi}{12}\tan\dfrac{\pi}{4}} = \tan\left(\dfrac{7\pi}{12} - \dfrac{\pi}{4}\right) = \tan\dfrac{\pi}{3} = \sqrt{3}$

19. $\sin 7x \cos 2x - \cos 7x \sin 2x = \sin(7x - 2x) = \sin 5x$

21. $\cos x \cos 2x + \sin x \sin 2x = \cos(x - 2x) = \cos(-x) = \cos x$

23. $\sin 7x \cos 3x - \cos 7x \sin 3x = \sin(7x - 3x) = \sin 4x$

25. $\cos 4x \cos(-2x) - \sin 4x \sin(-2x) = \cos 4x \cos 2x + \sin 4x \sin 2x$

$\qquad\qquad = \cos(4x - 2x)$

$\qquad\qquad = \cos 2x$

27. $\sin\dfrac{x}{3}\cos\dfrac{2x}{3} + \cos\dfrac{x}{3}\sin\dfrac{2x}{3} = \sin\left(\dfrac{x}{3} + \dfrac{2x}{3}\right) = \sin x$

29. $\dfrac{\tan 3x + \tan 4x}{1 - \tan 3x \tan 4x} = \tan(3x + 4x) = \tan 7x$

31. $\tan\alpha = -\dfrac{4}{3}, \ \sin\alpha = \dfrac{4}{5}, \ \cos\alpha = -\dfrac{3}{5}, \ \tan\beta = \dfrac{15}{8}, \ \sin\beta = -\dfrac{15}{17}, \ \cos\beta = -\dfrac{8}{17}$

a. $\sin(\alpha - \beta) = \sin\alpha\cos\beta - \cos\alpha\sin\beta$

$= \left(\dfrac{4}{5}\right)\left(-\dfrac{8}{17}\right) - \left(-\dfrac{3}{5}\right)\left(-\dfrac{15}{17}\right)$

$= -\dfrac{32}{85} - \dfrac{45}{85} = -\dfrac{77}{85}$

b. $\cos(\alpha + \beta) = \cos\alpha\cos\beta - \sin\alpha\sin\beta$

$= \left(-\dfrac{3}{5}\right)\left(-\dfrac{8}{17}\right) - \left(\dfrac{4}{5}\right)\left(-\dfrac{15}{17}\right)$

$= \dfrac{24}{85} + \dfrac{60}{85} = \dfrac{84}{85}$

c. $\tan(\alpha - \beta) = \dfrac{\tan\alpha - \tan\beta}{1 + \tan\alpha\tan\beta}$

$= \dfrac{-\dfrac{4}{3} - \dfrac{15}{8}}{1 + \left(-\dfrac{4}{3}\right)\left(\dfrac{15}{8}\right)} = \dfrac{-\dfrac{4}{3} - \dfrac{15}{8}}{1 - \dfrac{60}{24}} \cdot \dfrac{24}{24}$

$= \dfrac{-32 - 45}{24 - 60} = \dfrac{77}{36}$

33. $\sin\alpha = \dfrac{3}{5}$, $\cos\alpha = \dfrac{4}{5}$, $\tan\alpha = \dfrac{3}{4}$, $\cos\beta = -\dfrac{5}{13}$, $\sin\beta = \dfrac{12}{13}$, $\tan\beta = -\dfrac{12}{5}$

a. $\sin(\alpha - \beta) = \sin\alpha\cos\beta - \cos\alpha\sin\beta$

$$= \frac{3}{5}\left(-\frac{5}{13}\right) - \frac{4}{5}\left(\frac{12}{13}\right)$$

$$= -\frac{15}{65} - \frac{48}{65}$$

$$= -\frac{63}{65}$$

b. $\cos(\alpha + \beta) = \cos\alpha\cos\beta - \sin\alpha\sin\beta$

$$= \frac{4}{5}\left(-\frac{5}{13}\right) - \frac{3}{5}\left(\frac{12}{13}\right)$$

$$= -\frac{20}{65} - \frac{36}{65}$$

$$= -\frac{56}{65}$$

c. $\tan(\alpha - \beta) = \dfrac{\tan\alpha - \tan\beta}{1 + \tan\alpha\tan\beta}$

$$= \frac{\frac{3}{4} - \left(-\frac{12}{5}\right)}{1 + \left(\frac{3}{4}\right)\left(-\frac{12}{5}\right)}$$

$$= \frac{\frac{3}{4} + \frac{12}{5}}{1 - \frac{36}{20}} \cdot \frac{20}{20}$$

$$= \frac{15 + 48}{20 - 36} = -\frac{63}{16}$$

35. $\sin\alpha = -\dfrac{4}{5}$, $\cos\alpha = -\dfrac{3}{5}$, $\tan\alpha = \dfrac{4}{3}$, $\cos\beta = -\dfrac{12}{13}$, $\sin\beta = \dfrac{5}{13}$, $\tan\beta = -\dfrac{5}{12}$

a. $\sin(\alpha - \beta) = \sin\alpha\cos\beta - \cos\alpha\sin\beta$

$$= \left(-\frac{4}{5}\right)\left(-\frac{12}{13}\right) - \left(-\frac{3}{5}\right)\frac{5}{13}$$

$$= \frac{48}{65} + \frac{15}{65}$$

$$= \frac{63}{65}$$

b. $\cos(\alpha + \beta) = \cos\alpha\cos\beta - \sin\alpha\sin\beta$

$$= \left(-\frac{3}{5}\right)\left(-\frac{12}{13}\right) - \left(-\frac{4}{5}\right)\frac{5}{13}$$

$$= \frac{36}{65} + \frac{20}{65}$$

$$= \frac{56}{65}$$

c. $\tan(\alpha + \beta) = \dfrac{\tan\alpha + \tan\beta}{1 - \tan\alpha\tan\beta}$

$$= \frac{\frac{4}{3} + \left(-\frac{5}{12}\right)}{1 - \left(\frac{4}{3}\right)\left(-\frac{5}{12}\right)}$$

$$= \frac{\frac{4}{3} - \frac{5}{12}}{1 + \frac{29}{36}} = \frac{\frac{4}{3} - \frac{5}{12}}{1 + \frac{20}{36}} \cdot \frac{36}{36}$$

$$= \frac{48 - 15}{36 + 20}$$

$$= \frac{33}{56}$$

37. $\cos\alpha = \dfrac{15}{17}$, $\sin\alpha = \dfrac{8}{17}$, $\tan\alpha = \dfrac{8}{15}$, $\sin\beta = -\dfrac{3}{5}$, $\cos\beta = -\dfrac{4}{5}$, $\tan\beta = \dfrac{3}{4}$

a. $\sin(\alpha + \beta) = \sin\alpha\cos\beta + \cos\alpha\sin\beta$

$$= \dfrac{8}{17}\left(-\dfrac{4}{5}\right) + \dfrac{15}{17}\left(-\dfrac{3}{5}\right)$$

$$= -\dfrac{32}{85} - \dfrac{45}{85} = -\dfrac{77}{85}$$

b. $\cos(\alpha - \beta) = \cos\alpha\cos\beta + \sin\alpha\sin\beta$

$$= \dfrac{15}{17}\left(-\dfrac{4}{5}\right) + \dfrac{8}{17}\left(-\dfrac{3}{5}\right)$$

$$= -\dfrac{60}{85} - \dfrac{24}{85}$$

$$= -\dfrac{84}{85}$$

c. $\tan(\alpha - \beta) = \dfrac{\tan\alpha - \tan\beta}{1 + \tan\alpha\tan\beta}$

$$= \dfrac{\dfrac{8}{15} - \dfrac{3}{4}}{1 + \dfrac{8}{15}\left(\dfrac{3}{4}\right)} = \dfrac{\dfrac{8}{15} - \dfrac{3}{4}}{1 + \dfrac{24}{60}} = \dfrac{\dfrac{8}{15} - \dfrac{3}{4}}{1 + \dfrac{24}{60}} \cdot \dfrac{60}{60}$$

$$= \dfrac{32 - 45}{60 + 24} = -\dfrac{13}{84}$$

39. $\cos\alpha = -\dfrac{3}{5}$, $\sin\alpha = -\dfrac{4}{5}$, $\tan\alpha = \dfrac{4}{3}$, $\sin\beta = \dfrac{5}{13}$, $\cos\beta = \dfrac{12}{13}$, $\tan\beta = \dfrac{5}{12}$

a. $\sin(\alpha - \beta) = \sin\alpha\cos\beta - \cos\alpha\sin\beta$

$$= \left(-\dfrac{4}{5}\right)\dfrac{12}{13} - \left(-\dfrac{3}{5}\right)\left(\dfrac{5}{13}\right)$$

$$= -\dfrac{48}{65} + \dfrac{15}{65}$$

$$= -\dfrac{33}{65}$$

b. $\cos(\alpha + \beta) = \cos\alpha\cos\beta - \sin\alpha\sin\beta$

$$= \left(-\dfrac{3}{5}\right)\dfrac{12}{13} - \left(-\dfrac{4}{5}\right)\dfrac{5}{13}$$

$$= -\dfrac{36}{65} + \dfrac{20}{65}$$

$$= -\dfrac{16}{65}$$

c. $\tan(\alpha + \beta) = \dfrac{\tan\alpha + \tan\beta}{1 - \tan\alpha\tan\beta}$

$$= \dfrac{\dfrac{4}{3} + \dfrac{5}{12}}{1 - \dfrac{4}{3}\left(\dfrac{5}{12}\right)} = \dfrac{\dfrac{4}{3} + \dfrac{5}{12}}{1 - \dfrac{20}{36}}$$

$$= \dfrac{\dfrac{4}{3} + \dfrac{5}{12}}{1 - \dfrac{20}{36}} \cdot \dfrac{36}{36} = \dfrac{48 + 15}{36 - 20}$$

$$= \dfrac{63}{16}$$

41. $\sin\alpha = \dfrac{3}{5}$, $\cos\alpha = \dfrac{4}{5}$, $\tan\alpha = \dfrac{3}{4}$, $\tan\beta = \dfrac{5}{12}$, $\sin\beta = -\dfrac{5}{13}$, $\cos\beta = -\dfrac{12}{13}$

a. $\sin(\alpha + \beta) = \sin\alpha\cos\beta + \cos\alpha\sin\beta$

$$= \left(\dfrac{3}{5}\right)\left(-\dfrac{12}{13}\right) + \left(\dfrac{4}{5}\right)\left(-\dfrac{5}{13}\right)$$

$$= -\dfrac{36}{65} - \dfrac{20}{65} = -\dfrac{56}{65}$$

b. $\cos(\alpha - \beta) = \cos\alpha\cos\beta + \sin\alpha\sin\beta$

$$= \left(\dfrac{4}{5}\right)\left(-\dfrac{12}{13}\right) + \left(\dfrac{3}{5}\right)\left(-\dfrac{5}{13}\right)$$

$$= -\dfrac{48}{65} - \dfrac{15}{65} = -\dfrac{63}{65}$$

c. $\tan(\alpha - \beta) = \dfrac{\tan\alpha - \tan\beta}{1 + \tan\alpha\tan\beta}$

$$= \dfrac{\dfrac{3}{4} - \dfrac{5}{12}}{1 + \left(\dfrac{3}{4}\right)\left(\dfrac{5}{12}\right)} = \dfrac{\dfrac{3}{4} - \dfrac{5}{12}}{1 + \dfrac{15}{48}} \cdot \dfrac{48}{48}$$

$$= \dfrac{36 - 20}{48 + 15} = \dfrac{16}{63}$$

43. $\cos\left(\dfrac{\pi}{2}-\theta\right)=\cos\dfrac{\pi}{2}\cos\theta+\sin\dfrac{\pi}{2}\sin\theta$

$\qquad\qquad= 0\cdot\cos\theta+1\cdot\sin\theta$

$\qquad\qquad=\sin\theta$

45. $\sin\left(\theta+\dfrac{\pi}{2}\right)=\sin\theta\cos\dfrac{\pi}{2}+\cos\theta\sin\dfrac{\pi}{2}$

$\qquad\qquad=\sin\theta(0)+\cos\theta(1)$

$\qquad\qquad=\cos\theta$

47. $\tan\left(\theta+\dfrac{\pi}{4}\right)=\dfrac{\tan\theta+\tan\dfrac{\pi}{4}}{1-\tan\theta\tan\dfrac{\pi}{4}}$

$\qquad\qquad=\dfrac{\tan\theta+1}{1-\tan\theta}$

49. $\cos\left(\dfrac{3\pi}{2}-\theta\right)=\cos\dfrac{3\pi}{2}\cos\theta+\sin\dfrac{3\pi}{2}\sin\theta$

$\qquad\qquad= 0(\cos\theta)+(-1)\sin\theta$

$\qquad\qquad=-\sin\theta$

51. $\cot\left(\dfrac{\pi}{2}-\theta\right)=\dfrac{\cos\left(\pi/2-\theta\right)}{\sin\left(\pi/2-\theta\right)}$

$\qquad\qquad=\dfrac{\left(\cos\dfrac{\pi}{2}\right)\cos\theta+\left(\sin\dfrac{\pi}{2}\right)\sin\theta}{\left(\sin\dfrac{\pi}{2}\right)\cos\theta-\left(\cos\dfrac{\pi}{2}\right)\sin\theta}$

$\qquad\qquad=\dfrac{(0)\cos\theta+(1)\sin\theta}{(1)\cos\theta-(0)\sin\theta}$

$\qquad\qquad=\dfrac{\sin\theta}{\cos\theta}$

$\qquad\qquad=\tan\theta$

53. $\csc(\pi-\theta)=\dfrac{1}{\sin(\pi-\theta)}$

$\qquad\qquad=\dfrac{1}{\sin\pi\cos\theta-\cos\pi\sin\theta}$

$\qquad\qquad=\dfrac{1}{(0)\cos\theta-(-1)\sin\theta}$

$\qquad\qquad=\dfrac{1}{\sin\theta}$

$\qquad\qquad=\csc\theta$

55. $\sin 6x\cos 2x-\cos 6x\sin 2x=\sin(6x-2x)$

$\qquad\qquad=\sin 4x$

$\qquad\qquad=\sin(2x+2x)$

$\qquad\qquad=\sin 2x\cos 2x+\cos 2x\sin 2x$

$\qquad\qquad=2\sin 2x\cos 2x$

57. $\cos(\alpha+\beta)+\cos(\alpha-\beta)=\cos\alpha\cos\beta-\sin\alpha\sin\beta+\cos\alpha\cos\beta+\sin\alpha\sin\beta$

$\qquad\qquad=2\cos\alpha\cos\beta$

59. $\sin(\alpha+\beta)+\sin(\alpha-\beta)=\sin\alpha\cos\beta+\cos\alpha\sin\beta+\sin\alpha\cos\beta-\cos\alpha\sin\beta$

$\qquad\qquad=2\sin\alpha\cos\beta$

61. $\dfrac{\cos(\alpha-\beta)}{\sin(\alpha+\beta)}=\dfrac{\cos\alpha\cos\beta+\sin\alpha\sin\beta}{\sin\alpha\cos\beta+\cos\alpha\sin\beta}$

$\qquad\qquad=\dfrac{\dfrac{\cos\alpha\cos\beta}{\sin\alpha\cos\beta}+\dfrac{\sin\alpha\sin\beta}{\sin\alpha\cos\beta}}{\dfrac{\sin\alpha\cos\beta}{\sin\alpha\cos\beta}+\dfrac{\cos\alpha\sin\beta}{\sin\alpha\cos\beta}}$

$\qquad\qquad=\dfrac{\cot\alpha+\tan\beta}{1+\cot\alpha\tan\beta}$

63. $\sin\left(\dfrac{\pi}{2}+\alpha-\beta\right)=\sin\left[\dfrac{\pi}{2}+(\alpha-\beta)\right]$

$\qquad\qquad=\sin\dfrac{\pi}{2}\cos(\alpha-\beta)+\cos\dfrac{\pi}{2}\sin(\alpha-\beta)$

$\qquad\qquad=(1)\cos(\alpha-\beta)+(0)\sin(\alpha-\beta)$

$\qquad\qquad=\cos(\alpha-\beta)$

$\qquad\qquad=\cos\alpha\cos\beta+\sin\alpha\sin\beta$

65. $\sin 3x=\sin(2x+x)$

$\qquad\quad=\sin 2x\cos x+\cos 2x\sin x$

$\qquad\quad=\sin(x+x)\cos x+\cos(x+x)\sin x$

$\qquad\quad=(\sin x\cos x+\cos x\sin x)\cos x+(\cos x\cos x-\sin x\sin x)\sin x$

$\qquad\quad=2\sin x\cos^2 x+\sin x\cos^2 x-\sin^3 x$

$\qquad\quad=3\sin x\cos^2 x-\sin^3 x$

$\qquad\quad=3\sin x(1-\sin^2 x)-\sin^3 x$

$\qquad\quad=3\sin x-3\sin^3 x-\sin^3 x$

$\qquad\quad=3\sin x-4\sin^3 x$

67. $\cos(\theta + 3\pi) = \cos\theta\cos 3\pi - \sin\theta\sin 3\pi$
$= (\cos\theta)(-1) - (\sin\theta)(0)$
$= -\cos\theta$

69. $\tan(\theta + \pi) = \dfrac{\tan\theta + \tan\pi}{1 - \tan\theta\tan\pi}$
$= \dfrac{\tan\theta + 0}{1 - (\tan\theta)(0)}$
$= \tan\theta$

71. $\sin(\theta + 2k\pi) = \sin\theta\cos(2k\pi) + \cos\theta\sin 2k\pi$
$= (\sin\theta)(1) + (\cos\theta)(0)$
$= \sin\theta$

73. $y = \sin\left(\dfrac{\pi}{2} - x\right)$ and $y = \cos x$ both have the following graph.

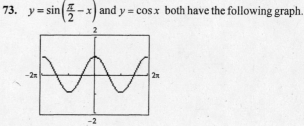

75. $y = \sin 7x\,\cos 2x - \cos 7x\,\sin 2x$ and $y = \sin 5x$ both have the following graph.

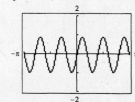

77. $\sin(x - y)\cdot\sin(x + y) = (\sin x\cos y - \cos x\sin y)(\sin x\cos y + \cos x\sin y)$
$= \sin^2 x\cos^2 y - \cos^2 x\sin^2 y$

79. $\cos(x + y + z) = \cos[x + (y + z)]$
$= \cos x\cos(y + z) - \sin x\sin(y + z)$
$= \cos x[\cos y\cos z - \sin y\sin z] - \sin x[\sin y\cos z + \cos y\sin z]$
$= \cos x\cos y\cos z - \cos x\sin y\sin z - \sin x\sin y\cos z - \sin x\cos y\sin z$

81. $\dfrac{\cos(x - y)}{\cos x\sin y} = \dfrac{\cos x\cos y + \sin x\sin y}{\cos x\sin y}$
$= \dfrac{\cos x\cos y}{\cos x\sin y} + \dfrac{\sin x\sin y}{\cos x\sin y}$
$= \cot y + \tan x$

83. $\dfrac{\cos(x + h) - \cos x}{h} = \dfrac{\cos x\cos h - \sin x\sin h - \cos x}{h}$
$= \dfrac{\cos x(\cos h - 1)}{h} - \dfrac{\sin x\sin h}{h}$
$= \cos x\dfrac{\cos h - 1}{h} - \sin x\dfrac{\sin h}{h}$

SECTION 3.3, Page 199

1. $2\sin 2\alpha\cos 2\alpha = \sin 2(2\alpha)$
$= \sin 4\alpha$

3. $1 - 2\sin^2 5\beta = \cos 2(5\beta)$
$= \cos 10\beta$

5. $\cos^2 3\alpha - \sin^2 3\alpha = \cos 2(3\alpha)$
$= \cos 6\alpha$

7. $\dfrac{2\tan 3\alpha}{1 - \tan^2 3\alpha} = \tan 2(3\alpha)$
$= \tan 6\alpha$

9. $\sin 75° = \sin\dfrac{150°}{2}$
$= +\sqrt{\dfrac{1 - \cos 150°}{2}}$
$= \sqrt{\dfrac{1 - (-\sqrt{3}/2)}{2}}$
$= \sqrt{\dfrac{2 + \sqrt{3}}{4}}$
$= \dfrac{\sqrt{2 + \sqrt{3}}}{2}$

11. $\tan 67.5° = \tan\dfrac{135°}{2}$
$= \dfrac{1 - \cos 135°}{\sin 135°}$
$= \dfrac{1 - (-\sqrt{2}/2)}{\sqrt{2}/2}$
$= \dfrac{2 + \sqrt{2}}{\sqrt{2}}\cdot\dfrac{\sqrt{2}}{\sqrt{2}}$
$= \dfrac{2\sqrt{2} + 2}{2}$
$= \sqrt{2} + 1$

13. $\cos 157.5° = \cos\dfrac{315°}{2}$

$\quad = -\sqrt{\dfrac{1+\cos 315°}{2}}$

$\quad = -\sqrt{\dfrac{1+(\sqrt{2}/2)}{2}}$

$\quad = -\sqrt{\dfrac{2+\sqrt{2}}{4}}$

$\quad = -\dfrac{\sqrt{2+\sqrt{2}}}{2}$

15. $\sin 22.5° = \sin\dfrac{45°}{2}$

$\quad = +\sqrt{\dfrac{1-\cos 45°}{2}}$

$\quad = \sqrt{\dfrac{1-\sqrt{2}/2}{2}}$

$\quad = \sqrt{\dfrac{2-\sqrt{2}}{4}}$

$\quad = \dfrac{\sqrt{2-\sqrt{2}}}{2}$

17. $\sin\dfrac{7\pi}{8} = \sin\dfrac{7\pi/4}{2}$

$\quad = +\sqrt{\dfrac{1-\cos 7\pi/4}{2}}$

$\quad = \sqrt{\dfrac{1-\sqrt{2}/2}{2}}$

$\quad = \sqrt{\dfrac{2-\sqrt{2}}{4}}$

$\quad = \dfrac{\sqrt{2-\sqrt{2}}}{2}$

19. $\cos\dfrac{5\pi}{12} = \cos\dfrac{5\pi/6}{2}$

$\quad = +\sqrt{\dfrac{1+\cos 5\pi/6}{2}}$

$\quad = \sqrt{\dfrac{1-\sqrt{3}/2}{2}}$

$\quad = \sqrt{\dfrac{2-\sqrt{3}}{4}}$

$\quad = \dfrac{\sqrt{2-\sqrt{3}}}{2}$

21. $\tan\dfrac{7\pi}{12} = \tan\dfrac{7\pi/6}{2}$

$\quad = \dfrac{1-\cos 7\pi/6}{\sin 7\pi/6}$

$\quad = \dfrac{1-(-\sqrt{3}/2)}{-1/2}$

$\quad = \dfrac{2+\sqrt{3}}{-1}$

$\quad = -2-\sqrt{3}$

23. $\cos\dfrac{\pi}{12} = \cos\dfrac{\pi/6}{2}$

$\quad = +\sqrt{\dfrac{1+\cos \pi/6}{2}}$

$\quad = \sqrt{\dfrac{1+\sqrt{3}/2}{2}}$

$\quad = \sqrt{\dfrac{2+\sqrt{3}}{4}}$

$\quad = \dfrac{\sqrt{2+\sqrt{3}}}{2}$

25. $\cos\theta = -\dfrac{4}{5}, \ \sin\theta = \sqrt{1-\left(\dfrac{4}{5}\right)^2} = \dfrac{3}{5}, \ \tan\theta = \dfrac{\sin\theta}{\cos\theta} = \dfrac{3/5}{-4/5} = -\dfrac{3}{4}$

$\sin 2\theta = 2\sin\theta\cos\theta$

$\quad = 2\left(\dfrac{3}{5}\right)\left(-\dfrac{4}{5}\right)$

$\quad = -\dfrac{24}{25}$

$\cos 2\theta = \cos^2\theta - \sin^2\theta$

$\quad = \left(-\dfrac{4}{5}\right)^2 - \left(\dfrac{3}{5}\right)^2$

$\quad = \dfrac{16}{25} - \dfrac{9}{25}$

$\quad = \dfrac{7}{25}$

$\tan 2\theta = \dfrac{2\tan\theta}{1-\tan^2\theta}$

$\quad = \dfrac{2\left(-\dfrac{3}{4}\right)}{1-\left(-\dfrac{3}{4}\right)^2} = \dfrac{-\dfrac{6}{4}}{1-\dfrac{9}{16}}$

$\quad = \dfrac{-\dfrac{6}{4}}{1-\dfrac{9}{16}}\cdot\dfrac{16}{16} = \dfrac{-24}{16-9}$

$\quad = -\dfrac{24}{7}$

27. $\sin\theta = \dfrac{8}{17}, \ \cos\theta = -\sqrt{1-\left(\dfrac{8}{17}\right)^2} = -\dfrac{15}{17}, \ \tan\theta = \dfrac{\sin\theta}{\cos\theta} = \dfrac{8/17}{-15/17} = -\dfrac{8}{15}$

$\sin 2\theta = 2\sin\theta\cos\theta$

$\quad = 2\left(\dfrac{8}{17}\right)\left(-\dfrac{15}{17}\right)$

$\quad = -\dfrac{240}{289}$

$\cos 2\theta = \cos^2\theta - \sin^2\theta$

$\quad = \left(-\dfrac{15}{17}\right)^2 - \left(\dfrac{8}{17}\right)^2$

$\quad = \dfrac{225}{289} - \dfrac{64}{289}$

$\quad = \dfrac{161}{289}$

$\tan 2\theta = \dfrac{2\tan\theta}{1-\tan^2\theta}$

$\quad = \dfrac{2\left(-\dfrac{8}{15}\right)}{1-\left(-\dfrac{8}{15}\right)^2} = \dfrac{-\dfrac{16}{15}}{1-\dfrac{64}{225}}$

$\quad = \dfrac{-\dfrac{16}{15}}{1-\dfrac{64}{225}}\cdot\dfrac{225}{225} = \dfrac{-240}{225-64}$

$\quad = -\dfrac{240}{161}$

29. $\tan\theta = -\dfrac{24}{7}$, $r = \sqrt{24^2 + 7^2} = \sqrt{576 + 49} = \sqrt{625} = 25$, $\sin\theta = -24/25$, $\cos\theta = 7/25$

$$\sin 2\theta = 2\sin\theta\cos\theta$$
$$= 2\left(-\frac{24}{25}\right)\left(\frac{7}{25}\right)$$
$$= -\frac{336}{625}$$

$$\cos 2\theta = \cos^2\theta - \sin^2\theta$$
$$= \left(\frac{7}{25}\right)^2 - \left(-\frac{24}{25}\right)^2$$
$$= \frac{49}{625} - \frac{576}{625}$$
$$= -\frac{527}{625}$$

$$\tan 2\theta = \frac{2\tan\theta}{1-\tan^2\theta}$$
$$= \frac{2\left(-\frac{24}{7}\right)}{1-\left(-\frac{24}{7}\right)^2}$$
$$= \frac{-\frac{48}{7}}{1-\frac{576}{49}} \cdot \frac{49}{49}$$
$$= \frac{-336}{49-576} = \frac{336}{527}$$

31. $\sin\theta = \dfrac{15}{17}$, $\cos\theta = \sqrt{1-\left(\dfrac{15}{17}\right)^2} = \sqrt{1-\dfrac{225}{289}} = \sqrt{\dfrac{289-225}{289}} = \sqrt{\dfrac{64}{289}} = \dfrac{8}{17}$, $\tan\theta = \dfrac{15/17}{8/17} = \dfrac{15}{8}$

$$\sin 2\theta = 2\sin\theta\cos\theta$$
$$= 2\left(\frac{15}{17}\right)\left(\frac{8}{17}\right)$$
$$= \frac{240}{289}$$

$$\cos 2\theta = \cos^2\theta - \sin^2\theta$$
$$= \left(\frac{8}{17}\right)^2 - \left(\frac{15}{17}\right)^2$$
$$= \frac{64}{289} - \frac{225}{289}$$
$$= -\frac{161}{289}$$

$$\tan 2\theta = \frac{2\tan\theta}{1-\tan^2\theta}$$
$$= \frac{2\left(\frac{15}{8}\right)}{1-\left(\frac{15}{8}\right)^2} = \frac{\frac{15}{4}}{1-\frac{225}{64}}$$
$$= \frac{\frac{15}{4}}{1-\frac{225}{64}} \cdot \frac{64}{64} = \frac{240}{64-225}$$
$$= -\frac{240}{161}$$

33. $\cos\theta = \dfrac{40}{41}$, $\sin\theta = -\sqrt{1-\left(\dfrac{40}{41}\right)^2} = -\sqrt{1-\dfrac{1600}{1681}} = -\dfrac{9}{41}$, $\tan\theta = \dfrac{-9/41}{40/41} = -\dfrac{9}{40}$

$$\sin 2\theta = 2\sin\theta\cos\theta$$
$$= 2\left(-\frac{9}{41}\right)\left(\frac{40}{41}\right)$$
$$= -\frac{720}{1681}$$

$$\cos 2\theta = \cos^2\theta - \sin^2\theta$$
$$= \left(\frac{40}{41}\right)^2 - \left(-\frac{9}{41}\right)^2$$
$$= \frac{1600}{1681} - \frac{81}{1681}$$
$$= \frac{1519}{1681}$$

$$\tan 2\theta = \frac{2\tan\theta}{1-\tan^2\theta}$$
$$= \frac{2\left(-\frac{9}{40}\right)}{1-\left(-\frac{9}{40}\right)^2} = \frac{-\frac{9}{20}}{1-\frac{81}{1600}}$$
$$= \frac{-\frac{9}{20}}{1-\frac{81}{1600}} \cdot \frac{1600}{1600}$$
$$= \frac{-720}{1600-81} = -\frac{720}{1519}$$

35. $\tan\theta = \dfrac{15}{8}$, $r = \sqrt{15^2 + 8^2} = \sqrt{225 + 64} = 17$, $\sin\theta = -\dfrac{15}{17}$, $\cos\theta = -\dfrac{8}{17}$

$$\sin 2\theta = 2\sin\theta\cos\theta$$
$$= 2\left(-\frac{15}{17}\right)\left(-\frac{8}{17}\right)$$
$$= \frac{240}{289}$$

$$\cos 2\theta = \cos^2\theta - \sin^2\theta$$
$$= \left(-\frac{8}{17}\right)^2 - \left(-\frac{15}{17}\right)^2$$
$$= \frac{64}{289} - \frac{225}{289}$$
$$= -\frac{161}{289}$$

$$\tan 2\theta = \frac{2\tan\theta}{1-\tan^2\theta}$$
$$= \frac{2\left(\frac{15}{8}\right)}{1-\left(\frac{15}{8}\right)^2} = \frac{\frac{15}{4}}{1-\frac{225}{64}}$$
$$= \frac{\frac{15}{4}}{1-\frac{225}{64}} \cdot \frac{64}{64} = \frac{120}{64-225}$$
$$= -\frac{120}{161}$$

37. $\sin\alpha = \dfrac{5}{13}$, $\cos\alpha = -\sqrt{1-\left(\dfrac{5}{13}\right)^2} = -\sqrt{1-\dfrac{25}{169}} = -\dfrac{12}{13}$.

$$\sin\frac{\alpha}{2} = \sqrt{\frac{1-\cos\alpha}{2}} \qquad\qquad \cos\frac{\alpha}{2} = \sqrt{\frac{1+\cos\alpha}{2}} \qquad\qquad \tan\frac{\alpha}{2} = \frac{1-\cos\alpha}{\sin\alpha}$$

$$= \sqrt{\frac{1-(-12/13)}{2}} \qquad\qquad = \sqrt{\frac{1-12/13}{2}} \qquad\qquad = \frac{1+\dfrac{12}{13}}{\dfrac{5}{13}}$$

$$= \sqrt{\frac{13+12}{26}} \qquad\qquad = \sqrt{\frac{13-12}{26}} \qquad\qquad = \frac{13+12}{5}$$

$$= \sqrt{\frac{25}{26}} = \frac{5}{\sqrt{26}} \qquad\qquad = \sqrt{\frac{1}{26}} = \frac{1}{\sqrt{26}} \qquad\qquad = 5$$

$$= \frac{5\sqrt{26}}{26} \qquad\qquad = \frac{\sqrt{26}}{26}$$

39. $\cos\alpha = -\dfrac{8}{17}$, $\sin\alpha = -\sqrt{1-\left(-\dfrac{8}{17}\right)^2} = -\sqrt{1-\dfrac{64}{289}} = -\dfrac{15}{17}$

$$\sin\frac{\alpha}{2} = \sqrt{\frac{1-\cos\alpha}{2}} \qquad\qquad \cos\frac{\alpha}{2} = -\sqrt{\frac{1+\cos\alpha}{2}} \qquad\qquad \tan\frac{\alpha}{2} = \frac{1-\cos\alpha}{\sin\alpha}$$

$$= \sqrt{\frac{1-(-8/17)}{2}} \qquad\qquad = -\sqrt{\frac{1-8/17}{2}} \qquad\qquad = \frac{1+\dfrac{18}{17}}{-\dfrac{15}{17}}$$

$$= \sqrt{\frac{17+8}{34}} \qquad\qquad = -\sqrt{\frac{17-8}{34}} \qquad\qquad = \frac{17+8}{-15}$$

$$= \sqrt{\frac{25}{34}} \qquad\qquad = -\sqrt{\frac{9}{34}} \qquad\qquad = -\frac{5}{3}$$

$$= \frac{5\sqrt{34}}{34} \qquad\qquad = -\frac{3\sqrt{34}}{34}$$

41. $\tan\alpha = \dfrac{4}{3}$, $r = \sqrt{3^2+4^2} = \sqrt{25} = 5$, $\sin\alpha = \dfrac{4}{5}$, $\cos = \dfrac{3}{5}$

$$\sin\frac{\alpha}{2} = \sqrt{\frac{1-\cos\alpha}{2}} \qquad\qquad \cos\frac{\alpha}{2} = \sqrt{\frac{1+\cos\alpha}{2}} \qquad\qquad \tan\frac{\alpha}{2} = \frac{1-\cos\alpha}{\sin\alpha}$$

$$= \sqrt{\frac{1-3/5}{2}} \qquad\qquad = \sqrt{\frac{1+3/5}{2}} \qquad\qquad = \frac{1-\dfrac{3}{5}}{\dfrac{4}{5}}$$

$$= \sqrt{\frac{5-3}{10}} \qquad\qquad = \sqrt{\frac{5+3}{10}} \qquad\qquad = \frac{5-3}{4}$$

$$= \sqrt{\frac{1}{5}} \qquad\qquad = \sqrt{\frac{4}{5}} \qquad\qquad = \frac{1}{2}$$

$$= \frac{\sqrt{5}}{5} \qquad\qquad = \frac{2\sqrt{5}}{5}$$

43. $\cos\alpha = \dfrac{24}{25}$, $\sin\alpha = -\sqrt{1-\left(\dfrac{24}{25}\right)^2} = -\sqrt{1-\dfrac{576}{625}} = -\dfrac{7}{25}$

$$\sin\frac{\alpha}{2} = \sqrt{\frac{1-\cos\alpha}{2}} \qquad\qquad \cos\frac{\alpha}{2} = -\sqrt{\frac{1+\cos\alpha}{2}} \qquad\qquad \tan\frac{\alpha}{2} = \frac{1-\cos\alpha}{\sin\alpha}$$

$$= \sqrt{\frac{1-24/25}{2}} \qquad\qquad = -\sqrt{\frac{1+24/25}{2}} \qquad\qquad = \frac{1-\dfrac{24}{25}}{-\dfrac{7}{25}} = \frac{25-24}{-7}$$

$$= \sqrt{\frac{25-24}{50}} = \sqrt{\frac{1}{50}} \qquad\qquad = -\sqrt{\frac{25+24}{50}} = -\sqrt{\frac{49}{50}} \qquad\qquad = -\frac{1}{7}$$

$$= \frac{\sqrt{2}}{10} \qquad\qquad = -\frac{7\sqrt{2}}{10}$$

45. $\sec\alpha = \dfrac{17}{15}, \ \cos\alpha = \dfrac{15}{17}, \ \sin\alpha = \sqrt{1-\left(\dfrac{15}{17}\right)^2} = \sqrt{1-\dfrac{225}{289}} = \dfrac{8}{17}$

$$
\begin{aligned}
\sin\frac{\alpha}{2} &= \sqrt{\frac{1-\cos\alpha}{2}} \\
&= \sqrt{\frac{1-15/17}{2}} \\
&= \sqrt{\frac{17-15}{34}} = \sqrt{\frac{1}{17}} \\
&= \frac{\sqrt{17}}{17}
\end{aligned}
\qquad
\begin{aligned}
\cos\frac{\alpha}{2} &= \sqrt{\frac{1+\cos\alpha}{2}} \\
&= \sqrt{\frac{1+15/17}{2}} \\
&= \sqrt{\frac{17+15}{34}} = \sqrt{\frac{16}{17}} \\
&= \frac{4}{\sqrt{17}} \\
&= \frac{4\sqrt{17}}{17}
\end{aligned}
\qquad
\begin{aligned}
\tan\frac{\alpha}{2} &= \frac{1-\cos\alpha}{\sin\alpha} \\
&= \frac{1-\frac{15}{17}}{\frac{8}{17}} = \frac{17-15}{8} \\
&= \frac{1}{4}
\end{aligned}
$$

47. $\cot\alpha = \dfrac{8}{15}, \ r = \sqrt{8^2+15^2} = \sqrt{64+225} = 17, \ \sin\alpha = -\dfrac{15}{17}, \ \cos\alpha = -\dfrac{8}{17}$

$$
\begin{aligned}
\sin\frac{\alpha}{2} &= \sqrt{\frac{1-\cos\alpha}{2}} \\
&= \sqrt{\frac{1-(-8/17)}{2}} \\
&= \sqrt{\frac{17+8}{34}} = \sqrt{\frac{25}{34}} \\
&= \frac{5\sqrt{34}}{34}
\end{aligned}
\qquad
\begin{aligned}
\cos\frac{\alpha}{2} &= -\sqrt{\frac{1+\cos\alpha}{2}} \\
&= -\sqrt{\frac{1+(-8/17)}{2}} \\
&= -\sqrt{\frac{17-8}{34}} = -\sqrt{\frac{9}{34}} \\
&= -\frac{3\sqrt{34}}{34}
\end{aligned}
\qquad
\begin{aligned}
\tan\frac{\alpha}{2} &= \frac{1-\cos\alpha}{\sin\alpha} \\
&= \frac{1-\left(-\frac{8}{17}\right)}{-\frac{15}{17}} \\
&= \frac{17+8}{-15} = \frac{25}{-15} \\
&= -\frac{5}{3}
\end{aligned}
$$

49. $\begin{aligned}[t] \sin 3x \cos 3x &= \frac{1}{2}(2\sin 3x \cos 3x) \\ &= \frac{1}{2}\sin 2(3x) \\ &= \frac{1}{2}\sin 6x \end{aligned}$

51. $\begin{aligned}[t] \sin^2 x + \cos 2x &= \sin^2 x + \cos^2 x - \sin^2 x \\ &= \cos^2 x \end{aligned}$

53. $\begin{aligned}[t] \frac{1+\cos 2x}{\sin 2x} &= \frac{1+2\cos^2 x - 1}{2\sin x \cos x} \\ &= \frac{2\cos^2 x}{2\sin x \cos x} \\ &= \cot x \end{aligned}$

55. $\begin{aligned}[t] \frac{\sin 2x}{1-\sin^2 x} &= \frac{2\sin x \cos x}{\cos^2 x} \\ &= 2\tan x \end{aligned}$

57. $\begin{aligned}[t] \frac{\cos 2x}{\cos^2 x} &= \frac{\cos^2 x - \sin^2 x}{\cos^2 x} \\ &= \frac{\cos^2 x}{\cos^2 x} - \frac{\sin^2 x}{\cos^2 x} \\ &= 1 - \tan^2 x \end{aligned}$

59. $\begin{aligned}[t] \sin 2x - \tan x &= 2\sin x \cos x - \frac{\sin x}{\cos x} \\ &= \frac{2\sin x \cos^2 x - \sin x}{\cos x} \\ &= \frac{\sin x(2\cos^2 x - 1)}{\cos x} \\ &= \tan x \cos 2x \end{aligned}$

61. $\begin{aligned}[t] \cos^4 x - \sin^4 x &= (\cos^2 x + \sin^2 x)(\cos^2 x - \sin^2 x) \\ &= \cos^2 x - \sin^2 x \\ &= \cos 2x \end{aligned}$

63. $\cos^2 x - 2\sin^2 x \cos^2 x - \sin^2 x + 2\sin^4 x = \cos^2 x(1 - 2\sin^2 x) - \sin^2 x(1 - 2\sin^2 x)$

$$= (1 - 2\sin^2 x)(\cos^2 x - \sin^2 x)$$
$$= \cos 2x \cos 2x$$
$$= \cos^2 2x$$

65. $\cos 4x = \cos 2(2x)$

$$= 2\cos^2 2x - 1$$
$$= 2(2\cos^2 x - 1)^2 - 1$$
$$= 2(4\cos^4 x - 4\cos^2 x + 1) - 1$$
$$= 8\cos^4 x - 8\cos^2 x + 1$$

67. $\cos 3x - \cos x = \cos(2x + x) - \cos x$

$$= \cos 2x \cos x - \sin 2x \sin x - \cos x$$
$$= (2\cos^2 x - 1)\cos x - 2\sin x \cos x \cdot \sin x - \cos x$$
$$= 2\cos^3 x - \cos x - 2\sin^2 x \cos x - \cos x$$
$$= 2\cos^3 x - 2\cos x - 2\sin^2 x \cos x$$
$$= 2\cos^3 x - 2\cos x - 2(1 - \cos^2 x)\cos x$$
$$= 2\cos^3 x - 2\cos x - 2\cos x + 2\cos^3 x$$
$$= 4\cos^3 x - 4\cos x$$

69. $\sin^3 x + \cos^3 x = (\sin x + \cos x)(\sin^2 x - \sin x \cos x + \cos^2 x)$

$$= (\sin x + \cos x)\left(\sin^2 x + \cos^2 x - \frac{2\sin x \cos x}{2}\right)$$
$$= (\sin x + \cos x)\left(1 - \frac{1}{2}\sin 2x\right)$$

71. $\sin^2 \dfrac{x}{2} = \left[\pm\sqrt{\dfrac{1 - \cos x}{2}}\right]^2$

$$= \frac{1 - \cos x}{2}$$
$$= \frac{1 - \cos x}{2} \cdot \frac{\sec x}{\sec x}$$
$$= \frac{\sec x - 1}{2\sec x}$$

73. $\tan \dfrac{x}{2} = \dfrac{1 - \cos x}{\sin x}$

$$= \frac{1}{\sin x} - \frac{\cos x}{\sin x}$$
$$= \csc x - \cot x$$

75. $2\sin\dfrac{x}{2}\cos\dfrac{x}{2} = \sin 2\left(\dfrac{x}{2}\right)$

$$= \sin x$$

77. $\left(\cos\dfrac{x}{2} + \sin\dfrac{x}{2}\right)^2 = \cos^2\dfrac{x}{2} + 2\sin\dfrac{x}{2}\cos\dfrac{x}{2} + \sin^2\dfrac{x}{2}$

$$= \cos^2\frac{x}{2} + \sin^2\frac{x}{2} + \sin 2\left(\frac{x}{2}\right)$$
$$= 1 + \sin x$$

79. $\sin^2\dfrac{x}{2}\sec x = \left(\pm\sqrt{\dfrac{1 - \cos x}{2}}\right)^2 \sec x$

$$= \frac{1 - \cos x}{2} \cdot \sec x$$
$$= \frac{1}{2}(\sec x - 1)$$

81. $\cos^2\dfrac{x}{2} - \cos x = \left(\pm\sqrt{\dfrac{1 + \cos x}{2}}\right)^2 - \cos x$

$$= \frac{1 + \cos x}{2} - \cos x$$
$$= \frac{1 + \cos x - 2\cos x}{2}$$
$$= \frac{1 - \cos x}{2}$$
$$= \sin^2\frac{x}{2}$$

83. $\sin^2\dfrac{x}{2} - \cos^2\dfrac{x}{2} = -\left(\cos^2\dfrac{x}{2} - \sin^2\dfrac{x}{2}\right)$

$\qquad = -\cos 2\left(\dfrac{x}{2}\right)$

$\qquad = -\cos x$

85. $\sin 2x - \cos x = 2\sin x\cos x - \cos x$

$\qquad = (\cos x)(2\sin x - 1)$

87. $\tan 2x = \dfrac{2\tan x}{1 - \tan^2 x}$

$\qquad = \dfrac{\dfrac{2\tan x}{\tan x}}{\dfrac{1}{\tan x} - \dfrac{\tan^2 x}{\tan x}}$

$\qquad = \dfrac{2}{\cot x - \tan x}$

89. $\dfrac{\sin^2 x + 1 - \cos^2 x}{\sin x(1 + \cos x)} = \dfrac{1 - \cos^2 x + 1 - \cos^2 x}{\sin x(1 + \cos x)}$

$\qquad = \dfrac{2(1 - \cos^2 x)}{\sin x(1 + \cos x)}$

$\qquad = \dfrac{2(1 - \cos x)(1 + \cos x)}{\sin x(1 + \cos x)}$

$\qquad = \dfrac{2(1 - \cos x)}{\sin x}$

$\qquad = 2\tan\dfrac{x}{2}$

91. $\csc 2x = \dfrac{1}{\sin 2x}$

$\qquad = \dfrac{1}{2\sin x\cos x}$

$\qquad = \tfrac{1}{2}\csc x\sec x$

93. $\cos\dfrac{x}{5} = \cos 2\left(\dfrac{x}{10}\right)$

$\qquad = 1 - 2\sin^2\dfrac{x}{10}$

95. $y = \sin^2 x + \cos 2x$ and $y = \cos^2 x$ both have the following graph.

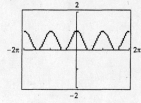

97. $y = 2\sin\dfrac{x}{2}\cos\dfrac{x}{2}$ and $y = \sin x$ both have the following graph.

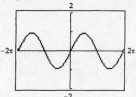

99. $\dfrac{\sin^3 x + \cos^3 x}{\sin x + \cos x} = \dfrac{(\sin x + \cos x)(\sin^2 x - \sin x\cos x + \cos^2 x)}{\sin x + \cos x}$

$\qquad = \sin^2 x + \cos^2 x - \dfrac{2\sin x\cos x}{2}$

$\qquad = 1 - \tfrac{1}{2}\sin 2x$

101. $\sin\dfrac{x}{2} - \cos\dfrac{x}{2} = \sqrt{\left(\sin\dfrac{x}{2} + \cos\dfrac{x}{2}\right)^2}$

$\qquad = \sqrt{\sin^2\dfrac{x}{2} - 2\sin\dfrac{x}{2}\cos\dfrac{x}{2} + \cos^2\dfrac{x}{2}}$

$\qquad = \sqrt{\sin^2\dfrac{x}{2} + \cos^2\dfrac{x}{2} - 2\sin\dfrac{x}{2}\cos\dfrac{x}{2}}$

$\qquad = \sqrt{1 - \sin x},\ 0° \le x \le 90°$

103. $x + y = 90°,\ y = 90° - x$

$\sin(x - y) = \sin(x - (90° - x))$

$\qquad = \sin(x - 90° + x) = \sin(2x - 90°)$

$\qquad = \sin 2x\cos 90° - \cos 2x\sin 90°$

$\qquad = \sin 2x(0) - \cos 2x(1)$

$\qquad = -\cos 2x$

105. $x + y = 180°,\ y = 180° - x$

$\sin(x - y) = \sin(x - 180° + x)$

$\qquad = \sin(2x - 180°)$

$\qquad = \sin 2x\cos 180° - \cos 2x\sin 180°$

$\qquad = \sin 2x(-1) - \cos 2x(0)$

$\qquad = -\sin 2x$

SECTION 3.4, Page 208

1. $2\sin x\cos 2x = 2\cdot\dfrac{1}{2}\big[\sin(x+2x)+\sin(x-2x)\big]$

$\qquad\qquad\quad = \sin 3x + \sin(-x)$

$\qquad\qquad\quad = \sin 3x - \sin x$

3. $\cos 6x\sin 2x = \dfrac{1}{2}\big[\sin(6x+2x)-\sin(6x-2x)\big]$

$\qquad\qquad\quad = \dfrac{1}{2}\big[\sin 8x - \sin 4x\big]$

5. $2\sin 5x\cos 3x = \sin(5x+3x)+\sin(5x-3x)$

$\qquad\qquad\quad = \sin 8x + \sin 2x$

7. $\sin x\cos 5x = \dfrac{1}{2}\big[\cos(x-5x)-\cos(x+5x)\big]$

$\qquad\qquad\quad = \dfrac{1}{2}\big[\cos(-4x)-\cos 6x\big]$

$\qquad\qquad\quad = \dfrac{1}{2}(\cos 4x - \cos 6x)$

9. $\cos 75^\circ\cos 15^\circ = \dfrac{1}{2}\big[\cos(75^\circ+15^\circ)+\cos(75^\circ-15^\circ)\big]$

$\qquad\qquad\quad = \dfrac{1}{2}(\cos 90^\circ + \cos 60^\circ)$

$\qquad\qquad\quad = \dfrac{1}{2}\left(0+\dfrac{1}{2}\right)$

$\qquad\qquad\quad = \dfrac{1}{4}$

11. $\cos 157.5^\circ\sin 22.5^\circ = \dfrac{1}{2}\big[\sin(157.5^\circ+22.5^\circ)-\sin(157.5^\circ-22.5^\circ)\big]$

$\qquad\qquad\quad = \dfrac{1}{2}(\sin 180^\circ - \sin 135^\circ)$

$\qquad\qquad\quad = \dfrac{1}{2}\left(0-\dfrac{\sqrt{2}}{2}\right)$

$\qquad\qquad\quad = -\dfrac{\sqrt{2}}{4}$

13. $\sin\dfrac{13\pi}{12}\cos\dfrac{\pi}{12} = \dfrac{1}{2}\left[\sin\left(\dfrac{13\pi}{12}+\dfrac{\pi}{12}\right)+\sin\left(\dfrac{13\pi}{12}-\dfrac{\pi}{12}\right)\right]$

$\qquad\qquad\quad = \dfrac{1}{2}\left(\sin\dfrac{7\pi}{6}+\sin\pi\right)$

$\qquad\qquad\quad = \dfrac{1}{2}\left(-\dfrac{1}{2}+0\right)$

$\qquad\qquad\quad = -\dfrac{1}{4}$

15. $\sin\dfrac{\pi}{12}\cos\dfrac{7\pi}{12} = \dfrac{1}{2}\left[\sin\left(\dfrac{\pi}{12}+\dfrac{7\pi}{12}\right)+\sin\left(\dfrac{\pi}{12}-\dfrac{7\pi}{12}\right)\right]$

$\qquad\qquad\quad = \dfrac{1}{2}\left[\sin\dfrac{2\pi}{3}+\sin\left(-\dfrac{\pi}{2}\right)\right]$

$\qquad\qquad\quad = \dfrac{1}{2}\left(\sin\dfrac{2\pi}{3}-\sin\dfrac{\pi}{2}\right)$

$\qquad\qquad\quad = \dfrac{1}{2}\left(\dfrac{\sqrt{3}}{2}-1\right)$

$\qquad\qquad\quad = \dfrac{\sqrt{3}-2}{4}$

17. $\sin 4\theta + \sin 2\theta = 2\sin\dfrac{4\theta+2\theta}{2}\cos\dfrac{4\theta-2\theta}{2}$

$\qquad\qquad\quad = 2\sin 3\theta\cos\theta$

19. $\cos 3\theta + \cos\theta = 2\cos\dfrac{3\theta+\theta}{2}\cos\dfrac{3\theta-\theta}{2}$

$\qquad\qquad\quad = 2\cos 2\theta\cos\theta$

21. $\cos 6\theta - \cos 2\theta = -2\sin\dfrac{6\theta+2\theta}{2}\sin\dfrac{6\theta-2\theta}{2}$

$\qquad\qquad\quad = -2\sin 4\theta\sin 2\theta$

23. $\cos\theta + \cos 7\theta = 2\cos\dfrac{\theta+7\theta}{2}\cos\dfrac{\theta-7\theta}{2}$

$\qquad\qquad\quad = 2\cos 4\theta\cos(-3\theta)$

$\qquad\qquad\quad = 2\cos 4\theta\cos 3\theta$

25. $\sin 5\theta + \sin 9\theta = 2\sin\dfrac{5\theta+9\theta}{2}\cos\dfrac{5\theta-9\theta}{2}$

$\qquad\qquad\quad = 2\sin 7\theta\cos(-2\theta)$

$\qquad\qquad\quad = 2\sin 7\theta\cos 2\theta$

27. $\cos 2\theta - \cos\theta = -2\sin\dfrac{2\theta+\theta}{2}\sin\dfrac{2\theta-\theta}{2}$

$\qquad\qquad\quad = -2\sin\dfrac{3}{2}\theta\sin\dfrac{1}{2}\theta$

29. $\cos\dfrac{\theta}{2} - \cos\theta = -2\sin\dfrac{\frac{\theta}{2}+\theta}{2}\sin\dfrac{\frac{\theta}{2}-\theta}{2}$

$\qquad\qquad\qquad = -2\sin\dfrac{3}{4}\theta\sin\left(-\dfrac{1}{4}\theta\right)$

$\qquad\qquad\qquad = 2\sin\dfrac{3}{4}\theta\sin\dfrac{1}{4}\theta$

31. $\sin\dfrac{\theta}{2} - \sin\dfrac{\theta}{3} = 2\cos\dfrac{\frac{\theta}{2}+\frac{\theta}{3}}{2}\sin\dfrac{\frac{\theta}{2}-\frac{\theta}{3}}{2}$

$\qquad\qquad\qquad = 2\cos\dfrac{5}{12}\theta\sin\dfrac{1}{12}\theta$

33. $\cos(\alpha+\beta) + \cos(\alpha-\beta) = \cos\alpha\cos\beta - \sin\alpha\sin\beta + \cos\alpha\cos\beta + \sin\alpha\sin\beta$

$\qquad\qquad\qquad\qquad\qquad = 2\cos\alpha\cos\beta$

35. $2\cos 3x\sin x = 2\cdot\dfrac{1}{2}\big[\sin(3x+x) - \sin(3x-x)\big]$

$\qquad\qquad = \sin 4x - \sin 2x$

$\qquad\qquad = 2\sin 2x\cos 2x - \sin 2x$

$\qquad\qquad = \sin 2x(2\cos 2x - 1)$

$\qquad\qquad = 2\sin x\cos x\Big[2(1-2\sin^2 x)-1\Big]$

$\qquad\qquad = 4\sin x\cos x - 8\sin^3 x\cos x - 2\sin x\cos x$

$\qquad\qquad = 2\sin x\cos x - 8\cos x\sin^3 x$

37. $2\cos 5x\cos 7x = 2\cdot\dfrac{1}{2}\big[\cos(5x+7x) + \cos(5x-7x)\big]$

$\qquad\qquad = \cos 12x + \cos(-2x)$

$\qquad\qquad = \cos 12x + \cos 2x$

$\qquad\qquad = \cos^2 6x - \sin^2 6x + 2\cos^2 x - 1$

39. $\sin 3x - \sin x = 2\cos\dfrac{3x+x}{2}\sin\dfrac{3x-x}{2}$

$\qquad\qquad = 2\cos 2x\sin x$

$\qquad\qquad = 2(1-2\sin^2 x)\sin x$

$\qquad\qquad = 2\sin x - 4\sin^3 x$

41. $\sin 2x + \sin 4x = 2\cos\dfrac{2x+4}{2}\cos\dfrac{2x-x4}{2}$

$\qquad\qquad = 2\cos 3x\cos(-x)$

$\qquad\qquad = 2\cos 3x\cos x$

$\qquad\qquad = 2\cos x\sin(2x+x)$

$\qquad\qquad = 2\cos x(\sin 2x\cos x + \cos 2x\sin x)$

$\qquad\qquad = 2\cos x[(2\sin x\cos x)\cos x + (2\cos^2 x - 1)\sin x]$

$\qquad\qquad = 2\cos x\sin x(4\cos^2 x - 1)$

43. $\dfrac{\sin 3x - \sin x}{\cos 3x - \cos x} = \dfrac{2\cos\frac{3x+x}{2}\sin\frac{3x-x}{2}}{-2\sin\frac{3x+x}{2}\sin\frac{3x-x}{2}}$

$\qquad\qquad = -\dfrac{\cos 2x}{\sin 2x}$

$\qquad\qquad = -\cot 2x$

45. $\dfrac{\sin 5x + \sin 3x}{4\sin x\cos^3 x - 4\sin^3 x\cos x} = \dfrac{2\sin\frac{5x+3x}{2}\cos\frac{5x-3x}{2}}{4\sin x\cos x(\cos^2 x - \sin^2 x)}$

$\qquad\qquad = \dfrac{\sin 4x\cos x}{2\sin x\cos x\cos 2x}$

$\qquad\qquad = \dfrac{2\sin 2x\cos 2x\cos x}{\sin 2x\cos 2x}$

$\qquad\qquad = 2\cos x$

47. $\sin(x+y)\cos(x-y) = \dfrac{1}{2}\big[\sin(x+y+x-y) + \sin(x+y-x+y)\big]$

$\qquad\qquad = \dfrac{1}{2}\big[\sin 2x + \sin 2y\big]$

$\qquad\qquad = \dfrac{1}{2}\big[2\sin x\cos x + 2\sin y\cos y\big]$

$\qquad\qquad = \sin x\cos x + \sin y\cos y$

124   **Chapter 3/Trigonometric Identities and Equations**

49. $a = -1$, $b = -1$, $k = \sqrt{(-1)^2 + (-1)^2} = \sqrt{2}$, α is a third quadrant angle.

$\sin \beta = \left| \dfrac{-1}{\sqrt{2}} \right| = \dfrac{1}{\sqrt{2}}$

$\beta = 45°$

$\alpha = -180° + 45° = -135°$

$y = \sqrt{2} \sin(x - 135°)$

51. $a = \dfrac{1}{2}$, $b = -\dfrac{\sqrt{3}}{2}$, $k = \sqrt{\left(\dfrac{1}{2}\right)^2 + \left(-\dfrac{\sqrt{3}}{2}\right)^2} = 1$, α is a fourth quadrant angle.

$\sin \beta = \left| \dfrac{-\frac{\sqrt{3}}{2}}{1} \right| = \dfrac{\sqrt{3}}{2}$

$\beta = 60°$

$\alpha = -60°$

$y = \sin(x - 60°)$

53. $a = \dfrac{1}{2}$, $b = -\dfrac{1}{2}$, $k = \sqrt{\left(\dfrac{1}{2}\right)^2 + \left(\dfrac{1}{2}\right)^2} = \dfrac{\sqrt{2}}{2}$, α is a fourth quadrant angle.

$\sin \beta = \left| \dfrac{-1/2}{\sqrt{2}/2} \right| = \dfrac{\sqrt{2}}{2}$

$\beta = 45°$

$\alpha = -45°$

$y = \dfrac{\sqrt{2}}{2} \sin(x - 45°)$

55. $a = -3$, $b = 3$, $k = \sqrt{(-3)^2 + 3^2} = 3\sqrt{2}$, α is a second quadrant angle.

$\sin \beta = \left| \dfrac{3}{3\sqrt{2}} \right| = \dfrac{1}{\sqrt{2}} - \dfrac{\sqrt{2}}{2}$

$\beta = 45°$

$\alpha = 180° - 45° = 135°$

$y = 3\sqrt{2} \sin(x + 135°)$

57. $a = \pi$, $b = -\pi$, $k = \sqrt{\pi^2 + (-\pi)^2} = \pi\sqrt{2}$, α is a fourth quadrant angle.

$\sin \beta = \left| \dfrac{-\pi}{\pi\sqrt{2}} \right| = \dfrac{1}{\sqrt{2}} = \dfrac{\sqrt{2}}{2}$

$\beta = 45°$

$\alpha = -45°$

$y = \pi\sqrt{2} \sin(x - 45°)$

59. $a = -1$, $b = 1$, $k = \sqrt{(-1)^2 + 1^2} = \sqrt{2}$, α is a second quadrant angle.

$\sin \beta = \left| \dfrac{1}{\sqrt{2}} \right| = \dfrac{1}{\sqrt{2}} = \dfrac{\sqrt{2}}{2}$

$\beta = \dfrac{\pi}{4}$

$\alpha = \pi - \dfrac{\pi}{4} = \dfrac{3\pi}{4}$

$y = \sqrt{2} \sin\left(x + \dfrac{3\pi}{4}\right)$

61. $a = \dfrac{\sqrt{3}}{2}$, $b = \dfrac{1}{2}$, $k = \sqrt{\left(\dfrac{\sqrt{3}}{2}\right)^2 + \left(\dfrac{1}{2}\right)^2} = 1$, α is a first quadrant angle.

$\sin\beta = \left|\dfrac{1/2}{1}\right| = \dfrac{1}{2}$

$\beta = \dfrac{\pi}{6}$

$\alpha = \dfrac{\pi}{6}$

$y = \sin\left(x + \dfrac{\pi}{6}\right)$

63. $a = -10$, $b = 10\sqrt{3}$, $k = \sqrt{(-10)^2 + (10\sqrt{3})^2} = 20$, α is a second quadrant angle.

$\sin\beta = \dfrac{10\sqrt{3}}{20} = \dfrac{\sqrt{3}}{2}$

$\beta = \dfrac{\pi}{3}$

$\alpha = \pi - \dfrac{\pi}{3} = \dfrac{2\pi}{3}$

$y = 20\sin\left(x + \dfrac{2\pi}{3}\right)$

65. $a = -5$, $b = 5$, $k = \sqrt{(-5)^2 + 5^2} = 5\sqrt{2}$, α is a second quadrant angle.

$\sin\beta = \left|\dfrac{5}{5\sqrt{2}}\right| = \dfrac{\sqrt{2}}{2}$

$\beta = \dfrac{\pi}{4}$

$\alpha = \pi - \dfrac{\pi}{4} = \dfrac{3\pi}{4}$

$y = 5\sqrt{2}\sin\left(x + \dfrac{3\pi}{4}\right)$

67. $y = -\sin x - \sqrt{3}\cos x$

$y = 2\sin\left(x - \dfrac{2\pi}{3}\right)$

69. $y = 2\sin x + 2\cos x$

$y = 2\sqrt{2}\sin\left(x + \dfrac{\pi}{4}\right)$

71. $y = -\sqrt{3}\sin x - \cos x$

$y = 2\sin\left(x - \dfrac{5\pi}{6}\right)$

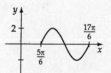

73. $y = -5\sin x + 5\sqrt{3}\cos x$

$y = 10\sin\left(x + \dfrac{2\pi}{3}\right)$

75. $y = 6\sqrt{3}\sin x - 6\cos x$

$y = 12\sin\left(x - \dfrac{\pi}{6}\right)$

77.

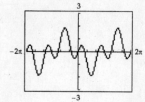

79.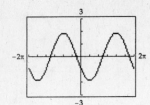

81.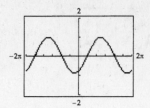

83. Let $x = \alpha + \beta$ and $y = \alpha - \beta$.

$x + y = \alpha + \beta + \alpha - \beta$ and $x - y = \alpha + \beta - (\alpha - \beta)$

$x + y = 2\alpha$ $\qquad\qquad\qquad$ $x - y = 2\beta$

$\qquad \alpha = \dfrac{x + y}{2}$ $\qquad\qquad\qquad$ $2\beta = \dfrac{x - y}{2}$

$\cos(\alpha - \beta) + \cos(\alpha + \beta) = 2\cos\alpha\cos\beta$

$\cos\left[\dfrac{x+y}{2} - \dfrac{x-y}{2}\right] + \cos\left[\dfrac{x+y}{2} + \dfrac{x-y}{2}\right] = 2\cos\dfrac{x+y}{2}\cos\dfrac{x-y}{2}$

$\cos y + \cos x = 2\cos\dfrac{x+y}{2}\cos\dfrac{x-y}{2}$

85. $x + y = 180°$

$\qquad y = 180° - x$

$\sin x + \sin y = \sin x + \sin(180° - x)$

$\qquad\qquad\quad = \sin x + \sin 180° \cos x - \cos 180° \sin x$

$\qquad\qquad\quad = \sin x + 0(\cos x) - (-1)\sin x$

$\qquad\qquad\quad = 2\sin x$

87. $\sin 2x + \sin 4x + \sin 6x = 2\sin\dfrac{2x+4x}{2}\cos\dfrac{2x-4x}{2} + 2\sin 3x \cos 3x$

$\qquad\qquad\qquad\qquad\qquad = 2\sin 3x \cos x + 2\sin 3x \cos 3x$

$\qquad\qquad\qquad\qquad\qquad = 2\sin 3x(\cos x + \cos 3x)$

$\qquad\qquad\qquad\qquad\qquad = 2\sin 3x\left(2\cos\dfrac{x+3x}{2}\cos\dfrac{x-3x}{2}\right)$

$\qquad\qquad\qquad\qquad\qquad = 4\sin 3x \cos 2x \cos x$

89. $\dfrac{\cos 10x + \cos 8x}{\sin 10x - \sin 8x} = \dfrac{2\cos\frac{10x+8x}{2}\cos\frac{10x-8x}{2}}{2\cos\frac{10x+8x}{2}\sin\frac{10x-8x}{2}}$

$\qquad\qquad\qquad\qquad = \dfrac{2\cos 9x \cos x}{2\cos 9x \sin x}$

$\qquad\qquad\qquad\qquad = \cot x$

91. $\dfrac{\sin 2x + \sin 4x + \sin 6x}{\cos 2x + \cos 4x + \cos 6x} = \dfrac{\sin 2x + \sin 6x + \sin 4x}{\cos 2x + \cos 6x + \cos 4x}$

$\qquad\qquad\qquad\qquad = \dfrac{2\sin\frac{2x+6x}{2}\cos\frac{2x-6x}{2} + \sin 4x}{2\cos\frac{2x+6x}{2}\cos\frac{2x-6x}{2} + \cos 4x}$

$\qquad\qquad\qquad\qquad = \dfrac{2\sin 4x \cos 2x + \sin 4x}{2\cos 4x \cos 2x + \cos 4x}$

$\qquad\qquad\qquad\qquad = \dfrac{\sin 4x(2\cos 2x + 1)}{\cos 4x(2\cos 2x + 1)}$

$\qquad\qquad\qquad\qquad = \dfrac{\sin 4x}{\cos 4x}$

$\qquad\qquad\qquad\qquad = \tan 4x$

93. $\cos^2 x - \sin^2 x = \cos x \cdot \cos x - \sin x \cdot \sin x$

$$= \frac{1}{2}\big[\cos(x+x) + \cos(x-x)\big] - \frac{1}{2}\big[\cos(x-x) - \cos(x+x)\big]$$

$$= \frac{1}{2}\cos 2x + \frac{1}{2}\cos 0 - \frac{1}{2}\cos 0 + \frac{1}{2}\cos 2x$$

$$= \cos 2x$$

95. Let $k = \sqrt{a^2 + b^2}$, $\tan \alpha = \dfrac{a}{b}$

$$a \sin x + b \cos x = \frac{\sqrt{a^2+b^2}}{\sqrt{a^2+b^2}}(a \sin x + b \cos x)$$

$$= \sqrt{a^2+b^2}\left(\frac{a}{\sqrt{a^2+b^2}}\sin x + \frac{b}{\sqrt{a^2+b^2}}\cos x\right)$$

$$= k(\sin\alpha \sin x + \cos\alpha \cos x) \text{ because } \sin\alpha = \frac{a}{\sqrt{a^2+b^2}} \text{ and } \cos\alpha = \frac{b}{\sqrt{a^2+b^2}}$$

$$= k(\cos x \cos\alpha + \sin x \sin\alpha) = k\cos(x-\alpha)$$

97. $y = \sin\dfrac{x}{2} - \cos\dfrac{x}{2}$

$y = \sqrt{2}\sin\left(\dfrac{x}{2} - \dfrac{\pi}{4}\right)$

amplitude $= \sqrt{2}$

phase shift $= \dfrac{\pi}{2}$

period $= 4\pi$

99. $y = \sqrt{3}\sin 2x - \cos 2x$

$y = 2\sin\left(2x - \dfrac{\pi}{6}\right)$

amplitude $= 2$

phase shift $= \dfrac{\pi}{12}$

period $= \pi$

101. $y = \sin\pi x + \sqrt{3}\cos\pi x$

$y = 2\sin\left(\pi x + \dfrac{\pi}{3}\right)$

amplitude $= 2$

phase shift $= -\dfrac{1}{3}$

period $= 2$

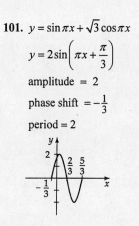

SECTION 3.5, Page 221

1. $y = \sin^{-1} 1$

$\sin y = 1$ with $-\dfrac{\pi}{2} \le y \le \dfrac{\pi}{2}$

$y = \dfrac{\pi}{2}$

3. $y = \cos^{-1}\left(-\dfrac{\sqrt{3}}{2}\right)$

$\cos y = -\dfrac{\sqrt{3}}{2}$ $0 \le y \le \pi$

$y = \dfrac{5\pi}{6}$

5. $y = \tan^{-1}(1)$

$\tan y = -1$ $-\dfrac{\pi}{2} < y < \dfrac{\pi}{2}$

$y = -\dfrac{\pi}{4}$

7. $y = \cot^{-1}\dfrac{\sqrt{3}}{3}$

$\cot y = \dfrac{\sqrt{3}}{3}$ $0 < y < \pi$

$y = \dfrac{\pi}{3}$

9. $y = \sec^{-1} 2$

$\sec y = 2$ $0 \le y \le \pi$

$y = \dfrac{\pi}{3}$

11. $y = \csc^{-1}\left(-\sqrt{2}\right)$

$\csc y = -\sqrt{2}$ $-\dfrac{\pi}{2} \le y \le \dfrac{\pi}{2}$

$y = -\dfrac{\pi}{4}$

13. $y = \sin^{-1}\left(-\dfrac{\sqrt{3}}{2}\right)$

$\sin y = -\dfrac{\sqrt{3}}{2}$ $-\dfrac{\pi}{2} \le y \le \dfrac{\pi}{2}$

$y = -\dfrac{\pi}{3}$

15. $y = \cos^{-1}\left(-\dfrac{1}{2}\right)$

$\cos y = -\dfrac{1}{2}$ $0 \le y \le \pi$

$y = \dfrac{2\pi}{3}$

17. $y = \tan^{-1}\dfrac{\sqrt{3}}{3}$

$\tan y = \dfrac{\sqrt{3}}{3}$ $-\dfrac{\pi}{2} < y < \dfrac{\pi}{2}$

$y = \dfrac{\pi}{6}$

19. $y = \cot^{-1}\sqrt{3}$

$\cot y = \sqrt{3} \quad 0 < y < \pi$

$y = \dfrac{\pi}{6}$

21. $y = \cos\left(\cos^{-1}\dfrac{1}{2}\right)$

$y = \cos\dfrac{\pi}{3}$

$y = \dfrac{1}{2}$

23. $y = \tan(\tan^{-1}2)$

$y = 2$

25. $y = \sin\left(\tan^{-1}\dfrac{3}{4}\right)$

$y = \dfrac{3}{5}$

27. $y = \tan\left(\sin^{-1}\dfrac{\sqrt{2}}{2}\right)$

$y = 1$

29. $y = \cos(\sec^{-1}2)$

$y = \dfrac{1}{2}$

31. $y = \sin^{-1}\left(\sin\dfrac{\pi}{6}\right)$

$= \sin^{-1}\dfrac{1}{2}$

$y = \dfrac{\pi}{6}$

33. $y = \cos^{-1}\left(\sin\dfrac{\pi}{4}\right)$

$= \cos^{-1}\dfrac{\sqrt{2}}{2}$

$y = \dfrac{\pi}{4}$

35. $y = \sin^{-1}\left(\tan\dfrac{\pi}{3}\right)$

$= \sin^{-1}\sqrt{3}$

y is not defined.

37. $y = \tan^{-1}\left(\sin\dfrac{\pi}{6}\right)$

$= \tan^{-1}\dfrac{1}{2}$

$y \approx 0.4636$

39. $y = \sin^{-1}\left(\cos\left[-\dfrac{2\pi}{3}\right]\right)$

$= \sin^{-1}\left(-\dfrac{1}{2}\right)$

$y = -\dfrac{\pi}{6}$

41. Let $\theta = \sin^{-1}\dfrac{1}{2}$ and find $y = \tan\theta$.

Then $\sin\theta = \dfrac{1}{2}$ and $-\dfrac{\pi}{2} \le \theta \le \dfrac{\pi}{2}$.

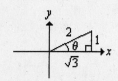

Thus $\tan\theta = \dfrac{1}{\sqrt{3}} = \dfrac{\sqrt{3}}{3}$.

$y = \dfrac{\sqrt{3}}{3}$

43. Let $\theta = \sin^{-1}\dfrac{1}{4}$ and find $y = \sec\theta$.

Then $\sin\theta = \dfrac{1}{4}$, and $-\dfrac{\pi}{2} \le \theta \le \dfrac{\pi}{2}$.

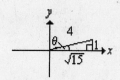

Thus $\sec\theta = \dfrac{4}{\sqrt{15}} = \dfrac{4\sqrt{15}}{15}$.

$y = \dfrac{4\sqrt{15}}{15}$

45. Let $\theta = \sin^{-1}\dfrac{7}{25}$ and find $y = \cos\theta$.

Then $\sin\theta = \dfrac{7}{25}$, and $-\dfrac{\pi}{2} \le \theta \le \dfrac{\pi}{2}$.

Thus $\cos\theta = \dfrac{24}{25}$.

$y = \dfrac{24}{25}$

47. Let $\theta = \tan^{-1}\dfrac{12}{5}$ and find $y = \sec\theta$.

Then $\tan\theta = \dfrac{12}{5}$, and $-\dfrac{\pi}{2} \le \theta \le \dfrac{\pi}{2}$.

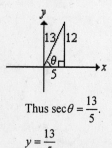

Thus $\sec\theta = \dfrac{13}{5}$.

$y = \dfrac{13}{5}$

49. Let $\alpha = \sin^{-1}\dfrac{\sqrt{2}}{2}$, $\alpha = \dfrac{\pi}{4}$, $\sin\alpha = \dfrac{\sqrt{2}}{2}$, $\cos\alpha = \dfrac{\sqrt{2}}{2}$.

$$y = \cos\left(2\sin^{-1}\dfrac{\sqrt{2}}{2}\right)$$
$$= \cos 2\alpha$$
$$= \cos^2\alpha - \sin^2\alpha$$
$$= \left(\dfrac{\sqrt{2}}{2}\right)^2 - \left(\dfrac{\sqrt{2}}{2}\right)^2$$
$$= 0$$

51. Let $\alpha = \sin^{-1}\dfrac{4}{5}$, $\sin\alpha = \dfrac{4}{5}$, $\cos\alpha = \sqrt{1-\left(\dfrac{4}{5}\right)^2} = \dfrac{3}{5}$.

$$y = \sin\left(2\sin^{-1}\dfrac{4}{5}\right)$$
$$= \sin 2\alpha = 2\sin\alpha\cos\alpha$$
$$= 2\left(\dfrac{4}{5}\right)\left(\dfrac{3}{5}\right) = \dfrac{24}{25}$$

53. $y = \sin\left(\sin^{-1}\dfrac{2}{3} + \cos^{-1}\dfrac{1}{2}\right)$

Let $\alpha = \sin^{-1}\dfrac{2}{3}$, $\sin\alpha = \dfrac{2}{3}$, $\cos\alpha = \sqrt{1-\left(\dfrac{2}{3}\right)^2} = \dfrac{\sqrt{5}}{3}$.

$\beta = \cos^{-1}\dfrac{1}{2}$, $\cos\beta = \dfrac{1}{2}$, $\sin\beta = \sqrt{1-\left(\dfrac{1}{2}\right)^2} = \dfrac{\sqrt{3}}{2}$.

$$y = \sin(\alpha+\beta)$$
$$= \sin\alpha\cos\beta + \cos\alpha\sin\beta$$
$$= \dfrac{2}{3}\left(\dfrac{1}{2}\right) + \dfrac{\sqrt{5}}{3}\left(\dfrac{\sqrt{3}}{2}\right)$$
$$= \dfrac{1}{3} + \dfrac{\sqrt{15}}{6} = \dfrac{2+\sqrt{15}}{6}$$

55. $y = \tan\left(\cos^{-1}\dfrac{1}{2} - \sin^{-1}\dfrac{3}{4}\right)$

Let $\alpha = \cos^{-1}\dfrac{1}{2}$, $\cos\alpha = \dfrac{1}{2}$, $\sin\alpha = \sqrt{1-\left(\dfrac{1}{2}\right)^2} = \dfrac{\sqrt{3}}{2}$, $\tan\alpha = \dfrac{\frac{\sqrt{3}}{2}}{\frac{1}{2}} = \sqrt{3}$.

$\beta = \sin^{-1}\dfrac{3}{4}$, $\sin\beta = \dfrac{3}{4}$, $\cos\beta = \sqrt{1-\left(\dfrac{3}{4}\right)^2} = \dfrac{\sqrt{7}}{4}$, $\tan\beta = \dfrac{\frac{3}{4}}{\frac{\sqrt{7}}{4}} = \dfrac{3}{\sqrt{7}} = \dfrac{3\sqrt{7}}{7}$.

$$y = \tan(\alpha-\beta)$$
$$= \dfrac{\tan\alpha - \tan\beta}{1+\tan\alpha\tan\beta}$$
$$= \dfrac{\sqrt{3} - \frac{3\sqrt{7}}{7}}{1+\sqrt{3}\cdot\frac{3\sqrt{7}}{7}} = \dfrac{\sqrt{3} - \frac{3\sqrt{7}}{7}}{1+\sqrt{3}\cdot\frac{3\sqrt{7}}{7}}\cdot\dfrac{7}{7} = \dfrac{7\sqrt{3}-3\sqrt{7}}{7+3\sqrt{21}} = \dfrac{7\sqrt{3}-3\sqrt{7}}{7+3\sqrt{21}}\cdot\dfrac{7-3\sqrt{21}}{7-3\sqrt{21}} = \dfrac{112\sqrt{3}-84\sqrt{7}}{-140} = \dfrac{3\sqrt{7}-4\sqrt{3}}{5} = \dfrac{1}{5}\left(3\sqrt{7}-4\sqrt{3}\right)$$

57. $\sin^{-1}x = \cos^{-1}\dfrac{5}{13}$

$\sin(\sin^{-1}x) = \sin\left(\cos^{-1}\dfrac{5}{13}\right)$

$x = \dfrac{12}{13}$

59. $\sin^{-1}(x-1) = \dfrac{\pi}{2}$

$(x-1) = \sin\dfrac{\pi}{2}$

$(x-1) = 1$

$x = 2$

61. $\tan^{-1}\left(x+\dfrac{\sqrt{2}}{2}\right) = \dfrac{\pi}{4}$

$\left(x+\dfrac{\sqrt{2}}{2}\right) = \tan\dfrac{\pi}{4}$

$x = 1 - \dfrac{\sqrt{2}}{2}$

$= \dfrac{2-\sqrt{2}}{2}$

63. $\sin^{-1}\dfrac{3}{5} + \cos^{-1}x = \dfrac{\pi}{4}$

$$\cos^{-1}x = \dfrac{\pi}{4} - \sin^{-1}\dfrac{3}{5}$$

$$x = \cos\left(\dfrac{\pi}{4} - \sin^{-1}\dfrac{3}{5}\right)$$

Let $\alpha = \sin^{-1}\dfrac{3}{5}$, $\sin\alpha = \dfrac{3}{5}$, $\cos\alpha = \dfrac{4}{5}$.

$$x = \cos\left(\dfrac{\pi}{4} - \alpha\right)$$

$$x = \cos\dfrac{\pi}{4}\cos\alpha + \sin\dfrac{\pi}{4}\sin\alpha$$

$$x = \dfrac{\sqrt{2}}{2}\cdot\dfrac{4}{5} + \dfrac{\sqrt{2}}{2}\cdot\dfrac{3}{5}$$

$$= \dfrac{4\sqrt{2}}{10} + \dfrac{3\sqrt{2}}{10} = \dfrac{7\sqrt{2}}{10}$$

65. $\sin^{-1}\dfrac{\sqrt{2}}{2} + \cos^{-1}x = \dfrac{2\pi}{3}$

$$\cos^{-1}x = \dfrac{2\pi}{3} - \sin^{-1}\dfrac{\sqrt{2}}{2}$$

$$x = \cos\left(\dfrac{2\pi}{3} - \sin^{-1}\dfrac{\sqrt{2}}{2}\right)$$

Let $\alpha = \sin^{-1}\dfrac{\sqrt{2}}{2}$, $\sin\alpha = \dfrac{\sqrt{2}}{2}$, $\cos\alpha = \dfrac{\sqrt{2}}{2}$.

$$x = \cos\left(\dfrac{2\pi}{3} - \alpha\right)$$

$$x = \cos\dfrac{2\pi}{3}\cos\alpha + \sin\dfrac{2\pi}{3}\sin\alpha$$

$$x = -\dfrac{1}{2}\cdot\dfrac{\sqrt{2}}{2} + \dfrac{\sqrt{3}}{2}\cdot\dfrac{\sqrt{2}}{2}$$

$$= -\dfrac{\sqrt{2}}{4} + \dfrac{\sqrt{6}}{4} = \dfrac{\sqrt{6}-\sqrt{2}}{4} \approx 0.2588$$

Note: Since $\sin\alpha = \dfrac{\sqrt{2}}{2}$ and $\cos\alpha = \dfrac{\sqrt{2}}{2}$, then $\alpha = \dfrac{\pi}{4}$.

Thus, $\cos\left(\dfrac{2\pi}{3} - \alpha\right) = \cos\left(\dfrac{2\pi}{3} - \dfrac{\pi}{4}\right) = \cos\left(\dfrac{5\pi}{12}\right) \approx 0.2588$.

67. $\cos(\sin^{-1}x) = \sqrt{1-x^2}$

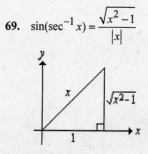

69. $\sin(\sec^{-1}x) = \dfrac{\sqrt{x^2-1}}{|x|}$

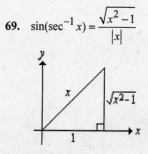

71. Let $\alpha = \sin^{-1}x$, $\sin\alpha = x$, $\cos\alpha = \sqrt{1-x^2}$.

Let $\beta = \sin^{-1}(-x)$, $\sin\beta = -x$, $\cos\beta = \sqrt{1-x^2}$.

$$\sin^{-1}x + \sin^{-1}(-x) = \alpha + \beta$$

$$= \sin^{-1}[\sin(\alpha+\beta)]$$

$$= \sin^{-1}(\sin\alpha\cos\beta + \cos\alpha\sin\beta)$$

$$= \sin^{-1}\left[x\sqrt{1-x^2} + \sqrt{1-x^2}(-x)\right]$$

$$= \sin^{-1}0$$

$$= 0$$

73. Let $\alpha = \tan^{-1}x$, $\tan\alpha = x$, $\beta = \tan^{-1}\dfrac{1}{x}$, $\tan\beta = \dfrac{1}{x}$.

$$\tan^{-1}x + \tan^{-1}\dfrac{1}{x} = \alpha + \beta$$

$$= \tan^{-1}[\tan(\alpha+\beta)]$$

$$= \tan^{-1}\left[\dfrac{\tan\alpha + \tan\beta}{1 - \tan\alpha\tan\beta}\right]$$

$$= \tan^{-1}\left[\dfrac{x + \dfrac{1}{x}}{1 - x\cdot\dfrac{1}{x}}\right]$$

$$= \tan^{-1}\dfrac{\dfrac{x^2+1}{x}}{1-1}, \text{ which is undefined}$$

Thus $x = \dfrac{\pi}{2}$

75. The graph of $y = \sin^{-1}(x) + 2$ (shown as a black graph) is the graph of $y = \sin^{-1} x$ (shown as a gray graph) moved two units up.

77. The graph of $y = \sin^{-1}(x+1) - 2$ (shown as a black graph) is the graph of $y = \sin^{-1} x$ (shown as a gray graph) moved one unit to the left and two units down.

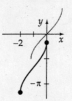

79. The graph of $y = 2\cos^{-1} x$ (shown as a black graph) is the graph of $y = \cos^{-1} x$ (shown as a gray graph) stretched.

81. The graph of $y = \tan^{-1}(x+1) - 2$ (shown as a black graph) is the graph of $y = \tan^{-1} x$ (shown as a gray graph) moved one unit to the left and two units down.

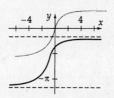

83. a.

$$\frac{A}{SD} = 1 - \frac{1}{2}\left[1 - \left(\frac{S}{D}\right)^2 + \frac{D}{S}\sin^{-1}\frac{S}{D}\right]$$

$$= 1 - \frac{1}{2}\left[1 - \left(\frac{1}{2}\right)^2 + 2\sin^{-1}\frac{1}{2}\right]$$

$$= 1 - \frac{1}{2}\left[\frac{3}{4} + \frac{\pi}{3}\right]$$

$$\approx 1 - \frac{1}{2}(1.79719755) \approx 1 - 0.8985$$

$$\frac{A}{SD} \approx 0.1014$$

b.

$$\frac{A}{SD} = 1 - \frac{1}{2}\left[1 - \left(\frac{S}{D}\right)^2 + \frac{D}{S}\sin^{-1}\frac{S}{D}\right]$$

$$\approx 1 - \frac{1}{2}\left[1 - 0.390625 + 1.6\sin^{-1}0.62\right]$$

$$\approx 1 - \frac{1}{2}\left[0.609375 + 1.0802105\right]$$

$$\approx 1 - \frac{1}{2}(1.689585) \approx 1 - 0.8448$$

$$\approx 0.1552$$

85.

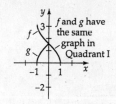

[Note: $f(x) = \cos^{-1} x$ is neither odd nor even.

$g(x) = \sin^{-1}\sqrt{1-x^2}$ is an even function.]

No, $f(x) \neq g(x)$ on the interval $[-1, 1]$.

87.

$y = \csc^{-1} 2x$

$\csc y = 2x$

$-\frac{\pi}{2} \leq y \leq \frac{\pi}{2}, \ y \neq 0$

$2x \leq -1$ or $2x \geq 1$

$x \leq -\frac{1}{2}$ or $x \geq \frac{1}{2}$

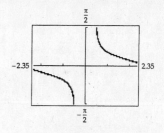

89.

$y = \sec^{-1}(x-1)$

$\sec y = x - 1$

$0 < y < \pi$

$y \neq \frac{\pi}{2}$

$x - 1 \leq -1 \quad x - 1 \geq 1$

$x \leq 0 \qquad x \geq 2$

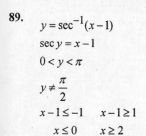
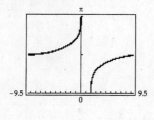

91.

$y = 2\tan^{-1} 2x$

$\frac{y}{2} = \tan^{-1} 2x$

$\tan y = 2x$

$-\frac{\pi}{2} < \frac{y}{2} < \frac{\pi}{2}$

$-\pi < y < \pi$

$-\infty < 2x < \infty$

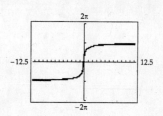

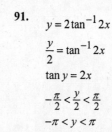

93.
$$y = \cot^{-1}\frac{x}{3}$$
$$\cot y = \frac{\pi}{3}$$
$$0 < y < \pi$$
$$-\infty < \frac{x}{3} < \infty$$

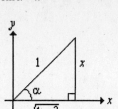

95. Let $\alpha = \sin^{-1}x$ $\cos\left(\sin^{-1}x\right) = \cos\alpha$
$$\sin\alpha = x$$
$$= \frac{\sqrt{1-x^2}}{1}$$
$$= \sqrt{1-x^2}$$

97. Let $\alpha = \csc^{-1}x$
$$\csc\alpha = x$$

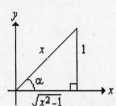

$$\tan(\csc^{-1}x) = \tan\alpha$$
$$= \frac{1}{\sqrt{x^2-1}}$$
$$= \frac{\sqrt{x^2-1}}{x^2-1}$$

99. $5x = \tan^{-1}3y$
$$\tan 5x = 3y$$
$$y = \frac{1}{3}\tan 5x$$

101. $x - \frac{\pi}{3} = \cos^{-1}(y-3)$
$$\cos\left(x - \frac{\pi}{3}\right) = y - 3$$
$$y = 3 + \cos\left(x - \frac{\pi}{3}\right)$$

SECTION 3.6, Page 232

1. $\sec x - \sqrt{2} = 0$
$$\sec x = \sqrt{2}$$
$$x = \frac{\pi}{4}, \frac{7\pi}{4}$$

3. $\tan x - \sqrt{3} = 0$
$$\tan x = \sqrt{3}$$
$$x = \frac{\pi}{3}, \frac{4\pi}{3}$$

5. $2\sin x\cos x = \sqrt{2}\cos x$
$$2\sin x\cos x - \sqrt{2}\cos x = 0$$
$$\cos x(2\sin x - \sqrt{2}) = 0$$

$\cos x = 0$ $2\sin x - \sqrt{2} = 0$
$$x = \frac{\pi}{2}, \frac{3\pi}{2} \qquad \sin x = \frac{\sqrt{2}}{2}$$
$$x = \frac{\pi}{4}, \frac{3\pi}{4}$$

The solutions are $\frac{\pi}{4}, \frac{\pi}{2}, \frac{3\pi}{4}, \frac{3\pi}{2}$.

7. $\sin^2 x - 1 = 0$
$$\sin^2 x = 1$$
$$\sin x = \pm\sqrt{1}$$
$$\sin x = \pm 1$$
$$x = \frac{\pi}{2}, \frac{3\pi}{2}$$

9. $4\sin x \cos x - 2\sqrt{3}\sin x - 2\sqrt{2}\cos x + \sqrt{6} = 0$

$2\sin x(2\cos x - \sqrt{3}) - \sqrt{2}(2\cos x - \sqrt{3}) = 0$

$(2\cos x - \sqrt{3})(2\sin x - \sqrt{2}) = 0$

$2\cos x - \sqrt{3} = 0$ $2\sin x - \sqrt{2} = 0$

$\cos x = \dfrac{\sqrt{3}}{2}$ $\sin x = \dfrac{\sqrt{2}}{2}$

$x = \dfrac{\pi}{6}, \dfrac{11\pi}{6}$ $x = \dfrac{\pi}{4}, \dfrac{3\pi}{4}$

The solutions are $\dfrac{\pi}{6}, \dfrac{\pi}{4}, \dfrac{3\pi}{4}, \dfrac{11\pi}{6}$.

11. $\csc x - \sqrt{2} = 0$

$\csc x = \sqrt{2}$

$\sin x = \dfrac{1}{\sqrt{2}} = \dfrac{\sqrt{2}}{2}$

$x = \dfrac{\pi}{4}, \dfrac{3\pi}{4}$

13. $2\sin^2 x + 1 = 3\sin x$

$2\sin^2 x - 3\sin x + 1 = 0$

$(2\sin x - 1)(\sin x - 1) = 0$

$2\sin x - 1 = 0$ $\sin x - 1 = 0$

$\sin x = \dfrac{1}{2}$ $\sin x = 1$

$x = \dfrac{\pi}{6}, \dfrac{5\pi}{6}$ $x = \dfrac{\pi}{2}$

The solutions are $\dfrac{\pi}{6}, \dfrac{\pi}{2}, \dfrac{5\pi}{6}$.

15. $4\cos^2 x - 3 = 0$

$\cos^2 x = \dfrac{3}{4}$

$\cos x = \pm\dfrac{\sqrt{3}}{2}$

$x = \dfrac{\pi}{6}, \dfrac{5\pi}{6}, \dfrac{7\pi}{6}, \dfrac{11\pi}{6}$

17. $2\sin^3 x = \sin x$

$2\sin^3 x - \sin x = 0$

$\sin x(2\sin^2 x - 1) = 0$

$\sin x = 0$ $2\sin^2 x = 1$

$x = 0, \pi$ $\sin x = \pm\dfrac{\sqrt{2}}{2}$

         $x = \dfrac{\pi}{4}, \dfrac{3\pi}{4}, \dfrac{5\pi}{4}, \dfrac{7\pi}{4}$

The solutions are $0, \dfrac{\pi}{4}, \dfrac{3\pi}{4}, \pi, \dfrac{5\pi}{4}, \dfrac{7\pi}{4}$.

19. $4\sin^2 x + 2\sqrt{3}\sin x - \sqrt{3} = 2\sin x$

$4\sin^2 x + 2\sqrt{3}\sin x - 2\sin x - \sqrt{3} = 0$

$2\sin x(2\sin x + \sqrt{3}) - (2\sin x + \sqrt{3}) = 0$

$(2\sin x + \sqrt{3})(2\sin x - 1) = 0$

$2\sin x + \sqrt{3} = 0$ $2\sin x - 1 = 0$

$\sin x = -\dfrac{\sqrt{3}}{2}$ $\sin x = \dfrac{1}{2}$

$x = \dfrac{4\pi}{3}, \dfrac{5\pi}{3}$ $x = \dfrac{\pi}{6}, \dfrac{5\pi}{6}$

The solutions are $\dfrac{\pi}{6}, \dfrac{5\pi}{6}, \dfrac{4\pi}{3}, \dfrac{5\pi}{3}$.

21. $\sin^4 x = \sin^2 x$

$\sin^4 x - \sin^2 x = 0$

$\sin^2 x(\sin^2 x - 1) = 0$

$\sin^2 x = 0$ $\sin^2 x - 1 = 0$

$\sin x = 0$ $\sin x = \pm 1$

$x = 0, \pi$ $x = \dfrac{\pi}{2}, \dfrac{3\pi}{2}$

The solutions are $0, \dfrac{\pi}{2}, \pi, \dfrac{3\pi}{2}$.

23. $\cos x - 0.75 = 0$

$\cos x = 0.75$

$x \approx 41.4°, 318.6°$

25. $3\sin x - 5 = 0$

$3\sin x = 5$

$\sin x = \dfrac{5}{3}$

no solution

27. $3\sec x - 8 = 0$

$3\sec x = 8$

$\sec x = \dfrac{8}{3}$

$\dfrac{1}{\cos x} = \dfrac{8}{3}$

$\cos x = \dfrac{3}{8}$

$x \approx 68.0°, \ 292.0°$

29. $\cos x + 3 = 0$

$\cos x = -3$

no solution

31. $3 - 5\sin x = 4\sin x + 1$

$-9\sin x = -2$

$\sin x = \dfrac{2}{9}$

$x \approx 12.8°, \ 167.2°$

33. $\dfrac{1}{2}\sin x + \dfrac{2}{3} = \dfrac{3}{4}\sin x + \dfrac{3}{5}$

$-\dfrac{1}{4}\sin x = \dfrac{3}{5} - \dfrac{2}{3}$

$-\dfrac{1}{4}\sin x = -\dfrac{1}{15}$

$\sin x = \dfrac{4}{15}$

$x \approx 15.5°, \ 164.5°$

35. $3\tan^2 x - 2\tan x = 0$

$\tan x(3\tan x - 2) = 0$

$\tan x = 0 \qquad\qquad 3\tan x - 2 = 0$

$x = 0, \ 180° \qquad\qquad \tan x = \dfrac{2}{3}$

$\qquad\qquad\qquad\qquad x \approx 33.7°, \ 213.7°$

The solutions are $0°, 33.7°, 180°, 213.7°$.

37. $3\cos x + \sec x = 0$

$3\cos x + \dfrac{1}{\cos x} = 0$

$3\cos^2 x + 1 = 0$

$\cos^2 x = -\dfrac{1}{3}$

no solution

39. $\tan^2 x = 3\sec^2 x - 2$

$\tan^2 x = 3(1 + \tan^2 x) - 2$

$\tan^2 x = 3 + 3\tan^2 x - 2$

$-2\tan^2 x = 1$

$\tan^2 x = -\dfrac{1}{2}$

no solution

41. $2\sin^2 x = 1 - \cos x$

$2(1 - \cos^2 x) = 1 - \cos x$

$2 - 2\cos^2 x = 1 - \cos x$

$0 = 2\cos^2 x - \cos x - 1$

$0 = (2\cos x + 1)(\cos x - 1)$

$2\cos x + 1 = 0 \qquad\qquad \cos x - 1 = 0$

$\cos x = -\dfrac{1}{2} \qquad\qquad \cos x = 1$

$x = 120°, \ 240° \qquad\qquad x = 0°$

The solutions are $0°, 120°, 240°$.

43. $3\cos^2 x + 5\cos x - 2 = 0$

$\cos x = \dfrac{-5 \pm \sqrt{5^2 - 4(3)(-2)}}{2 \cdot 3}$

$= \dfrac{-5 \pm \sqrt{49}}{6} = \dfrac{-5 \pm 7}{6}$

$\cos x = \dfrac{1}{3} \qquad\qquad \cos x = -2$

$x \approx 70.5°, 289.5° \qquad$ no solution

The solutions are $70.5°, 289.5°$.

45. $2\tan^2 x - \tan x - 10 = 0$

$(\tan x + 2)(2\tan x - 5) = 0$

$\tan x + 2 = 0 \qquad\qquad 2\tan x - 5 = 0$

$\tan x = -2 \qquad\qquad \tan x = \dfrac{5}{2}$

$x \approx 116.6°, \ 296.6° \quad x \approx 68.2°, \ 248.2°$

The solutions are $68.2°, 116.6°, 248.2°, 296.6°$.

47. $3\sin x\cos x - \cos x = 0$

$\cos x(3\sin x - 1) = 0$

$\cos x = 0 \qquad\qquad 3\sin x - 1 = 0$

$x = 90°, \ 270° \qquad\qquad \sin x = \dfrac{1}{3}$

$\qquad\qquad\qquad\qquad x \approx 19.5°, \ 160.5°$

The solutions are $19.5°, 90°, 160.5°, 270°$.

49. $2\sin x \cos x - \sin x - 2\cos x + 1 = 0$

$\sin x(2\cos x - 1) - (2\cos x - 1) = 0$

$(2\cos x - 1)(\sin x - 1) = 0$

$2\cos x - 1 = 0 \qquad \sin x - 1 = 0$

$\cos x = \dfrac{1}{2} \qquad \sin x = 1$

$x = 60°, \ 300° \qquad x = 90°$

The solutions are $60°, \ 90°, \ 300°$.

51. $2\sin x - \cos x = 1$

$2\sin x - 1 = \cos x$

$(2\sin x - 1)^2 = (\cos x)^2$

$4\sin^2 x - 4\sin x + 1 = \cos^2 x$

$4\sin^2 x - 4\sin x + 1 = 1 - \sin^2 x$

$5\sin^2 x - 4\sin x = 0$

$\sin x(5\sin x - 4) = 0$

$\sin x = 0 \qquad\qquad\qquad 5\sin x - 4 = 0$

$x = 180° \qquad\qquad\qquad \sin x = \dfrac{4}{5}$

$x \approx 53.1°$ or $126.9°$

$126.9°$ does not check.

The solutions are $53.1°, \ 180°$.

53. $2\sin x - 3\cos x = 1$

$2\sin x = 3\cos x + 1$

$(2\sin x)^2 = (3\cos x + 1)^2$

$4\sin^2 x = 9\cos^2 x + 6\cos x + 1$

$4(1 - \cos^2 x) = 9\cos^2 x + 6\cos x + 1$

$0 = 13\cos^2 x + 6\cos x - 3$

55. $3\sin^2 x - \sin x - 1 = 0$

$\sin x = \dfrac{1 \pm \sqrt{(-1)^2 - 4(3)(-1)}}{2(3)}$

$= \dfrac{1 \pm \sqrt{13}}{6}$

$\sin x = 0.7676 \qquad\qquad \sin x = -0.4343$

$x = 50.1°, \ 129.9° \qquad x = 205.7°, \ 334.3°$

The solutions are $50.1°, \ 129.9°, \ 205.7°, \ 334.3°$.

57. $2\cos x - 1 + 3\sec x = 0$

$2\cos x - 1 + \dfrac{3}{\cos x} = 0$

$2\cos^2 x - \cos x + 3 = 0$

$\cos x = \dfrac{1 \pm \sqrt{(-1)^2 - 4(2)(3)}}{2(2)}$

$= \dfrac{1 \pm \sqrt{-23}}{4}$

no solution

59. $\cos^2 x - 3\sin x + 2\sin^2 x = 0$

$1 - \sin^2 x - 3\sin x + 2\sin^2 x = 0$

$\sin^2 x - 3\sin x + 1 = 0$

$\sin x = \dfrac{3 \pm \sqrt{(-3)^2 - 4(1)(1)}}{2(1)}$

$= \dfrac{3 \pm \sqrt{5}}{2}$

$\sin x = 2.6180 \qquad\qquad \sin x = 0.3820$

no solution $\qquad\qquad\qquad x = 22.5°, \ 157.5°$

The solutions are $22.5°, \ 157.5°$.

61. $\tan 2x - 1 = 0$

$\tan 2x = 1$

$2x = \dfrac{\pi}{4} + k\pi$

$x = \dfrac{\pi}{8} + \dfrac{k\pi}{2}$, where k is an integer

63. $\sin 5x = 1$

$5x = \dfrac{\pi}{2} + 2k\pi$

$x = \dfrac{\pi}{10} + \dfrac{2}{5}k\pi$, where k is an integer

65. $\sin 2x - \sin x = 0$

$2 \sin x \cos x - \sin x = 0$

$\sin x (2 \cos x - 1) = 0$

$\sin x = 0$ $2 \cos x - 1 = 0$

$\quad x = 0 + 2k\pi$ $\cos x = \dfrac{1}{2}$

$\qquad$ or

$\quad x = \pi + 2k\pi$ $x = \dfrac{\pi}{3} + 2k\pi$

$\qquad\qquad$ or

$x = \dfrac{5\pi}{3} + 2k\pi$

The solutions are $0 + 2k\pi$, $\dfrac{\pi}{3} + 2k\pi$, $\pi + 2k\pi$, $\dfrac{5\pi}{3} + 2k\pi$

where k is an integer.

67. $\sin\left(2x + \dfrac{\pi}{6}\right) = -\dfrac{1}{2}$

$2x + \dfrac{\pi}{6} = \dfrac{7\pi}{6} + 2k\pi$ or $2x + \dfrac{\pi}{6} = \dfrac{11\pi}{6} + 2k\pi$

$2x = \pi + 2k\pi$ $2x = \dfrac{5\pi}{3} + 2k\pi$

$x = \dfrac{\pi}{2} + k\pi$ $x = \dfrac{5\pi}{6} + k\pi$

The solutions are $\dfrac{\pi}{2} + k\pi$, $\dfrac{5\pi}{6} + k\pi$ where k is an integer.

69. $\sin^2 \dfrac{x}{2} + \cos x = 1$

$\left(\pm\sqrt{\dfrac{1 - \cos x}{2}}\right)^2 + \cos x = 1$

$\dfrac{1 - \cos x}{2} + \cos x = 1$

$1 - \cos x + 2 \cos x = 2$

$\cos x = 1$

$x = 0 + 2k\pi$ where k is an integer

71. $\qquad \cos 2x - 1 - 3 \sin x$

$1 - 2 \sin^2 x = 1 - 3 \sin x$

$0 = 2 \sin^2 x - 3 \sin x$

$0 = \sin x (2 \sin x - 3)$

$\sin x = 0 \qquad\qquad 2 \sin x - 3 = 0$

$\quad x = 0, \pi \qquad\qquad \sin x = \dfrac{3}{2}$

$\qquad\qquad\qquad\qquad$ no solution.

The solutions are 0, π.

73. $\qquad \sin 4x - \sin 2x = 0$

$2 \sin 2x \cos 2x - \sin 2x = 0$

$\sin 2x (2 \cos 2x - 1) = 0$

$\sin 2x = 0$ $2 \cos 2x - 1 = 0$

$2x = 0 + 2k\pi$ $\cos 2x = \dfrac{1}{2}$

$\qquad$ or

$2x = \pi + 2k\pi$ $2x = \dfrac{\pi}{3} + 2k\pi$

$\qquad\qquad\qquad\qquad\qquad$ or

$2x = \dfrac{5\pi}{3} + 2k\pi$

$x = 0 + k\pi$, $\dfrac{\pi}{2} + k\pi$, $\dfrac{\pi}{6} + k\pi$, $\dfrac{5\pi}{6} + k\pi$

The solutions are 0, $\dfrac{\pi}{6}$, $\dfrac{\pi}{2}$, $\dfrac{5\pi}{6}$, π, $\dfrac{7\pi}{6}$, $\dfrac{3\pi}{2}$, $\dfrac{11\pi}{6}$.

75. $\tan \dfrac{\pi}{2} = \sin x$

$\dfrac{1 - \cos x}{\sin x} = \sin x$

$1 - \cos x = \sin^2 x$

$1 - \cos x = 1 - \cos^2 x$

$\cos^2 x - \cos x = 0$

$\cos x (\cos x - 1) = 0$

$\cos x = 0 \qquad\qquad \cos x = 1$

$x = \dfrac{\pi}{2}, \dfrac{3\pi}{2} \qquad\qquad x = 0$

The solutions are 0, $\dfrac{\pi}{2}$, $\dfrac{3\pi}{2}$.

77. $\sin 2x \cos x + \cos 2x \sin x = 0$

$\sin(2x + x) = 0$

$\sin 3x = 0$

$3x = 0 + 2k\pi$ or $3x = \pi + 2k\pi$

$x = 0 + \dfrac{2}{3}k\pi$ or $x = \dfrac{\pi}{3} + \dfrac{2}{3}k\pi$

The solutions are 0, $\dfrac{\pi}{3}$, $\dfrac{2}{3}\pi$, π, $\dfrac{4}{3}\pi$, $\dfrac{5}{3}\pi$.

79. $\sin x \cos 2x - \cos x \sin 2x = \dfrac{\sqrt{3}}{2}$

$\sin(x - 2x) = \dfrac{\sqrt{3}}{2}$

$\sin(-x) = \dfrac{\sqrt{3}}{2}$

$\sin x = -\dfrac{\sqrt{3}}{2}$

$x = \dfrac{4\pi}{3}, \dfrac{5\pi}{3}$

81. $\sin 3x - \sin x = 0$

$2\cos\dfrac{3x+x}{2}\sin\dfrac{3x-2}{2} = 0$

$2\cos 2x \sin x = 0$

$2(1 - 2\sin^2 x)\sin x = 0$

$\begin{array}{ll} \sin x = 0 & 1 - 2\sin^2 x = 0 \\ x = 0,\ \pi & \sin^2 x = \dfrac{1}{2} \\ & \sin x = \pm\dfrac{\sqrt{2}}{2} \end{array}$

$x = \dfrac{\pi}{4},\ \dfrac{3\pi}{4},\ \dfrac{5\pi}{4},\ \dfrac{7\pi}{4}$

The solutions are $0,\ \dfrac{\pi}{4},\ \dfrac{3\pi}{4},\ \pi,\ \dfrac{5\pi}{4},\ \dfrac{7\pi}{4}$.

83. $2\sin x \cos x + 2\sin x - \cos x - 1 = 0$

$2\sin x(\cos x + 1) - (\cos x + 1) = 0$

$(\cos x + 1)(2\sin x - 1) = 0$

$\begin{array}{ll} \cos x + 1 = 0 & 2\sin x - 1 = 0 \\ \cos x = -1 & \sin x = \dfrac{1}{2} \\ x = \pi & \\ & x = \dfrac{\pi}{6},\ \dfrac{5\pi}{6} \end{array}$

The solutions are $\dfrac{\pi}{6},\ \dfrac{5\pi}{6},\ \pi$.

85. 0.7391 **87.** $-3.2957,\ 3.2957$ **89.** 1.16

91. Set your graphing utility to "degree" mode and graph $d = \dfrac{(288)^2}{16}\sin\theta\cos\theta$ and $d = 1295$ for $0° \le \theta \le 90°$, $-500 \le d \le 3000$.

Use the TRACE or INTERSECT feature of your graphing utility to determine the intersection of the two graphs.

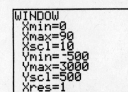 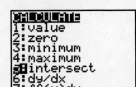 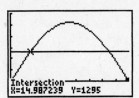 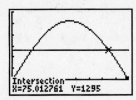

Thus, $d = 1295$ for $\theta \approx 14.99°$ and $\theta \approx 75.01°$.

The sine regression functions in Exercises 93–98 were obtained on a TI-83 calculator by using an iteration factor of 16. The use of a different iteration factor may produce a sine regression function that varies from the regression functions listed below.

93. a.

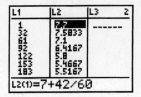 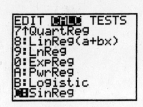 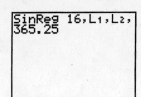

$f(x) = y \approx 1.121306016\sin(0.0159513906x + 1.836184752) + 6.625736057$

b. Set your graphing utility to "radian" mode.

$f(71) \approx 1.121306016\sin(0.0159513906(71) + 1.836184752) + 6.625736057$

≈ 6.818600251 hours

$\approx 6 + .818600251(60) \to 6:49$

95. a.

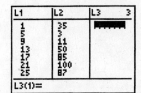 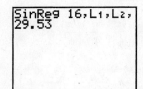

$f(x) = y \approx 49.51489157\sin(0.2127412216x - 2.91197818) + 53.18502398$

b. Set your graphing utility to "radian" mode.

$f(31) \approx 49.51489157\sin(0.2127412216(31) - 2.91197818) + 53.18502398$

$\approx 28\%$

97. a.

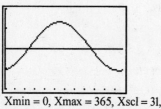

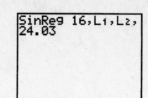

$$f(x) = y \approx 35.32348722\sin(0.3023630568x - 2.142585551) + 1.720109382$$

b. Set your graphing utility to "radian" mode.

$$f\left(9\frac{25}{60}\right) \approx 35.32348722\sin\left(0.3023630568\left(9\frac{25}{60}\right) - 2.142585551\right) + 1.720109382$$

$$\approx 24.6°$$

99. $d(t) \geq 12$ for $74 \leq t \leq 269$, so Mexico City will have at least 12 hours of daylight about $269 - 74 = 195$ days of the year.

Xmin = 0, Xmax = 365, Xscl = 31,
Ymin = 10, Ymax = 14, Yscl = 1

101. $\sqrt{3}\sin x + \cos x = \sqrt{3}$

$a = \sqrt{3}$, $b = 1$

$k = \sqrt{(\sqrt{3})^2 + 1} = 2$

α is in first quadrant

$\tan \beta = \dfrac{1}{\sqrt{3}}$

$\beta = \dfrac{\pi}{6}$

$\alpha = \dfrac{\pi}{6}$

$2\sin\left(x + \dfrac{\pi}{6}\right) = \sqrt{3}$

$\sin\left(x + \dfrac{\pi}{6}\right) = \dfrac{\sqrt{3}}{2}$

$x + \dfrac{\pi}{6} = \dfrac{\pi}{3}$ $x + \dfrac{\pi}{6} = \dfrac{2\pi}{3}$

$x = \dfrac{\pi}{6}$ $x = \dfrac{\pi}{2}$

103. $-\sin x + \sqrt{3}\cos x = \sqrt{3}$

$k = \sqrt{(-1)^2 + (\sqrt{3})^2} = 2$

$\tan \beta = \left|\dfrac{\sqrt{3}}{-1}\right| = \sqrt{3}$

$\beta = \dfrac{\pi}{3}$

$\alpha = \dfrac{2\pi}{3}$ second quadrant

$2\sin\left(x + \dfrac{2\pi}{3}\right) = \sqrt{3}$

$\sin\left(x + \dfrac{2\pi}{3}\right) = \dfrac{\sqrt{3}}{2}$

$x + \dfrac{2\pi}{3} = \dfrac{\pi}{3}$ $x + \dfrac{2\pi}{3} = \dfrac{2\pi}{3}$

$x = -\dfrac{\pi}{3}$ $x = 0$

$-\dfrac{\pi}{3} + 2\pi = \dfrac{5\pi}{3}$

The solutions are 0 and $\dfrac{5\pi}{3}$.

105.
$$\cos 5x - \cos 3x = 0$$
$$-2\sin\frac{5x+3x}{2}\sin\frac{5x-3x}{2} = 0$$
$$-2\sin 4x \sin x = 0$$

$\sin 4x = 0$ $\sin x = 0$

$4x = 0 + 2k\pi$ $4x = \pi + 2k\pi$ $x = 0, \pi$

$x = 0 + \dfrac{1}{2}k\pi$ $x = \dfrac{\pi}{4} + \dfrac{1}{2}k\pi$

$x = 0, \dfrac{\pi}{2}, \pi, \dfrac{3\pi}{2}$ $x = \dfrac{\pi}{4}, \dfrac{3\pi}{4}, \dfrac{5\pi}{4}, \dfrac{7\pi}{4}$

The solutions are $0, \dfrac{\pi}{4}, \dfrac{\pi}{2}, \dfrac{3\pi}{4}, \pi, \dfrac{5\pi}{4}, \dfrac{3\pi}{2}, \dfrac{7\pi}{4}$.

107.
$$\sin 3x + \sin x = 0$$
$$2\sin\frac{3x+x}{2}\cos\frac{3x-x}{2} = 0$$
$$2\sin 2x \cos x = 0$$

$\sin 2x = 0$ $\cos x = 0$

$2x = 0 + 2k\pi$ $2x = \pi + 2k\pi$ $x = \dfrac{\pi}{2}, \dfrac{3\pi}{2}$

$x = 0 + k\pi$

$x = 0, \pi$ $x = \dfrac{\pi}{2} + k\pi$

$x = \dfrac{\pi}{2}, \dfrac{3\pi}{2}$

The solutions are $0, \dfrac{\pi}{2}, \pi, \dfrac{3\pi}{2}$.

109. $\cos 4x + \cos 2x = 0$

$$2\cos\frac{4x+x}{2}\cos\frac{4x-x}{2} = 0$$

$$2\cos 3x \cos x = 0$$

$\cos 3x = 0$ $\qquad\qquad\qquad\qquad$ $\cos x = 0$

$3x = \dfrac{\pi}{2} + 2k\pi$ $\qquad$ $3x = \dfrac{3\pi}{2} + 2k\pi$ $\qquad$ $x = \dfrac{\pi}{2}, \dfrac{3\pi}{2}$

$x = \dfrac{\pi}{6} + \dfrac{2}{3}k\pi$ $\qquad$ $x = \dfrac{\pi}{2} + \dfrac{2}{3}k\pi$

$x = \dfrac{\pi}{6}, \dfrac{5\pi}{6}, \dfrac{3\pi}{2}$ $\qquad$ $x = \dfrac{\pi}{2}, \dfrac{7\pi}{6}, \dfrac{11\pi}{6}$

The solutions are $\dfrac{\pi}{6}, \dfrac{\pi}{2}, \dfrac{5\pi}{6}, \dfrac{7\pi}{6}, \dfrac{3\pi}{2}, \dfrac{11\pi}{6}$.

111. $\theta = 20°$

$$x = \sqrt{(4+18\cot 20°)^2 + 100} - (4+18\cot 20°)$$

$x \approx 0.93$ ft.

$\theta = 30°$

$$x = \sqrt{(4+18\cot 30°)^2 + 100} - (4+18\cot 30°)$$

$x \approx 1.39$ ft.

EXPLORING CONCEPTS WITH TECHNOLOGY, Page 237

1. $y = f_1 = x + \dfrac{x^3}{2\cdot 3}$ $\quad$ where $-1 \le x \le 1$

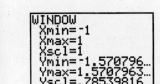 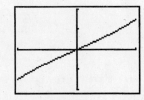

$y = f_2 = x + \dfrac{x^3}{2\cdot 3} + \dfrac{1\cdot 3 x^5}{2\cdot 4\cdot 5}$ $\quad$ where $-1 \le x \le 1$

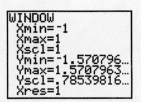 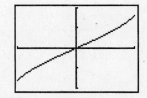

$y = f_3 = x + \dfrac{x^3}{2\cdot 3} + \dfrac{1\cdot 3 x^5}{2\cdot 4\cdot 5} + \dfrac{1\cdot 3\cdot 5 x^7}{2\cdot 4\cdot 6\cdot 7}$ $\quad$ where $-1 \le x \le 1$

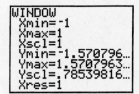 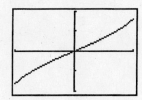

$y = f_4 = x + \dfrac{x^3}{2\cdot 3} + \dfrac{1\cdot 3 x^5}{2\cdot 4\cdot 5} + \dfrac{1\cdot 3\cdot 5 x^7}{2\cdot 4\cdot 6\cdot 7} + \dfrac{1\cdot 3\cdot 5\cdot 7 x^9}{2\cdot 4\cdot 6\cdot 8\cdot 9}$

where $-1 \le x \le 1$

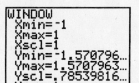 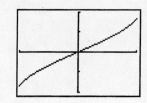

3. $\left| f_4(x) - \sin^{-1} x \right| < 0.001$ for $-0.7186 < x < 0.7186$

5. $f_6(1) = 1 + \dfrac{1}{6} + \dfrac{3}{40} + \dfrac{5}{112} + \dfrac{35}{1152} + \dfrac{63}{2816} + \dfrac{231}{13312}$

$\approx 1 + 0.167 + 0.075 + 0.045 + 0.030 + 0.022 + 0.017$

Each term is much smaller than the previous term.

CHAPTER 3 TRUE/FALSE EXERCISES, Page 239

1. False; if $\alpha = 45°$ and $\beta = 60°$, then $\dfrac{\tan\alpha}{\tan\beta} = \dfrac{\tan 45°}{\tan 60°} = \dfrac{1}{\sqrt{3}/2} = \dfrac{2\sqrt{3}}{3} \neq \dfrac{45°}{60°}$

3 False; if $x = 1$, then $\sin^{-1} x = \sin^{-1} 1 = \dfrac{\pi}{2}$ but $\csc x^{-1} = \csc 1^{-1} = \csc 1 = \dfrac{1}{\sin 1} \approx 1.188$.

5. False; if $\alpha = \beta = \dfrac{\pi}{4}$, we get $\sin\dfrac{\pi}{2} = 1 \neq \dfrac{\sqrt{2}}{2} + \dfrac{\sqrt{2}}{2} = \sqrt{2}$.

7. False; $\tan 45° = \tan 225°$ but $45° \neq 225°$.

9. False; $\cos\left(\cos^{-1} 2\right) \neq 2$ because $\cos^{-1} 2$ is undefined.

11. False; if $\theta = 30°$, then $\cos 30° = \dfrac{\sqrt{3}}{2} \neq \dfrac{1}{2} = \sin 150°$.

CHAPTER 3 REVIEW EXERCISES, Page 239

1. $\cos(45° + 30°) = \cos 45° \cos 30° - \sin 45° \sin 30°$

$$= \frac{\sqrt{2}}{2} \cdot \frac{\sqrt{3}}{2} - \frac{\sqrt{2}}{2} \cdot \frac{1}{2}$$

$$= \frac{\sqrt{6}}{4} - \frac{\sqrt{2}}{4}$$

$$= \frac{\sqrt{6} - \sqrt{2}}{4}$$

3. $\sin\left(\frac{2\pi}{3} + \frac{\pi}{4}\right) = \sin\frac{2\pi}{3}\cos\frac{\pi}{4} + \cos\frac{2\pi}{3}\sin\frac{\pi}{4}$

$$= \frac{\sqrt{3}}{2} \cdot \frac{\sqrt{2}}{2} - \frac{1}{2} \cdot \frac{\sqrt{2}}{2}$$

$$= \frac{\sqrt{6}}{4} - \frac{\sqrt{2}}{4}$$

$$= \frac{\sqrt{6} - \sqrt{2}}{4}$$

5. $\sin(60° - 135°) = \sin 60° \cos 135° - \cos 60° \sin 135°$

$$= \frac{\sqrt{3}}{2}\left(-\frac{\sqrt{2}}{2}\right) - \frac{1}{2} \cdot \frac{\sqrt{2}}{2}$$

$$= -\frac{\sqrt{6}}{4} - \frac{\sqrt{2}}{4} = -\frac{\sqrt{6} + \sqrt{2}}{4}$$

7. $\sin 22.5° = \sin\frac{45°}{2}$

$$= \sqrt{\frac{1 - \cos 45°}{2}}$$

$$= \sqrt{\frac{1 - \frac{\sqrt{2}}{2}}{2}}$$

$$= \sqrt{\frac{2 - \sqrt{2}}{4}}$$

$$= \frac{\sqrt{2 - \sqrt{2}}}{2}$$

9. $\tan 67.5° = \tan\frac{135°}{2}$

$$= \frac{1 - \cos 135°}{\sin 135°}$$

$$= \frac{1 - \left(-\frac{1}{\sqrt{2}}\right)}{\frac{1}{\sqrt{2}}}$$

$$= \frac{1 + \frac{1}{\sqrt{2}}}{\frac{1}{\sqrt{2}}}$$

$$= \sqrt{2} + 1$$

11. $\sin\alpha = \frac{1}{2}, \cos\alpha = \frac{\sqrt{3}}{2}$, quadrant I, $\tan\alpha = \frac{\sqrt{3}}{3}$

$\cos\beta = \frac{1}{2}, \sin\beta = -\frac{\sqrt{3}}{2}$, quadrant IV, $\tan\beta = -\sqrt{3}$

a. $\cos(\alpha - \beta) = \cos\alpha\cos\beta + \sin\alpha\sin\beta$

$$= \frac{\sqrt{3}}{2} \cdot \frac{1}{2} + \frac{1}{2}\left(-\frac{\sqrt{3}}{2}\right)$$

$$= \frac{\sqrt{3}}{4} - \frac{\sqrt{3}}{4} = 0$$

b. $\tan 2\alpha = \frac{2\tan\alpha}{1 - \tan^2\alpha}$

$$= \frac{2\left(\frac{\sqrt{3}}{3}\right)}{1 - \left(\frac{\sqrt{3}}{3}\right)^2} = \frac{\frac{2\sqrt{3}}{3}}{1 - \frac{1}{3}} \cdot \frac{3}{3}$$

$$= \frac{2\sqrt{3}}{3 - 1} = \sqrt{3}$$

c. $\sin\frac{\beta}{2} = \sqrt{\frac{1 - \cos\beta}{2}}$

$$= \sqrt{\frac{1 - \frac{1}{2}}{2}}$$

$$= \sqrt{\frac{1}{4}}$$

$$= \frac{1}{2}$$

13. $\sin\alpha = -\dfrac{1}{2}$, $\cos\alpha = \dfrac{\sqrt{3}}{2}$, quadrant IV, $\tan\alpha = -\dfrac{\sqrt{3}}{3}$

$\cos\beta = -\dfrac{\sqrt{3}}{2}$, $\sin\beta = -\dfrac{1}{2}$, quadrant III

a. $\sin(\alpha-\beta) = \sin\alpha\cos\beta - \cos\alpha\sin\beta$

$\qquad = -\dfrac{1}{2}\left(-\dfrac{\sqrt{3}}{2}\right) - \dfrac{\sqrt{3}}{2}\left(-\dfrac{1}{2}\right)$

$\qquad = \dfrac{\sqrt{3}}{4} + \dfrac{\sqrt{3}}{4} = \dfrac{2\sqrt{3}}{4}$

$\qquad = \dfrac{\sqrt{3}}{2}$

b. $\tan 2\alpha = \dfrac{2\tan\alpha}{1-\tan^2\alpha}$

$\qquad = \dfrac{2\left(-\dfrac{\sqrt{3}}{3}\right)}{1-\left(-\dfrac{\sqrt{3}}{3}\right)^2}$

$\qquad = \dfrac{-\dfrac{2\sqrt{3}}{3}}{1-\dfrac{1}{3}} \cdot \dfrac{3}{3}$

$\qquad = \dfrac{-2\sqrt{3}}{3-1}$

$\qquad = -\sqrt{3}$

c. $\cos\dfrac{\beta}{2} = -\sqrt{\dfrac{1+\cos\beta}{2}}$

$\qquad = -\sqrt{\dfrac{1+\left(-\dfrac{\sqrt{3}}{2}\right)}{2}}$

$\qquad = -\sqrt{\dfrac{2-\sqrt{3}}{4}}$

$\qquad = -\dfrac{\sqrt{2-\sqrt{3}}}{2}$

15. $2\sin 3x\cos 3x = \sin 2(3x)$

$\qquad\qquad\qquad = \sin 6x$

17. $\sin 4x\cos x - \cos 4x\sin x = \sin(4x-x)$

$\qquad\qquad\qquad\qquad\qquad = \sin 3x$

19. $1-2\sin^2\dfrac{\beta}{2} = \cos 2\left(\dfrac{\beta}{2}\right)$

$\qquad\qquad\quad = \cos\beta$

21. $\sin 47°\sin 22° = \dfrac{1}{2}\left[\cos(47°-22°) - \cos(47°+22°)\right]$

$\qquad\qquad\quad = \dfrac{1}{2}(\cos 25° - \cos 69°)$

$\qquad\qquad\quad \approx \dfrac{1}{2}(0.9063 - 0.3584)$

$\qquad\qquad\quad \approx 0.2740$

23. $2\sin\dfrac{\pi}{3}\cos\dfrac{2\pi}{3} = \sin\left(\dfrac{\pi}{3}+\dfrac{2\pi}{3}\right) + \sin\left(\dfrac{\pi}{3}-\dfrac{2\pi}{3}\right)$

$\qquad\qquad\qquad = \sin x + \sin\left(-\dfrac{\pi}{3}\right)$

$\qquad\qquad\qquad = 0 - \sin\dfrac{\pi}{3}$

$\qquad\qquad\qquad = -\dfrac{\sqrt{3}}{2}$

25. $\cos 2\theta - \cos 4\theta = -2\sin\dfrac{2\theta+4\theta}{2}\sin\dfrac{2\theta-4\theta}{2}$

$\qquad\qquad\qquad = -2\sin 3\theta\sin(-\theta)$

$\qquad\qquad\qquad = 2\sin 3\theta\sin\theta$

27. $\sin 6\theta + \sin 2\theta = 2\sin\dfrac{6\theta+2\theta}{2}\cos\dfrac{6\theta-2\theta}{2}$

$\qquad\qquad\qquad = 2\sin 4\theta\cos 2\theta$

29. $\dfrac{1}{\sin x-1} + \dfrac{1}{\sin x+1} = \dfrac{(\sin x+1)+(\sin x-1)}{(\sin x-1)(\sin x+1)}$

$\qquad\qquad\qquad\qquad = \dfrac{2\sin x}{\sin^2 x-1}$

$\qquad\qquad\qquad\qquad = \dfrac{2\sin x}{-\cos^2 x}$

$\qquad\qquad\qquad\qquad = -2\tan x\sec x$

31. $\dfrac{1+\sin x}{\cos^2 x} = \dfrac{1}{\cos^2 x} + \dfrac{\sin x}{\cos^2 x}$

$\qquad\qquad = \sec^2 x + \tan x\sec x$

$\qquad\qquad = \tan^2 x + 1 + \tan x\sec x$

33. $\dfrac{1}{\cos x} - \cos x = \dfrac{1-\cos^2 x}{\cos x}$

$\qquad\qquad\qquad = \dfrac{\sin^2 x}{\cos x}$

$\qquad\qquad\qquad = \tan x\sin x$

35. $\sin\left(\dfrac{\pi}{4}-\alpha\right)=\sin\dfrac{\pi}{4}\cos\alpha-\cos\dfrac{\pi}{4}\sin\alpha$

$\qquad\qquad\quad =\dfrac{\sqrt{2}}{2}\cos\alpha-\dfrac{\sqrt{2}}{2}\sin\alpha$

$\qquad\qquad\quad =\dfrac{\sqrt{2}}{2}(\cos\alpha-\sin\alpha)$

37. $\dfrac{\sin 4x-\sin 2x}{\cos 4x-\cos 2x}=\dfrac{2\cos\frac{4x+2x}{2}\sin\frac{4x-2x}{2}}{-2\sin\frac{4x+2x}{2}\sin\frac{4x-2x}{2}}$

$\qquad\qquad\qquad =-\dfrac{\cos 3x\sin x}{\sin 3x\sin x}$

$\qquad\qquad\qquad =-\cot 3x$

39. $\sin x-\cos 2x=\sin x-(1-2\sin^2 x)$

$\qquad\qquad\quad =2\sin^2 x+\sin x-1$

$\qquad\qquad\quad =(2\sin x-1)(\sin x+1)$

41. $\tan 4x=\dfrac{2\tan 2x}{1-\tan^2 2x}$

$\qquad\quad =\dfrac{2\left(\dfrac{2\tan x}{1-\tan^2 x}\right)}{1-\left(\dfrac{2\tan x}{1-\tan^2 x}\right)^2}$

$\qquad\quad =\dfrac{\dfrac{4\tan x}{1-\tan^2 x}}{\dfrac{(1-\tan^2 x)^2-(2\tan x)^2}{(1-\tan^2 x)^2}}\cdot\dfrac{(1-\tan^2 x)^2}{(1-\tan^2 x)^2}$

$\qquad\quad =\dfrac{4\tan x(1-\tan^2 x)}{(1-\tan^2 x)^2-4\tan^2 x}$

$\qquad\quad =\dfrac{4\tan x-4\tan^3 x}{1-2\tan^2 x+\tan^4 x-4\tan^2 x}$

$\qquad\quad =\dfrac{4\tan x-4\tan^3 x}{1-6\tan^2 x+\tan^4 x}$

43. $2\sin 3x\cos 3x-2\sin x\cos x=\sin 6x-\sin 2x$

$\qquad\qquad\qquad\qquad\quad =2\cos\dfrac{6x+2x}{2}\sin\dfrac{6x-2x}{2}$

$\qquad\qquad\qquad\qquad\quad =2\cos 4x\sin 2x$

45. $\cos(x+y)\cos(x-y)=\dfrac{1}{2}\big[\cos(x+y+x-y)+\cos(x+y-x+y)\big]$

$\qquad\qquad\qquad\qquad =\dfrac{1}{2}(\cos 2x+\cos 2y)$

$\qquad\qquad\qquad\qquad =\dfrac{1}{2}\big[2\cos^2 x-1+2\cos^2 y-1\big]$

$\qquad\qquad\qquad\qquad =\cos^2 x+\cos^2 y-1$

47. $y=\sec\left(\sin^{-1}\dfrac{12}{13}\right)$, $\alpha=\sin^{-1}\dfrac{12}{13}$, $\sin\alpha=\dfrac{12}{13}$, $\cos\alpha=\dfrac{5}{13}$, $\sec\alpha=\dfrac{13}{5}$

$\qquad y=\sec\alpha=\dfrac{13}{5}$

49. $2\sin^{-1}(x-1)=\dfrac{\pi}{3}$

$\qquad\sin^{-1}(x-1)=\dfrac{\pi}{6}$

$\qquad\qquad x-1=\sin\dfrac{\pi}{6}$

$\qquad\qquad x-1=\dfrac{1}{2}$

$\qquad\qquad\quad x=\dfrac{3}{2}$

51. $\sin^{-1}x+\cos^{-1}\dfrac{4}{5}=\dfrac{\pi}{2}$ $\qquad\alpha=\cos^{-1}\dfrac{4}{5}$

$\qquad\sin^{-1}x+\alpha=\dfrac{\pi}{2}$ $\qquad\qquad\cos\alpha=\dfrac{4}{5}$

$\qquad\quad\sin^{-1}x=\dfrac{\pi}{2}-\alpha$ $\qquad\qquad\sin\alpha=\dfrac{3}{5}$

$\qquad\qquad\quad x=\sin\left(\dfrac{\pi}{2}-\alpha\right)$

$\qquad\qquad\qquad =\sin\dfrac{\pi}{2}\cos\alpha-\cos\dfrac{\pi}{2}\sin\alpha$

$\qquad\qquad\qquad =1\cdot\dfrac{4}{5}-0\cdot\dfrac{3}{5}$

$\qquad\qquad\qquad =\dfrac{4}{5}$

53.

$$4\sin^2 x + 2\sqrt{3}\sin x - 2\sin x - \sqrt{3} = 0$$

$$2\sin x\left(2\sin x + \sqrt{3}\right) - \left(2\sin x + \sqrt{3}\right) = 0$$

$$\left(2\sin x + \sqrt{3}\right)\left(2\sin x - 1\right) = 0$$

$2\sin x + \sqrt{3} = 0$	$2\sin x - 1 = 0$
$\sin x = -\dfrac{\sqrt{3}}{2}$	$\sin x = \dfrac{1}{2}$
$x = 240°, 300°$	$x = 30°, 150°$

The solutions are $30°, 150°, 240°, 300°$.

55.

$$3\cos^2 x + \sin x = 1$$

$$3(1 - \sin^2 x) + \sin x = 1$$

$$0 = 3\sin^2 x - \sin x - 2$$

$$0 = (3\sin x + 2)(\sin x - 1)$$

$3\sin x + 2 = 0$	$\sin x - 1 = 0$
$\sin x = -\dfrac{2}{3}$	$\sin x = 1$
$x = 3.8713 \text{ or } 5.553$	$x = \dfrac{\pi}{2}$

The solutions are $\dfrac{\pi}{2} + 2k\pi,\ 3.8713 + 2k\pi,\ 5.553 + 2k\pi$ where k is an integer.

57.

$$\sin 3x \cos x - \cos 3x \sin x = \frac{1}{2}$$

$$\sin(3x - x) = \frac{1}{2}$$

$$\sin 2x = \frac{1}{2}$$

$2x = \dfrac{\pi}{6} + 2k\pi$	$2x = \dfrac{5\pi}{6} + 2k\pi$
$x = \dfrac{\pi}{12} + k\pi$	$x = \dfrac{5\pi}{12} + k\pi$

The solutions are $\dfrac{\pi}{12},\ \dfrac{5\pi}{12},\ \dfrac{13\pi}{12},\ \dfrac{17\pi}{12}$.

59.

$$f(x) = \sqrt{3}\sin x + \cos x$$

$$f(x) = 2\sin\left(x + \frac{\pi}{6}\right)$$

amplitude $= 2$

phase shift $= -\dfrac{\pi}{6}$

61.

$$f(x) = -\sin x - \sqrt{3}\cos x$$

$$f(x) = 2\sin\left(x + \frac{4\pi}{3}\right)$$

amplitude $= 2$

phase shift $= -\dfrac{4\pi}{3}$

63. $f(x) = 2\cos^{-1} x$

65. $f(x) = \sin^{-1}\dfrac{x}{2}$

CHAPTER 3 TEST, Page 241

1.

$$1 + \sin^2 x \sec^2 x = 1 + \sin^2 x \frac{1}{\cos^2 x}$$

$$= 1 + \tan^2 x$$

$$= \sec^2 x$$

2.

$$\frac{1}{\sec x - \tan x} - \frac{1}{\sec x + \tan x} = \frac{\sec x + \tan x - \sec x + \tan x}{\sec^2 x - \tan^2 x}$$

$$= \frac{2\tan x}{1}$$

$$= 2\tan x$$

3.

$$\cos^3 x + \cos x \sin^2 x = \cos x(\cos^2 x + \sin^2 x)$$

$$= \cos x$$

4.

$$\csc x - \cot x = \frac{1}{\sin x} - \frac{\cos x}{\sin x}$$

$$= \frac{1 - \cos x}{\sin x}$$

5. $\sin 195° = \sin(150° + 45°)$

$\qquad = \sin 150° \cos 45° + \cos 150° \sin 45°$

$\qquad = \dfrac{1}{2}\left(\dfrac{\sqrt{2}}{2}\right) + \left(-\dfrac{\sqrt{3}}{2}\right)\left(\dfrac{\sqrt{2}}{2}\right)$

$\qquad = \dfrac{\sqrt{2} - \sqrt{6}}{4}$

6. $\sin\alpha = -\dfrac{3}{5}, \cos\alpha = -\dfrac{4}{5}, \cos\beta = -\dfrac{\sqrt{2}}{2}, \sin\beta\,\dfrac{\sqrt{2}}{2}$

$\sin(\alpha + \beta) = \sin\alpha\cos\beta + \cos\alpha\sin\beta$

$\qquad = \left(-\dfrac{3}{5}\right)\left(-\dfrac{\sqrt{2}}{2}\right) + \left(-\dfrac{4}{5}\right)\left(\dfrac{\sqrt{2}}{2}\right)$

$\qquad = \dfrac{3\sqrt{2} - 4\sqrt{2}}{10}$

$\qquad = -\dfrac{\sqrt{2}}{10}$

7. $\sin\left(\theta - \dfrac{3\pi}{2}\right) = \sin\theta\cos\dfrac{3\pi}{2} - \cos\theta\sin\dfrac{3\pi}{2}$

$\qquad = \sin\theta\,(0) - \cos\theta\,(-1)$

$\qquad = \cos\theta$

8. $\cos 6x \sin 3x + \sin 6x \cos 3x = \sin(6x + 3x)$

$\qquad = \sin 9x$

9. $\sin\theta = \dfrac{4}{5}, \cos\theta = -\dfrac{3}{5}$

$\cos 2\theta = 2\cos^2\theta - 1$

$\qquad = 2\left(-\dfrac{3}{5}\right)^2 - 1$

$\qquad = \dfrac{18}{25} - 1 = -\dfrac{7}{25}$

10. $\tan\dfrac{\theta}{2} + \dfrac{\cos\theta}{\sin\theta} = \dfrac{1 - \cos\theta}{\sin\theta} + \dfrac{\cos\theta}{\sin\theta}$

$\qquad = \dfrac{1 - \cos\theta + \cos\theta}{\sin\theta}$

$\qquad = \csc\theta$

11. $\sin^2 2x + 4\cos^4 x = (2\sin x\cos x)^2 + 4\cos^4 x$

$\qquad = 4\sin^2 x\cos^2 x + 4\cos^4 x$

$\qquad = 4\cos^2 x(\sin^2 x + \cos^2 x)$

$\qquad = 4\cos^2 x$

12. $\sin 15° \cos 75° = \dfrac{1}{2}\left[\sin(15° + 75°) + \sin(15° - 75°)\right]$

$\qquad = \dfrac{1}{2}(\sin 90° - \sin 60°)$

$\qquad = \dfrac{1}{2}\left(1 - \dfrac{\sqrt{3}}{2}\right)$

$\qquad = \dfrac{1}{2} - \dfrac{\sqrt{3}}{4}$

$\qquad = \dfrac{2 - \sqrt{3}}{4}$

13. $y = -\dfrac{\sqrt{3}}{2}\sin x + \dfrac{1}{2}\cos x$

$a = -\dfrac{\sqrt{3}}{2},\ b = \dfrac{1}{2}$

$k = \sqrt{\left(-\dfrac{\sqrt{3}}{2}\right)^2 + \left(\dfrac{1}{2}\right)^2} = 1$

$\sin\beta = \left|\dfrac{\frac{1}{2}}{1}\right| = \dfrac{1}{2}$

$\beta = \dfrac{\pi}{6}$

$\alpha = \pi - \dfrac{\pi}{6}$

$\qquad = \dfrac{5\pi}{6}$

$y = \sin\left(x + \dfrac{5\pi}{6}\right)$

14. $\theta = \cos^{-1}(0.7644)$

$\theta = 0.701$

15. $\sin\left(\cos^{-1}\dfrac{12}{13}\right)$

Let $\theta = \cos^{-1}\dfrac{12}{13}$ and find $\sin\theta$.

Then $\cos\theta = \dfrac{12}{13}$ and $0 \le \theta \le \pi$.

$\sin\theta = \dfrac{5}{13}$

$\sin\left(\cos^{-1}\dfrac{12}{13}\right) = \dfrac{5}{13}$

16. The graph of $y = \sin^{-1}(x+2)$ is the graph of
$y = \sin^{-1}x$ moved two units to the left.

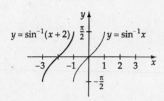

17. $3\sin x - 2 = 0$

$$\sin x = \frac{2}{3}$$

$$x = 41.8°, \ 138.2°$$

18. $\sin x \cos x - \dfrac{\sqrt{3}}{2}\sin x = 0$

$$\sin x\left(\cos x - \frac{\sqrt{3}}{2}\right) = 0$$

$\sin x = 0$ $\qquad\qquad \cos x - \dfrac{\sqrt{3}}{2} = 0$

$x = 0, \ \pi$

$$\cos x = \frac{\sqrt{3}}{2}$$

$$x = \frac{\pi}{6}, \ \frac{11\pi}{6}$$

The solutions are $0, \ \dfrac{\pi}{6}, \ \pi, \ \dfrac{11\pi}{6}$.

19. $\qquad \sin 2x + \sin x - 2\cos x - 1 = 0$

$$2\sin x \cos x + \sin x - 2\cos x - 1 = 0$$

$$\sin x(2\cos x + 1) - (2\cos x + 1) = 0$$

$$(2\cos x + 1)(\sin x - 1) = 0$$

$\cos x = -\dfrac{1}{2}$ $\qquad\qquad \sin x = 1$

$x = \dfrac{2\pi}{3}, \ \dfrac{4\pi}{3}$ $\qquad\qquad x = \dfrac{\pi}{2}$

The solutions are $\dfrac{\pi}{2}, \ \dfrac{2\pi}{3}, \ \dfrac{4\pi}{3}$.

20. a.

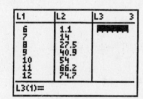

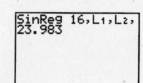

```
EDIT ░CALC░ TESTS
7↑QuartReg
8:LinReg(a+bx)
9:LnReg
0:ExpReg
A:PwrReg
B:Logistic
█:SinReg
```

```
SinReg 16,L₁,L₂,
23.983
```

```
SinReg
y=a*sin(bx+c)+d
a=38.96121154
b=.400329557
c=2.920598682
d=33.05340949
```

$f(x) = y \approx 38.96121154\sin(0.400329557x + 2.920598682) + 33.05340949$

b. $f\left(10\dfrac{40}{60}\right) \approx 38.96121154\sin\left(0.400329557\left(10\dfrac{40}{60}\right) + 2.920598682\right) + 33.05340949$

$\qquad \approx 63.8°$

SECTION 4.1, Page 249

1.

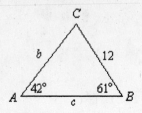

$C = 180° - 42° - 61°$

$C = 77°$

$$\frac{b}{\sin B} = \frac{a}{\sin A}$$

$$\frac{b}{\sin 61^0} = \frac{12}{\sin 42^0}$$

$$b = \frac{12\sin 61^0}{\sin 42^0}$$

$$b \approx 16$$

$$\frac{c}{\sin C} = \frac{a}{\sin A}$$

$$\frac{c}{\sin 77^0} = \frac{12}{\sin 42^0}$$

$$c = \frac{12\sin 77^0}{\sin 42^0}$$

$$c \approx 17$$

3.

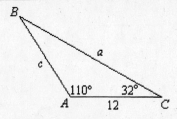

$B = 180° - 110° - 32°$

$B = 38°$

$$\frac{b}{\sin B} = \frac{a}{\sin A}$$

$$\frac{12}{\sin 38^0} = \frac{a}{\sin 110^0}$$

$$a = \frac{12\sin 110^0}{\sin 38^0}$$

$$a \approx 18$$

$$\frac{c}{\sin C} = \frac{b}{\sin B}$$

$$\frac{c}{\sin 32^0} = \frac{12}{\sin 38^0}$$

$$c = \frac{12\sin 32^0}{\sin 38^0}$$

$$c \approx 10$$

5.

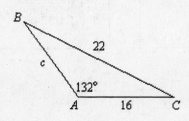

$$\frac{a}{\sin A} = \frac{b}{\sin B}$$

$$\frac{22}{\sin 132^0} = \frac{16}{\sin B}$$

$$\sin B = \frac{16\sin 132^0}{22}$$

$$\sin B \approx 0.5405$$

$$B \approx 33^0$$

$C \approx 180° - 33° - 132°$

$C \approx 15°$

$$\frac{a}{\sin A} = \frac{c}{\sin C}$$

$$\frac{22}{\sin 132^0} \approx \frac{c}{\sin 15^0}$$

$$c \approx \frac{22\sin 15^0}{\sin 132^0}$$

$$c \approx 7.7$$

7.

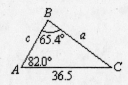

$C = 180° - 65.4° - 82.0°$

$C = 32.6°$

$$\frac{c}{\sin C} = \frac{b}{\sin B}$$

$$\frac{c}{\sin 32.6°} = \frac{36.5}{\sin 65.4°}$$

$$c = \frac{36.5\sin 32.6°}{\sin 65.4°}$$

$$c \approx 21.6$$

$$\frac{a}{\sin A} = \frac{b}{\sin B}$$

$$\frac{a}{\sin 82.0°} = \frac{36.5}{\sin 65.4°}$$

$$a = \frac{36.5\sin 82.0°}{\sin 65.4°}$$

$$a \approx 39.8$$

9.

$B = 180° - 98.5° - 33.8°$

$B = 47.7°$

$$\frac{a}{\sin A} = \frac{c}{\sin C}$$

$$\frac{a}{\sin 33.8°} = \frac{102}{\sin 98.5°}$$

$$a = \frac{102\sin 33.8°}{\sin 98.5°}$$

$$a \approx 57.4$$

$$\frac{b}{\sin B} = \frac{c}{\sin C}$$

$$\frac{b}{\sin 47.7°} = \frac{102}{\sin 98.5°}$$

$$b = \frac{102\sin 47.7°}{\sin 98.5°}$$

$$b \approx 76.3$$

11.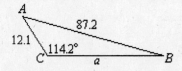

$$\frac{c}{\sin C} = \frac{b}{\sin B}$$

$$\frac{87.2}{\sin 114.2^\circ} = \frac{12.1}{\sin B}$$

$$\sin B = \frac{12.1 \sin 114.2^\circ}{87.2} \approx 0.1266$$

$$B \approx 7.3^\circ$$

$$\frac{c}{\sin C} = \frac{a}{\sin A}$$

$$\frac{87.2}{\sin 114.2^\circ} \approx \frac{a}{\sin 58.5^\circ}$$

$$a = \frac{87.2 \sin 58.5^\circ}{\sin 114.2^\circ} \approx 81.5$$

$$A \approx 180^\circ - 114.2^\circ - 7.3^\circ$$

$$A \approx 58.5^\circ$$

13.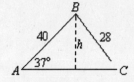

$$\sin 37^\circ = \frac{h}{40}$$

$$h = 40 \sin 37^\circ$$

$$h \approx 24$$

Since $h < 28$, two triangles exist.

$$\frac{c}{\sin C} = \frac{a}{\sin A}$$

$$\frac{40}{\sin C} = \frac{28}{\sin 37^\circ}$$

$$\sin C = \frac{40 \sin 37^\circ}{28} = 0.8597$$

$$C \approx 59^\circ \text{ or } 121^\circ$$

$$C = 59^\circ$$

$$B = 180^\circ - 37^\circ - 59^\circ = 84^\circ$$

$$\frac{b}{\sin 84^\circ} = \frac{28}{\sin 37^\circ}$$

$$b = \frac{28 \sin 84^\circ}{\sin 37^\circ} \approx 46$$

$$C = 121^\circ$$

$$B = 180^\circ - 121^\circ - 37^\circ = 22^\circ$$

$$\frac{b}{\sin 22^\circ} = \frac{28}{\sin 37^\circ}$$

$$b = \frac{28 \sin 22^\circ}{\sin 37^\circ} \approx 17$$

15.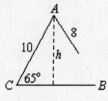

$$\sin 65^\circ = \frac{h}{10}$$

$$h = 10 \sin 65^\circ$$

$$h \approx 9.06$$

Since $h > 8$, no triangle is formed.

17.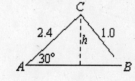

$$\sin 30^\circ = \frac{h}{2.4}$$

$$h = 2.4 \sin 30^\circ$$

$$h \approx 1.2$$

Since $h > 1$, no triangle is formed.

19.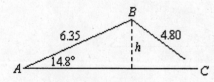

$$\sin 14.8^\circ = \frac{h}{6.35}$$

$$h = 6.35 \sin 14.8^\circ$$

$$h \approx 1.62$$

Since $h < 4.80$, two solutions exist.

$$\frac{c}{\sin C} = \frac{a}{\sin A}$$

$$\frac{6.35}{\sin C} = \frac{4.80}{\sin 14.8^\circ}$$

$$\sin C = \frac{6.35 \sin 14.8^\circ}{4.80}$$

$$\sin C = 0.3379$$

$$C \approx 19.8^\circ \text{ or } 160.2^\circ$$

$$C = 19.8^\circ$$

$$B = 180^\circ - 19.8^\circ - 14.8^\circ$$

$$= 145.4^\circ$$

$$\frac{b}{\sin 145.4^\circ} = \frac{4.80}{\sin 14.8^\circ}$$

$$b = \frac{4.80 \sin 145.4^\circ}{\sin 14.8^\circ}$$

$$b \approx 10.7$$

$$C = 160.2^\circ$$

$$B = 180^\circ - 160.2^\circ - 14.8^\circ$$

$$= 5.0^\circ$$

$$\frac{b}{\sin 5.0^\circ} = \frac{4.80}{\sin 14.8^\circ}$$

$$b = \frac{4.80 \sin 5.0^\circ}{\sin 14.8^\circ}$$

$$b \approx 1.64$$

21.

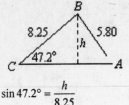

$$\sin 47.2° = \frac{h}{8.25}$$
$$h = 8.25 \sin 47.2°$$
$$h \approx 6.05$$

Since $h > 5.80$, no triangle is formed.

23.

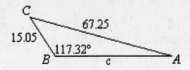

Since $b > a$, one triangle exists.

$$\frac{a}{\sin A} = \frac{b}{\sin B}$$
$$\frac{15.05}{\sin A} = \frac{67.25}{\sin 117.32°}$$
$$\sin A = \frac{15.04 \sin 117.32°}{67.25} \approx 0.1988$$
$$A = 11.47°$$

$$C = 180° - 11.47° - 117.32°$$
$$C = 51.21°$$

$$\frac{c}{\sin C} = \frac{b}{\sin B}$$
$$\frac{c}{\sin 51.21°} = \frac{67.25}{\sin 117.32°}$$
$$c = \frac{67.25 \sin 51.21°}{\sin 117.32°}$$
$$c \approx 59.00$$

25.

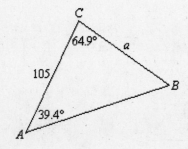

$$\angle B = 180° - (39.4° + 64.9°)$$
$$\angle B = 75.7°$$

$$\frac{a}{\sin A} = \frac{b}{\sin B}$$
$$\frac{a}{\sin 39.4°} = \frac{105}{\sin 75.7°}$$
$$a = \frac{105 \sin 39.4°}{\sin 75.7°}$$
$$a \approx 68.8 \text{ miles}$$

27.

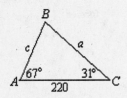

$$B = 180° - 67° - 31°$$
$$B = 82°$$

$$\frac{c}{\sin C} = \frac{b}{\sin B}$$
$$\frac{c}{\sin 31°} = \frac{220}{\sin 82°}$$
$$c = \frac{220 \sin 31°}{\sin 82°}$$
$$c \approx 110 \text{ feet}$$

29.

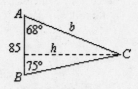

$$C = 180° - 68° - 75°$$
$$C = 37°$$

$$\frac{b}{\sin B} = \frac{c}{\sin C}$$
$$\frac{b}{\sin 75°} = \frac{85}{\sin 37°}$$
$$b = \frac{85 \sin 75°}{\sin 37°}$$
$$b \approx 136.4267957$$

$$\sin 68° = \frac{h}{b}$$
$$h = b \sin 68°$$
$$h \approx 130 \text{ yards}$$

31.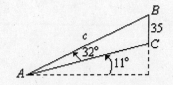

$A = 32° - 11°$
$A = 21°$
$B = 180° - 90° - 32°$
$B = 58°$
$C = 180° - 58° - 21°$
$C = 101°$

$\dfrac{c}{\sin C} = \dfrac{a}{\sin A}$

$\dfrac{c}{\sin 101°} = \dfrac{35}{\sin 21°}$

$c = \dfrac{35\sin 101°}{\sin 21°}$

$c \approx 96$ feet

33.

$A = 5°$
$B = 180° - 90° - 75°$
$B = 15°$
$C = 180° - 15° - 5°$
$C = 160°$

$\dfrac{b}{\sin B} = \dfrac{a}{\sin A}$

$\dfrac{b}{\sin 15°} = \dfrac{12}{\sin 5°}$

$b = \dfrac{12\sin 15°}{\sin 5°}$

$\sin 70° = \dfrac{h}{b}$

$h = b\sin 70°$

$h = \left(\dfrac{12\sin 15°}{\sin 5°}\right)\sin 70°$

$h \approx 33$ feet

35.

$\phi = 360° - 332°$
$\phi = 28°$
$\alpha = 28°$

$C = 82° - 28°$
$C = 54°$
$A = 28° + 36°$
$A = 64°$
$B = 180° - 64° - 54°$
$B = 62°$

$\dfrac{a}{\sin A} = \dfrac{b}{\sin B}$

$\dfrac{a}{\sin 64°} = \dfrac{8.0}{\sin 62°}$

$a = \dfrac{8.0\sin 64°}{\sin 62°}$

$a \approx 8.1$ miles

37.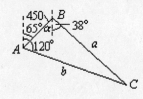

$A = 120° - 65°$
$A = 55°$
$\alpha = 65°$
$B = 38° + 65°$
$B = 103°$
$C = 180° - 103° - 55°$
$C = 22°$

$\dfrac{b}{\sin B} = \dfrac{c}{\sin C}$

$\dfrac{b}{\sin 103°} = \dfrac{450}{\sin 22°}$

$b = \dfrac{450\sin 103°}{\sin 22°}$

$b \approx 1200$ miles

39.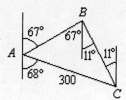

41. $\dfrac{a}{\sin A} = \dfrac{b}{\sin B}$

$\dfrac{a}{b} = \dfrac{\sin A}{\sin B}$

$\dfrac{a}{b} + 1 = \dfrac{\sin A}{\sin B} + 1$

$\dfrac{a+b}{b} = \dfrac{\sin A + \sin B}{\sin B}$

$A = 180° - 67° - 68°$
$A = 45°$
$B = 67° + 11°$
$B = 78°$
$C = 180° - 45° - 78°$
$C = 57°$

$\dfrac{c}{\sin C} = \dfrac{b}{\sin B}$

$\dfrac{c}{\sin 57°} = \dfrac{300}{\sin 78°}$

$c = \dfrac{300\sin 57°}{\sin 78°}$

$c \approx 260$ meters

43.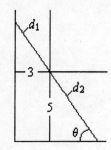

$\dfrac{3}{d_1} = \cos\theta, \ \dfrac{5}{d_2} = \sin\theta$

$d_1 = \dfrac{3}{\cos\theta}, \ d_2 = \dfrac{5}{\sin\theta}$

$L = d_1 + d_2$

$L(\theta) = \dfrac{3}{\cos\theta} + \dfrac{5}{\sin\theta}$

The graph of L is shown.
The minimum value of L is approximately 11.19 meters

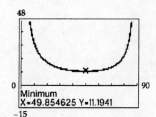

SECTION 4.2, Page 258

1.
$$c^2 = a^2 + b^2 - 2ab\cos C$$
$$c^2 = 12^2 + 18^2 - 2(12)(18)\cos 44°$$
$$c^2 = 468 - 432\cos 44°$$

$$c = \sqrt{468 - 432\cos 44°}$$
$$c \approx 13$$

3.
$$b^2 = a^2 + c^2 - 2ac\cos B$$
$$b^2 = 120^2 + 180^2 - 2(120)(180)\cos 56°$$
$$b^2 = 46,800 - 43,200\cos 56°$$
$$b = \sqrt{46,800 - 43,200\cos 56°}$$
$$b \approx 150$$

5.
$$a^2 = b^2 + c^2 - 2bc\cos A$$
$$a^2 = 60^2 + 84^2 - 2(60)(84)\cos 13°$$
$$a^2 = 10,656 - 10,080\cos 13°$$
$$a = \sqrt{10,656 - 10,080\cos 13°}$$
$$a \approx 29$$

7.
$$c^2 = a^2 + b^2 - 2ab\cos C$$
$$c^2 = 9.0^2 + 7.0^2 - 2(9.0)(7.0)\cos 72°$$
$$c^2 = 130 - 126\cos 72°$$
$$c = \sqrt{130 - 126\cos 72°}$$
$$c \approx 9.5$$

9.
$$c^2 = a^2 + b^2 - 2ab\cos C$$
$$c^2 = 4.6^2 + 7.2^2 - 2(4.6)(7.2)\cos 124°$$
$$c^2 = 73 - 66.24\cos 124°$$
$$c = \sqrt{73 - 66.24\cos 124°}$$
$$c \approx 10$$

11.
$$b^2 = a^2 + c^2 - 2ac\cos B$$
$$b^2 = 25.9^2 + 33.4^2 - 2(25.9)(33.4)\cos 84°$$
$$b^2 = 1786.37 - 1730.12\cos 84°$$
$$b = \sqrt{1786.37 - 1730\cos 84°}$$
$$b \approx 40.1$$

13.
$$b^2 = a^2 + c^2 - 2ac\cos B$$
$$b^2 = 122^2 + 55.9^2 - 2(122)(55.9)\cos 44.2°$$
$$b^2 = 18,008.81 - 13,639.6\cos 44.2°$$
$$b = \sqrt{18,008.81 - 13,639.6\cos 44.2°}$$
$$b \approx 90.7$$

15.
$$\cos A = \frac{b^2 + c^2 - a^2}{2bc}$$
$$\cos A = \frac{32^2 + 40^2 - 25^2}{2(32)(40)}$$
$$\cos A = \frac{1999}{2560}$$
$$A = \cos^{-1}\left(\frac{1999}{2560}\right) \approx 39°$$

17.
$$\cos C = \frac{a^2 + b^2 - c^2}{2ab}$$
$$\cos C = \frac{8.0^2 + 9.0^2 - 12^2}{2(8.0)(9.0)}$$
$$\cos C = \frac{1}{144}$$
$$C = \cos^{-1}\left(\frac{1}{144}\right) \approx 90°$$

19.
$$\cos B = \frac{a^2 + c^2 - b^2}{2ac}$$
$$\cos B = \frac{80^2 + 124^2 - 92^2}{2(80)(124)}$$
$$\cos B = \frac{13,312}{19,840}$$
$$B = \cos^{-1}\left(\frac{13312}{19840}\right) \approx 47.9°$$

21.
$$\cos C = \frac{a^2 + b^2 - c^2}{2ab}$$
$$\cos C = \frac{1025^2 + 625^2 - 1402^2}{2(1025)(625)}$$
$$\cos C = \frac{-575,150}{1,281,250}$$
$$C = \cos^{-1}\left(\frac{-575,150}{1,281,250}\right) \approx 116.67°$$

23.
$$\cos B = \frac{a^2 + c^2 - b^2}{2ac}$$
$$\cos B = \frac{32.5^2 + 29.6^2 - 40.1^2}{2(32.5)(29.6)}$$
$$\cos B = \frac{324.4}{1924}$$
$$B = \cos^{-1}\left(\frac{324.4}{1924}\right)$$
$$B \approx 80.3°$$

25. $K = \frac{1}{2}bc\sin A$

$K = \frac{1}{2}(12)(24)\sin 105°$

$K \approx 140$ square units

27. $C = 180° - 42° - 76°$
$C = 62°$

$K = \frac{c^2 \sin A \sin B}{2\sin C}$

$K = \frac{12^2 \sin 42° \sin 76°}{2\sin 62°}$

$K \approx 53$ square units

29. $s = \frac{1}{2}(a+b+c)$

$s = \frac{1}{2}(16+12+14)$

$s = 21$

$K = \sqrt{s(s-a)(s-b)(s-c)}$

$K = \sqrt{21(21-16)(21-12)(21-14)}$

$K = \sqrt{21(5)(9)(7)}$

$K \approx 81$ square units

31. $\frac{a}{\sin A} = \frac{b}{\sin B}$

$\frac{22.4}{\sin A} = \frac{26.9}{\sin 54.3°}$

$\sin A = \frac{22.4\sin 54.3°}{26.9}$

$\sin A \approx 0.6762$

$A \approx 42.5°$

$C = 180° - 42.5° - 54.3°$

$C = 83.2°$

$K = \frac{1}{2}ab\sin C$

$K = \frac{1}{2}(22.4)(26.9)\sin 83.2°$

$K \approx 299$ square units

33. $C = 180° - 116° - 34°$
$C = 30°$

$K = \frac{c^2 \sin A \sin B}{2\sin C}$

$K = \frac{8.5^2 \sin 116° \sin 34°}{2\sin 30°}$

$K \approx 36$ square units

35. $s = \frac{1}{2}(a+b+c)$

$s = \frac{1}{2}(3.6+4.2+4.8)$

$s = 6.3$

$K = \sqrt{s(s-a)(s-b)(s-c)}$

$K = \sqrt{6.3(6.3-3.6)(6.3-4.2)(6.3-4.8)}$

$K = \sqrt{6.3(2.7)(2.1)(1.5)}$

$K \approx 7.3$ square units

37.

$\alpha = 32°$
$\beta = 72°$
$B = 72° + 32°$
$B = 104°$

$b^2 = a^2 + c^2 - 2ac\cos 104°$
$b^2 = 320^2 + 560^2 - 2(320)(560)\cos 104°$
$b^2 = 416,000 - 358,400\cos 104°$
$b = \sqrt{416,000 - 358,400\cos 104°}$
$b \approx 710$ miles

39.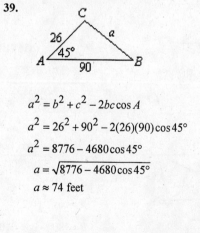

$a^2 = b^2 + c^2 - 2bc\cos A$
$a^2 = 26^2 + 90^2 - 2(26)(90)\cos 45°$
$a^2 = 8776 - 4680\cos 45°$
$a = \sqrt{8776 - 4680\cos 45°}$
$a \approx 74$ feet

41.

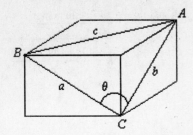

Let a = the length of the diagonal on the front of the box.
Let b = the length of the diagonal on the right side of the box.
Let c = the length of the diagonal on the top of the box.

$$a^2 = (4.75)^2 + (6.50)^2 = 64.8125$$
$$a = \sqrt{64.8125}$$
$$b^2 = (3.25)^2 + (4.75)^2 = 33.125$$
$$b = \sqrt{33.125}$$
$$c^2 = (6.50)^2 + (3.25)^2 = 52.8125$$
$$\theta = C$$
$$\cos C = \frac{a^2 + b^2 - c^2}{2ab}$$
$$\cos\theta = \frac{64.8125 + 33.125 - 52.8125}{2\sqrt{64.8125}\sqrt{33.125}}$$
$$\cos\theta = \frac{45.125}{2\sqrt{64.8125}\sqrt{33.125}}$$
$$\theta = \cos^{-1}\left(\frac{45.125}{2\sqrt{64.8125}\sqrt{33.125}}\right)$$
$$\theta \approx 60.9°$$

43.

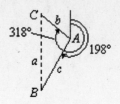

$$b = (18 \text{ mph})(10 \text{ hours}) = 180 \text{ miles}$$
$$c = (22 \text{ mph})(10 \text{ hours}) = 220 \text{ miles}$$
$$A = 318° - 198°$$
$$A = 120°$$

$$a^2 = b^2 + c^2 - 2bc \cos A$$
$$a^2 = 180^2 + 220^2 - 2(180)(220)\cos 120°$$
$$a^2 = 120,400$$
$$a \approx 350 \text{ miles}$$

45.

$$A = \frac{360°}{6}$$
$$A = 60°$$

$$a^2 = 40^2 + 40^2 - 2(40)(40)\cos 60°$$
$$a^2 = 1600$$
$$a = 40 \text{ cm}$$

47.

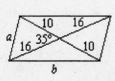

$$a^2 = 16^2 + 10^2 - 2(16)(10)\cos 35°$$
$$a^2 = 356 - 320\cos 35°$$
$$a = \sqrt{356 - 320\cos 35°}$$
$$a \approx 9.7 \text{ inches}$$

$$b^2 = 10^2 + 16^2 - 2(10)(16)\cos 145°$$
$$b^2 = 356 - 320\cos 145°$$
$$b = \sqrt{356 - 320\cos 145°}$$
$$b \approx 25 \text{ inches}$$

49.

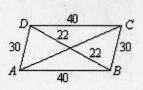

$$\cos A = \frac{30^2 + 40^2 - 44^2}{2(30)(40)}$$
$$\cos A = 0.235$$
$$A = \cos^{-1}(0.235)$$
$$B = 180° - \cos^{-1}(0.235)$$

$$(AC)^2 = 40^2 + 30^2 - 2(40)(30)\cos(180° - \cos^{-1}(0.235))$$
$$AC = \sqrt{40^2 + 30^2 - 2(40)(30)\cos(180° - \cos^{-1}(0.235))}$$
$$\approx 55 \text{ centimeters}$$

51.

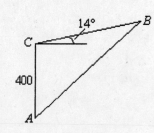

$C = 90° + 14°$

$C = 104°$

$a = \dfrac{180(5280)}{3600} \cdot 10$

$a = 2640$ feet

$c^2 = 2640^2 + 400^2 - 2(2640)(400)(\cos 104°)$

$c^2 \approx 7,640,539$

$c \approx 2800$ feet

53.

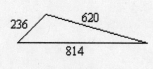

$s = \dfrac{1}{2}(a + b + c)$

$s = \dfrac{1}{2}(236 + 620 + 814)$

$s = 835$

$K = \sqrt{s(s-a)(s-b)(s-c)}$

$K = \sqrt{835(835 - 236)(835 - 620)(835 - 814)}$

$K = \sqrt{835(599)(215)(21)}$

$K \approx \sqrt{2,258,240,000}$

$K \approx 47,500$ square meters

55.

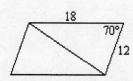

$K = 2\left[\dfrac{1}{2}(12)(18)\sin 70°\right]$

$K \approx 203$ square meters

57.

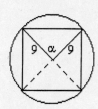

$\alpha = 90°$

$K = 4\left[\dfrac{1}{2}(9)(9)\sin 90°\right]$

$K = 162$ in^2

59.

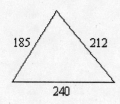

$s = \dfrac{1}{2}(185 + 212 + 240)$

$s = 318.5$

$K = \sqrt{318.5(318.5 - 185)(318.5 - 212)(318.5 - 240)}$

$K \approx 18,854$ ft^2

cost $= 2.20(18,854)$

cost $\approx \$41,000$

61.

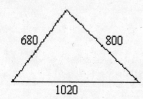

$s = \dfrac{1}{2}(680 + 800 + 1020)$

$s = 1250$

$K = \sqrt{1250(1250 - 680)(1250 - 800)(1250 - 1020)}$

$K \approx 271,558$ ft^2

Acres $= \dfrac{271,558}{43,560}$

Acres ≈ 6.23

63. a.

$A = (\text{side})^2$

$= 8^2$

$= 64$ square units

$s = \dfrac{8 + 8 + 8 + 8}{2}$

$= \dfrac{32}{2}$

$= 16$

$K = \sqrt{(16-8)(16-8)(16-8)(16-8)}$

$= \sqrt{8 \cdot 8 \cdot 8 \cdot 8}$

$= 8 \cdot 8$

$= 64$ square units

b.

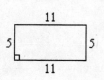

$A = LW$

$= 11(5)$

$= 55$ square units

$s = \dfrac{11 + 5 + 11 + 5}{2}$

$= \dfrac{32}{2}$

$= 16$

$K = \sqrt{(16-11)(16-5)(16-11)(16-5)}$

$= \sqrt{5 \cdot 11 \cdot 5 \cdot 11}$

$= 5 \cdot 11$

$= 55$ square units

65.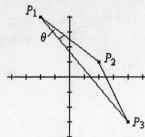

$$d(P_1, P_2) = \sqrt{[2-(-2)]^2 + (1-4)^2} = 5$$

$$d(P_1, P_3) = \sqrt{(-2-4)^2 + (4-(-3))^2} = \sqrt{85}$$

$$d(P_2, P_3) = \sqrt{(2-4)^2 + (1-(-3))^2} = 2\sqrt{5}$$

$$\cos\theta = \frac{5^2 + \left(\sqrt{85}\right)^2 - \left(2\sqrt{5}\right)^2}{2 \cdot 5 \cdot \sqrt{85}}$$

$$\cos\theta \approx 0.9762$$

$$\theta \approx 12.5°$$

67.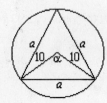

$$\alpha = \frac{360}{3}$$

$$= 120°$$

$$a^2 = 10^2 + 10^2 - 2(10)(10)\cos 120°$$

$$a^2 = 300$$

$$a = \sqrt{300}$$

$$\text{perimeter} = 3a$$

$$= 3\sqrt{300}$$

$$\approx 52.0 \text{ cm}$$

69. $\cos A = \dfrac{b^2 + c^2 - a^2}{2bc} = \dfrac{b^2 + 2bc + c^2 - a^2 - 2bc}{2bc} = \dfrac{(b+c)^2 - a^2}{2bc} - \dfrac{2bc}{2bc} = \dfrac{(b+c-a)(b+c+a)}{2bc} - 1$

71.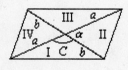

$$\alpha + C = 180°$$

$$\alpha = 180° - C$$

$$\sin\alpha = \sin C$$

$$K_I = K_{III} = \frac{1}{2}ab\sin C$$

$$K_{II} = K_{IV} = \frac{1}{2}ab\sin\alpha = \frac{1}{2}ab\sin C$$

$$K = K_I + K_{II} + K_{III} + K_{IV}$$

$$= \frac{1}{2}ab\sin C + \frac{1}{2}ab\sin C + \frac{1}{2}ab\sin C + \frac{1}{2}ab\sin C$$

$$K = 2ab\sin C$$

73. $V = 18K$, where $K =$ area of triangular base

$$V = \frac{1}{2}(4)(4)(\sin 72°)(18)$$

$$V \approx 140 \text{ in}^3$$

SECTION 4.3, Page 274

1. $a = 4 - (-3) = 7$
$b = -1 - 0 = -1$

A vector equivalent to $\mathbf{P_1P_2}$ is $\mathbf{v} = \langle 7, -1 \rangle$.

3. $a = -3 - 4 = -7$
$b = -3 - 2 = -5$

A vector equivalent to $\mathbf{P_1P_2}$ is $\mathbf{v} = \langle -7, -5 \rangle$.

5. $a = 2 - 2 = 0$
$b = 3 - (-5) = 8$

A vector equivalent to $\mathbf{P_1P_2}$ is $\mathbf{v} = \langle 0, 8 \rangle$.

7. $\|\mathbf{v}\| = \sqrt{(-3)^2 + 4^2}$
$\|\mathbf{v}\| = \sqrt{9 + 16}$
$\|\mathbf{v}\| = 5$

$\alpha = \tan^{-1}\left|\dfrac{4}{-3}\right| = \tan^{-1}\dfrac{4}{3}$

$\alpha \approx 53.1°$
$\theta = 180° - \alpha$
$\theta \approx 180° - 53.1°$
$\theta \approx 126.9°$

$$\mathbf{u} = \left\langle \frac{-3}{5}, \frac{4}{5} \right\rangle$$

A unit vector in the direction of $\mathbf{v}$ is $\mathbf{u} = \left\langle -\frac{3}{5}, \frac{4}{5} \right\rangle$.

9. $\|\mathbf{v}\| = \sqrt{20^2 + (-40)^2}$ $\alpha = \tan^{-1}\left|\dfrac{-40}{20}\right| = \tan^{-1}2$

$\|\mathbf{v}\| = \sqrt{400 + 1600}$ $\alpha \approx 63.4°$

$\|\mathbf{v}\| = \sqrt{2000} = 20\sqrt{5}$ $\theta = 360° - \alpha$

≈ 44.7 $\theta \approx 360° - 63.4°$

$\theta \approx 296.6°$

$\mathbf{u} = \left\langle \dfrac{20}{20\sqrt{5}}, \dfrac{-40}{20\sqrt{5}} \right\rangle = \left\langle \dfrac{\sqrt{5}}{5}, \dfrac{-2\sqrt{5}}{5} \right\rangle$

A unit vector in the direction of $\mathbf{v}$ is $\mathbf{u} = \left\langle \dfrac{\sqrt{5}}{5}, -\dfrac{2\sqrt{5}}{5} \right\rangle$.

11. $\|\mathbf{v}\| = \sqrt{2^2 + (-4)^2}$ $\alpha = \tan^{-1}\left|\dfrac{-4}{2}\right| = \tan^{-1}2$

$\|\mathbf{v}\| = \sqrt{4 + 16}$ $\alpha \approx 63.4°$

$\|\mathbf{v}\| = \sqrt{20} = 2\sqrt{5}$ $\theta = 360° - \alpha$

≈ 4.5 $\theta \approx 360° - 63.4°$

$\theta \approx 296.6°$

$\mathbf{u} = \left\langle \dfrac{2}{2\sqrt{5}}, \dfrac{-4}{2\sqrt{5}} \right\rangle = \left\langle \dfrac{\sqrt{5}}{5}, -\dfrac{2\sqrt{5}}{5} \right\rangle$

A unit vector in the direction of $\mathbf{v}$ is $\mathbf{u} = \left\langle \dfrac{\sqrt{5}}{5}, -\dfrac{2\sqrt{5}}{5} \right\rangle$.

13. $\|\mathbf{v}\| = \sqrt{42^2 + (-18)^2}$ $\alpha = \tan^{-1}\left|\dfrac{-18}{42}\right| = \tan^{-1}\dfrac{3}{7}$

$\|\mathbf{v}\| = \sqrt{1764 + 324}$ $\alpha \approx 23.2°$

$\|\mathbf{v}\| = \sqrt{2088}$ $\theta = 360° - \alpha$

$= 6\sqrt{58}$ $\theta \approx 360° - 23.2°$

≈ 45.7 $\theta \approx 336.8°$

$\mathbf{u} = \left\langle \dfrac{42}{6\sqrt{58}}, \dfrac{-18}{6\sqrt{58}} \right\rangle = \left\langle \dfrac{7\sqrt{58}}{58}, -\dfrac{3\sqrt{58}}{58} \right\rangle$

A unit vector in the direction of $\mathbf{v}$ is $\mathbf{u} = \left\langle \dfrac{7\sqrt{58}}{58}, -\dfrac{3\sqrt{58}}{58} \right\rangle$.

15. $3\mathbf{u} = 3\langle -2, 4 \rangle = \langle -6, 12 \rangle$

17. $2\mathbf{u} - \mathbf{v} = 2\langle -2, 4 \rangle - \langle -3, -2 \rangle$

$= \langle -4, 8 \rangle - \langle -3, -2 \rangle$

$= \langle -1, 10 \rangle$

19. $\dfrac{2}{3}\mathbf{u} + \dfrac{1}{6}\mathbf{v} = \dfrac{2}{3}\langle -2, 4 \rangle + \dfrac{1}{6}\langle -3, -2 \rangle$

$= \left\langle -\dfrac{4}{3}, \dfrac{8}{3} \right\rangle + \left\langle -\dfrac{1}{2}, -\dfrac{1}{3} \right\rangle$

$= \left\langle -\dfrac{11}{6}, \dfrac{7}{3} \right\rangle$

21. $\|\mathbf{u}\| = \sqrt{(-2)^2 + 4^2} = \sqrt{20} = 2\sqrt{5}$

23. $3\mathbf{u} - 4\mathbf{v} = 3\langle -2, 4 \rangle - 4\langle -3, -2 \rangle$

$= \langle -6, 12 \rangle - \langle -12, -8 \rangle$

$= \langle 6, 20 \rangle$

$\|3\mathbf{u} - 4\mathbf{v}\| = \sqrt{6^2 + 20^2}$

$= \sqrt{436}$

$= 2\sqrt{109}$

25. $4\mathbf{v} = 4(-2\mathbf{i} + 3\mathbf{j})$

$= -8\mathbf{i} + 12\mathbf{j}$

27. $6\mathbf{u} + 2\mathbf{v} = 6(3\mathbf{i} - 2\mathbf{j}) + 2(-2\mathbf{i} + 3\mathbf{j})$

$= (18\mathbf{i} - 12\mathbf{j}) + (-4\mathbf{i} + 6\mathbf{j})$

$= (18 - 4)\mathbf{i} + (-12 + 6)\mathbf{j}$

$= 14\mathbf{i} - 6\mathbf{j}$

29. $\dfrac{2}{3}\mathbf{v} + \dfrac{3}{4}\mathbf{u} = \dfrac{2}{3}(-2\mathbf{i} + 3\mathbf{j}) + \dfrac{3}{4}(3\mathbf{i} - 2\mathbf{j})$

$= \left(-\dfrac{4}{3}\mathbf{i} + 2\mathbf{j} \right) + \left(\dfrac{9}{4}\mathbf{i} - \dfrac{3}{2}\mathbf{j} \right)$

$= \left(-\dfrac{4}{3} + \dfrac{9}{4} \right)\mathbf{i} + \left(2 - \dfrac{3}{2} \right)\mathbf{j}$

$= \dfrac{11}{12}\mathbf{i} + \dfrac{1}{2}\mathbf{j}$

31. $\mathbf{u} - 2\mathbf{v} = (3\mathbf{i} - 2\mathbf{j}) - 2(-2\mathbf{i} + 3\mathbf{j})$

$= (3\mathbf{i} - 2\mathbf{j}) - (-4\mathbf{i} + 6\mathbf{j})$

$= (3 + 4)\mathbf{i} + (-2 - 6)\mathbf{j}$

$= 7\mathbf{i} - 8\mathbf{j}$

$\|\mathbf{u} - 2\mathbf{v}\| = \sqrt{7^2 + (-8)^2} = \sqrt{113}$

33. $a_1 = 5\cos 27° \approx 4.5$

$a_2 = 5\sin 27° \approx 2.3$

$\mathbf{v} = a_1\mathbf{i} + a_2\mathbf{j} \approx 4.5\mathbf{i} + 2.3\mathbf{j}$

35. $a_1 = 4\cos\dfrac{\pi}{4} \approx 2.8$

$a_2 = 4\sin\dfrac{\pi}{4} \approx 2.8$

$\mathbf{v} = a_1\mathbf{i} + a_2\mathbf{j} \approx 2.8\mathbf{i} + 2.8\mathbf{j}$

37.   heading = 124° ⇒   wind from the west ⇒
          direction angle = –34°   direction angle = 0°

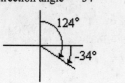

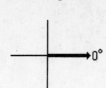

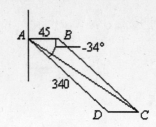

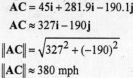

$\mathbf{AB} = 45\mathbf{i}$

$\mathbf{AD} = 340\cos(-34°)\mathbf{i} + 340\sin(-34°)\mathbf{j}$

$\mathbf{AD} \approx 281.9\mathbf{i} - 190.1\mathbf{j}$

$\mathbf{AC} = \mathbf{AB} + \mathbf{AD}$

$\mathbf{AC} = 45\mathbf{i} + 281.9\mathbf{i} - 190.1\mathbf{j}$

$\mathbf{AC} \approx 327\mathbf{i} - 190\mathbf{j}$

$\|\mathbf{AC}\| = \sqrt{327^2 + (-190)^2}$

$\|\mathbf{AC}\| \approx 380 \text{ mph}$

The ground speed of the plane is approximately 380 mph.

39.   heading = 96° ⇒   heading = 37° ⇒
          direction angle = –6°   direction angle = 53°

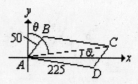

$\mathbf{AB} = 50\cos 53°\mathbf{i} + 50\sin 53°\mathbf{j}$

$\qquad \approx 30.1\mathbf{i} + 39.9\mathbf{j}$

$\mathbf{AD} = 225\cos(-6°)\mathbf{i} + 225\sin(-6°)\mathbf{j}$

$\qquad \approx 223.8\mathbf{i} - 23.5\mathbf{j}$

$\mathbf{AC} = \mathbf{AB} + \mathbf{AD}$

$\qquad \approx 30.1\mathbf{i} + 39.9\mathbf{j} + 223.8\mathbf{i} - 23.5\mathbf{j}$

$\qquad \approx 253.9\mathbf{i} + 16.4\mathbf{j}$

$\|\mathbf{AC}\| = \sqrt{(253.9)^2 + (16.4)^2} \approx 250$

$\alpha = \tan^{-1}\left|\dfrac{16.4}{253.9}\right|$

$\quad = \tan^{-1}\dfrac{16.4}{253.9}$

$\alpha \approx 4°$

$\theta = 90° - \alpha$

$\theta \approx 90° - 4°$

$\theta \approx 86°$

The ground speed of the plane is about 250 mph at a heading of approximately 86°.

41.

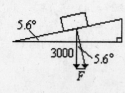

$\sin 5.6° = \dfrac{F}{3000}$

$F = 3000 \sin 5.6°$

$F \approx 293 \text{ lb}$

43.

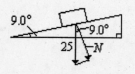

$\cos 9.0° = \dfrac{N}{25}$

$N = 25\cos 9.0°$

$N \approx 24.7 \text{ lb}$

45. $\mathbf{v} \cdot \mathbf{w} = \langle 3, -2\rangle \cdot \langle 1, 3\rangle$

$\qquad = 3(1) + (-2)3$

$\qquad = 3 - 6$

$\qquad = -3$

47. $\mathbf{v} \cdot \mathbf{w} = \langle 4, 1\rangle \cdot \langle -1, 4\rangle$

$\qquad = 4(-1) + 1(4)$

$\qquad = -4 + 4$

$\qquad = 0$

49. $\mathbf{v} \cdot \mathbf{w} = (\mathbf{i} + 2\mathbf{j}) \cdot (-\mathbf{i} + \mathbf{j})$

$\qquad = 1(-1) + 2(1)$

$\qquad = -1 + 2$

$\qquad = 1$

51. $\mathbf{v} \cdot \mathbf{w} = (6\mathbf{i} - 4\mathbf{j}) \cdot (-2\mathbf{i} - 3\mathbf{j})$

$\qquad = 6(-2) + (-4)(-3)$

$\qquad = -12 + 12$

$\qquad = 0$

53. $\cos\theta = \dfrac{\mathbf{v} \cdot \mathbf{w}}{\|\mathbf{v}\|\,\|\mathbf{w}\|}$

$\cos\theta = \dfrac{\langle 2, -1\rangle \cdot \langle 3, 4\rangle}{\sqrt{2^2 + (-1)^2}\,\sqrt{3^2 + 4^2}}$

$\cos\theta = \dfrac{2(3) + (-1)4}{\sqrt{5}\,\sqrt{25}}$

$\cos\theta = \dfrac{2}{5\sqrt{5}} \approx 0.1789$

$\theta \approx 79.7°$

55. $\cos\theta = \dfrac{\mathbf{v} \cdot \mathbf{w}}{\|\mathbf{v}\|\,\|\mathbf{w}\|}$

$\cos\theta = \dfrac{\langle 0, 3\rangle \cdot \langle 2, 2\rangle}{\sqrt{0^2 + 3^2}\,\sqrt{2^2 + 2^2}}$

$\cos\theta = \dfrac{0(2) + 3(2)}{\sqrt{9}\,\sqrt{8}}$

$\cos\theta = \dfrac{6}{6\sqrt{2}} \approx 0.7071$

$\theta = 45°$

57. $\cos\theta = \dfrac{\mathbf{v}\cdot\mathbf{w}}{\|\mathbf{v}\|\,\|\mathbf{w}\|}$

$\cos\theta = \dfrac{(5\mathbf{i}-2\mathbf{j})\cdot(2\mathbf{i}+5\mathbf{j})}{\sqrt{5^2+(-2)^2}\sqrt{2^2+5^2}}$

$\cos\theta = \dfrac{5(2)+(-2)(5)}{\sqrt{29}\sqrt{29}}$

$\cos\theta = \dfrac{0}{\sqrt{29}\sqrt{29}} = 0$

$\theta = 90^{\circ}$

Thus, the vectors are orthogonal.

59. $\cos\theta = \dfrac{\mathbf{v}\cdot\mathbf{w}}{\|\mathbf{v}\|\,\|\mathbf{w}\|}$

$\cos\theta = \dfrac{(5\mathbf{i}+2\mathbf{j})\cdot(-5\mathbf{i}-2\mathbf{j})}{\sqrt{5^2+2^2}\sqrt{(-5)^2+(-2)^2}}$

$\cos\theta = \dfrac{5(-5)+2(-2)}{\sqrt{29}\sqrt{29}}$

$\cos\theta = \dfrac{-29}{\sqrt{29}\sqrt{29}} = -1$

$\theta = 180^{\circ}$

61. $\text{proj}_{\mathbf{w}}\mathbf{v} = \dfrac{\mathbf{v}\cdot\mathbf{w}}{\|\mathbf{w}\|}$

$\text{proj}_{\mathbf{w}}\mathbf{v} = \dfrac{\langle 6,\ 7\rangle\cdot\langle 3,\ 4\rangle}{\sqrt{3^2+4^2}}$

$= \dfrac{18+28}{\sqrt{25}}$

$= \dfrac{46}{5}$

63. $\text{proj}_{\mathbf{w}}\mathbf{v} = \dfrac{\mathbf{v}\cdot\mathbf{w}}{\|\mathbf{w}\|}$

$\text{proj}_{\mathbf{w}}\mathbf{v} = \dfrac{\langle -3,\ 4\rangle\cdot\langle 2,\ 5\rangle}{\sqrt{2^2+5^2}} = \dfrac{-6+20}{\sqrt{29}} = \dfrac{14}{\sqrt{29}} = \dfrac{14\sqrt{29}}{29} \approx 2.6$

65. $\text{proj}_{\mathbf{w}}\mathbf{v} = \dfrac{\mathbf{v}\cdot\mathbf{w}}{\|\mathbf{w}\|}$

$\text{proj}_{\mathbf{w}}\mathbf{v} = \dfrac{(2\mathbf{i}+\mathbf{j})\cdot(6\mathbf{i}+3\mathbf{j})}{\sqrt{6^2+3^2}} = \dfrac{12+3}{\sqrt{45}} = \dfrac{5}{\sqrt{5}} = \sqrt{5} \approx 2.2$

67. $\text{proj}_{\mathbf{w}}\mathbf{v} = \dfrac{\mathbf{v}\cdot\mathbf{w}}{\|\mathbf{w}\|}$

$\text{proj}_{\mathbf{w}}\mathbf{v} = \dfrac{(3\mathbf{i}-4\mathbf{j})\cdot(3\mathbf{i}-4\mathbf{j})}{\sqrt{(-6)^2+12^2}} = \dfrac{-18-48}{\sqrt{180}} = -\dfrac{11}{\sqrt{5}}$

$= -\dfrac{11\sqrt{5}}{5} \approx -4.9$

69. $W = \mathbf{F}\cdot\mathbf{s}$

$W = \|\mathbf{F}\|\,\|\mathbf{s}\|\cos\alpha$

$W = (75)(15)(\cos 32^{\circ})$

$W \approx 954$ foot-pounds

71.

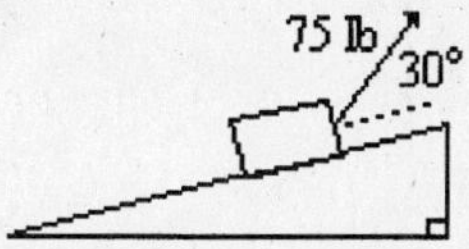

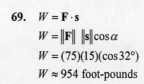

$W = \mathbf{F}\cdot\mathbf{s}$

$W = \|\mathbf{F}\|\,\|\mathbf{s}\|\cos\alpha$

$W = (75)(12)(\cos 30^{\circ})$

$W \approx 779$ foot-pounds

73.

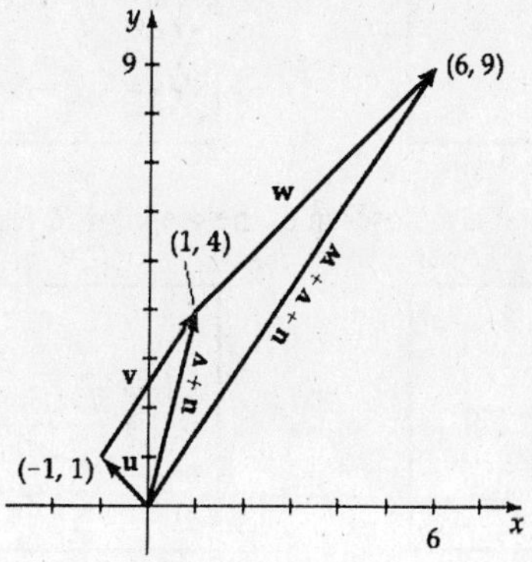

Thus, the sum is $\langle 6.\ 9\rangle$.

75.

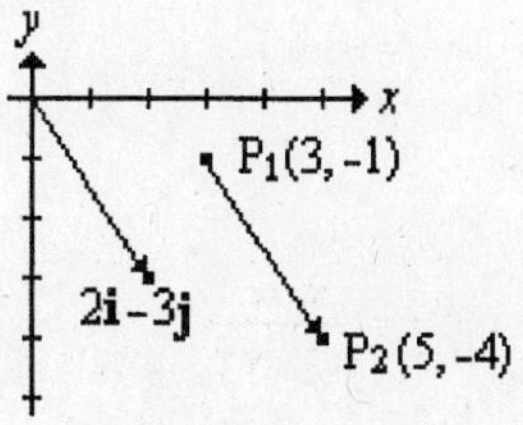

The vector from $P_1(3,-1)$ to $P_2(5,-4)$ is equivalent to $2\mathbf{i}-3\mathbf{j}$.

77.

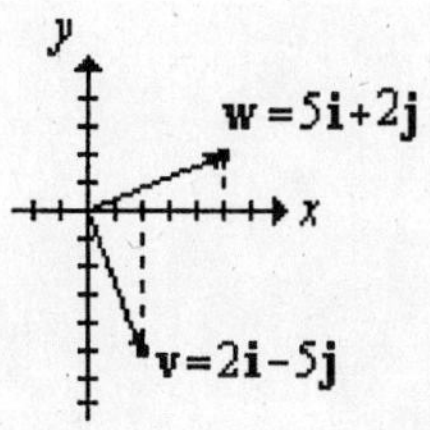

$\mathbf{v}\cdot\mathbf{w} = (2\mathbf{i}-5\mathbf{j})\cdot(5\mathbf{i}+2\mathbf{j})$

$= 10-10$

$= 0$

The two vectors are perpendicular.

79. $\mathbf{v} = \langle -2,\ 7\rangle$

$\langle -2,\ 7\rangle\cdot\langle a,\ b\rangle = 0$

$-2a+7b = 0$

$a = \dfrac{7}{2}b$

Let $b = 2$

$a = 7$

Thus, $\mathbf{u} = \langle 7,\ 2\rangle$ is one example.

81.

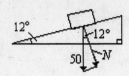

$$\cos 12° = \frac{N}{50}$$
$$N = 50\cos 12°$$
$$F\mu = 0.13N$$
$$\quad = 0.13(50\cos 12°)$$
$$F\mu = 6.4 \text{ pounds}$$

85. Let $\mathbf{v} = \langle a,b \rangle$ and $\mathbf{w} = \langle d,e \rangle$

$$c\mathbf{v} = \langle ca, ab \rangle$$

$$c(\mathbf{v} \cdot \mathbf{w}) = c\langle a,b \rangle \cdot \langle d,e \rangle = c(ad + be) = cad + cbe$$
$$(c\mathbf{v} \cdot \mathbf{w}) = \langle ca, ab \rangle \cdot \langle d,e \rangle = cad + cbe$$

Therefore, $c(\mathbf{v} \cdot \mathbf{w}) = (c\mathbf{v}) \cdot \mathbf{w}$.

83. Let $\mathbf{u} = c i + b\mathbf{j}$, $\mathbf{v} = c i + d\mathbf{j}$, and $\mathbf{w} = e i + f\mathbf{j}$.

$$(\mathbf{u} \cdot \mathbf{v}) \cdot \mathbf{w} = \left[(a i + b\mathbf{j}) \cdot (c i + d\mathbf{j}) \right] \cdot (e i + f\mathbf{j}).$$
$$= (ac + bd) \cdot (e i + f\mathbf{j})$$

$ac + bd$ is a scalar quantity. The product of a scalar and a vector is not defined. Therefore, no, $(\mathbf{u} \cdot \mathbf{v}) \cdot \mathbf{w}$ does not equal $\mathbf{u} \cdot (\mathbf{v} \cdot \mathbf{w})$.

87. Neither. If the force and the distance are the same, the work will be the same.

EXPLORING CONCEPTS WITH TECHNOLOGY, Page 276

The following graph is the graph of R as given in Equation (2) with $a = 8$ cm, $b = 4$ cm, $r_1 = 0.4$ cm, and $r_2 = 0.3$ cm.

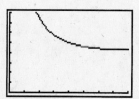

The following graph is a close-up of the graph of R, for $71° \leq \theta \leq 74°$.

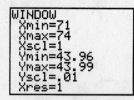

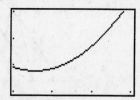

According to this graph R is a minimum when $\theta = 72°$ (to the nearest degree).

Using Equation (3) yields

$$\cos\theta = \left(\frac{\frac{3}{4}r_1}{r_1} \right)^4$$

$$\cos\theta = \frac{81}{256}$$

$$\theta = \cos^{-1}\left(\frac{81}{256} \right) \approx 72°$$

CHAPTER 4 TRUE/FALSE EXERCISES, Page 277

1. False, it cannot be used to solve a triangle given the angle opposite one of the given sides.

3. True   **5.** True   **7.** True   **9.** True

11. False, let $\mathbf{v} = i + \mathbf{j}, \mathbf{w} = i - \mathbf{j}$. Then $\mathbf{v} \cdot \mathbf{w} = 1 - 1 = 0$

CHAPTER 4 REVIEW EXERCISES, Page 278

1.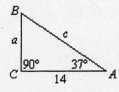

$B = 180° - 90° - 37°$
$B = 53°$

$\tan A = \dfrac{a}{4}$
$a = 14 \tan 37°$
$a \approx 11$

$\cos A = \dfrac{14}{c}$
$c = \dfrac{14}{\cos 37°}$
$c \approx 18$

3.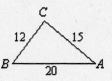

$\cos B = \dfrac{12^2 + 20^2 - 15^2}{2(12)(20)}$
$\cos B \approx 0.6646$
$B \approx 48°$

$\cos C = \dfrac{12^2 + 15^2 - 20^2}{2(12)(15)}$
$\cos C \approx -0.0861$
$C \approx 95°$

$A = 180° - 48° - 95°$
$A = 37°$

5.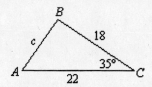

$c^2 = 22^2 + 18^2 - 2(22)(18)\cos 35°$
$c^2 \approx 159$
$c = \sqrt{159}$
$c \approx 13$

$\dfrac{18}{\sin A} = \dfrac{\sqrt{159}}{\sin 35°}$
$\sin A = \dfrac{18 \sin 35°}{\sqrt{159}}$
$\sin A \approx 0.8188$
$A \approx 55°$

$B \approx 180 - 35° - 55°$
$B \approx 90°$

7.

$\dfrac{10}{\sin C} = \dfrac{8}{\sin 105°}$
$\sin C = \dfrac{10 \sin 105°}{8}$
$\sin C \approx 1.207$
No triangle is formed.

9.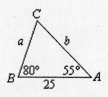

$C = 180° - 80° - 55°$
$C = 45°$

$\dfrac{25}{\sin 45°} = \dfrac{a}{\sin 55°}$
$a = \dfrac{25 \sin 55°}{\sin 45°}$
$a \approx 29$

$\dfrac{25}{\sin 45°} = \dfrac{b}{\sin 80°}$
$b = \dfrac{25 \sin 80°}{\sin 45°}$
$b \approx 35$

11. $s = \dfrac{1}{2}(a+b+c)$

$s = \dfrac{1}{2}(24 + 30 + 36)$

$s = 45$

$K = \sqrt{s(s-a)(s-b)(s-c)}$
$K = \sqrt{45(45-24)(45-30)(45-36)}$
$K = \sqrt{127,575}$
$K \approx 360$ square units

13.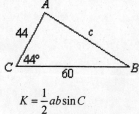

$K = \dfrac{1}{2}ab \sin C$
$K = \dfrac{1}{2}(60)(44)\sin 44°$
$K \approx 920$ square units

15.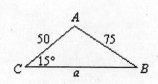

$\dfrac{50}{\sin B} = \dfrac{75}{\sin 15°}$
$\sin B = \dfrac{50 \sin 15°}{75}$
$\sin B \approx 0.1725$
$B \approx 10°$

$A \approx 180° - 10° - 15°$
$A \approx 155°$

$K \approx \dfrac{1}{2}(50)(75)\sin 155°$
$K \approx 790$ square units

160   **Chapter 4/Applications of Trigonometry**

17.

$$\frac{15}{\sin B} = \frac{32}{\sin 110°}$$

$$\sin B = \frac{15\sin 110°}{32}$$

$$\sin B \approx 0.4405$$

$$B \approx 26°$$

$C \approx 180° - 110° - 26°$
$C \approx 44°$

$K \approx \frac{1}{2}(15)(32)\sin 44°$
$K \approx 170$ square units

19. Let $\mathbf{P_1P_2} = a_1\mathbf{i} + a_2\mathbf{j}$.

$a_1 = 3 - (-2) = 5$
$a_2 = 7 - 4 = 3$

A vector equivalent to $\mathbf{P_1P_2}$ is $\mathbf{v} = \langle 5, \, 3 \rangle$.

21. $\|\mathbf{v}\| = \sqrt{(-4)^2 + 2^2}$

$\|\mathbf{v}\| = \sqrt{16 + 4}$

$\|\mathbf{v}\| \approx 4.5$

$\alpha \approx \tan^{-1}\left|\frac{2}{-4}\right| = \tan^{-1}\frac{1}{2}$
$\alpha \approx 26.6°$
$\theta \approx 180° - 26.6°$
$\theta \approx 153.4°$

23. $\|\mathbf{u}\| = \sqrt{(-2)^2 + 3^2}$

$\|\mathbf{u}\| = \sqrt{4 + 9}$

$\|\mathbf{u}\| \approx 3.6$

$\alpha = \tan^{-1}\left|\frac{3}{-2}\right| = \tan^{-1}\frac{3}{2}$
$\alpha \approx 56.3°$
$\theta \approx 180° - 56.3°$
$\theta \approx 123.7°$

25. $\|\mathbf{w}\| = \sqrt{(-8)^2 + 5^2}$

$\|\mathbf{w}\| = \sqrt{89}$

$\|\mathbf{u}\| = \left\langle \frac{-8}{\sqrt{89}}, \, \frac{5}{\sqrt{89}} \right\rangle = \left\langle -\frac{8\sqrt{89}}{89}, \, \frac{5\sqrt{89}}{89} \right\rangle$

A unit vector in the direction of $\|\mathbf{w}\|$ is $\|\mathbf{u}\| = \left\langle -\frac{8\sqrt{89}}{89}, \, \frac{5\sqrt{89}}{89} \right\rangle$.

27. $\|\mathbf{v}\| = \sqrt{5^2 + 1^2}$

$\|\mathbf{v}\| = \sqrt{26}$

$\mathbf{u} = \frac{5}{\sqrt{26}}\mathbf{i} + \frac{1}{\sqrt{26}}\mathbf{j} = \frac{5\sqrt{26}}{26}\mathbf{i} + \frac{\sqrt{26}}{26}\mathbf{j}$

A unit vector in the direction of $\mathbf{v}$ is $\mathbf{u} = \frac{5\sqrt{26}}{26}\mathbf{i} + \frac{\sqrt{26}}{26}\mathbf{j}$.

29. $\mathbf{v} - \mathbf{u} = \langle -4, -1 \rangle - \langle 3, 2 \rangle$

$= \langle -7, -3 \rangle$

31. $-\mathbf{u} + \frac{1}{2}\mathbf{v} = -(10\mathbf{i} + 6\mathbf{j}) + \frac{1}{2}(8\mathbf{i} - 5\mathbf{j})$

$= (-10\mathbf{i} - 6\mathbf{j}) + \left(4\mathbf{i} - \frac{5}{2}\mathbf{j}\right)$

$= (-10 + 4)\mathbf{i} + \left(-6 - \frac{5}{2}\right)\mathbf{j}$

$= -6\mathbf{i} - \frac{17}{2}\mathbf{j}$

33. $\mathbf{v} = 400\sin 204°\mathbf{i} + 400\cos 204°\mathbf{j}$

$\mathbf{v} \approx -162.7\mathbf{i} - 365.4\mathbf{j}$

$\mathbf{w} = -45\mathbf{i}$

$\mathbf{R} = \mathbf{v} + \mathbf{w}$

$\mathbf{R} \approx -162.7\mathbf{i} - 365.4\mathbf{j} - 45\mathbf{i}$

$\mathbf{R} \approx -207.7\mathbf{i} - 365.4\mathbf{j}$

$\|\mathbf{R}\| \approx \sqrt{(-207.7)^2 + (-365.4)^2}$

$\|\mathbf{R}\| \approx 420$ mph

$\alpha = \tan^{-1}\left|\frac{-365.4}{-207.7}\right| = \tan^{-1}\frac{365.4}{207.7}$

$\alpha \approx 60°$
$\theta \approx 180° + 60°$
$\theta \approx 240°$

The ground speed is approximately 420 mph at a heading of 240°

35. $\mathbf{u} \cdot \mathbf{v} = \langle 3, \, 7 \rangle \cdot \langle -1, \, 3 \rangle$

$= (3)(-1) + (7)(3)$

$= 18$

37. $\mathbf{v} \cdot \mathbf{u} = (-4\mathbf{i} - \mathbf{j}) \cdot (2\mathbf{i} + \mathbf{j})$

$= (-4)(2) + (-1)(1)$

$= -9$

39. $\cos\alpha = \dfrac{\langle 7, -4 \rangle \cdot \langle 2, 3 \rangle}{\sqrt{7^2 + (-4)^2}\sqrt{2^2 + 3^2}}$

$\cos\alpha = \dfrac{14 + (-12)}{\sqrt{65}\sqrt{13}}$

$\cos\alpha \approx 0.0688$

$\alpha \approx 86°$

41. $\cos\alpha = \dfrac{(6\mathbf{i}-11\mathbf{j})\cdot(2\mathbf{i}-4\mathbf{j})}{\sqrt{6^2+(-11)^2}\sqrt{2^2+4^2}}$

$\cos\alpha = \dfrac{12-44}{\sqrt{157}\sqrt{20}}$

$\cos\alpha \approx -0.5711$

$\cos\alpha \approx 125°$

43. $\text{proj}_{\mathbf{W}}\mathbf{v} = \dfrac{\mathbf{v}\cdot\mathbf{w}}{\|\mathbf{w}\|}$

$\text{proj}_{\mathbf{W}}\mathbf{v} = \dfrac{\langle -2,5\rangle\cdot\langle 5,4\rangle}{\sqrt{5^2+4^2}}$

$= \dfrac{-10+20}{\sqrt{41}}$

$= \dfrac{10}{\sqrt{41}}$

$= \dfrac{10\sqrt{41}}{41}$

45. $\mathbf{w} = \|\mathbf{F}\|\ \|\mathbf{S}\|\cos\theta$

$\mathbf{w} = 60\cdot14\cos38°$

$\mathbf{w} \approx 662$ foot-pounds

CHAPTER 4 TEST, Page 278

1.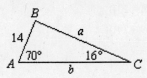

$B = 180° - 70° - 16°$

$B = 94°$

$\dfrac{c}{\sin C} = \dfrac{a}{\sin A}$ $\dfrac{c}{\sin C} = \dfrac{b}{\sin B}$

$a = \dfrac{14\sin 70^0}{\sin 16^0}$ $b = \dfrac{14\sin 94^0}{\sin 16^0}$

$a \approx 48$ $b \approx 51$

2.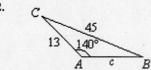

$\dfrac{b}{\sin B} = \dfrac{a}{\sin A}$

$B = \sin^{-1}\left(\dfrac{b\sin A}{a}\right)$

$B = \sin^{-1}\left(\dfrac{13\sin 140°}{45}\right)$

$B \approx 11°$

3.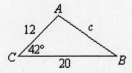

$c^2 = a^2 + b^2 - 2ab\cos C$

$c^2 = 20^2 + 12^2 - 2(20)(12)\cos 42°$

$c \approx 14$

4.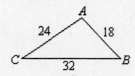

$\cos B = \dfrac{a^2+c^2-b^2}{2ac}$

$B = \cos^{-1}\left(\dfrac{32^2+18^2-24^2}{2(32)(18)}\right)$

$B \approx 48°$

5.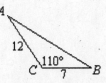

$K = \dfrac{1}{2}ab\sin C$

$K = \dfrac{1}{2}(7)(12)(\sin 110^0)$

$K \approx 39$ square units

6.

$A = 180° - 42° - 75°$

$A = 63°$

$K = \dfrac{b^2\sin A\sin C}{2\sin B}$

$K = \dfrac{12^2\sin 63°\sin 75°}{2\sin 42°}$

$K \approx 93$ square units

7.

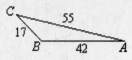

$$s = \frac{1}{2}(a+b+c)$$

$$s = \frac{1}{2}(17+55+42) = 57$$

$$K = \sqrt{s(s-a)(s-b)(s-c)}$$

$$K = \sqrt{57(57-17)(57-55)(57-45)}$$

$K \approx 260$ square units

8. $\mathbf{v} = -2\mathbf{i} + 3\mathbf{j}$

$$\|\mathbf{v}\| = \sqrt{\mathbf{v} \cdot \mathbf{v}}$$

$$= \sqrt{(-2\mathbf{i}+3\mathbf{j}) \cdot (-2\mathbf{i}+3\mathbf{j})}$$

$$= \sqrt{(-2)(-2)+(3)(3)}$$

$$= \sqrt{4+9}$$

$$= \sqrt{13}$$

9. $a_1 = 12\cos 220° \approx -9.193$

$a_2 = 12\sin 220° \approx -7.713$

$\mathbf{v} = a_1\mathbf{i} + a_2\mathbf{j}$

$\mathbf{v} = -9.193\mathbf{i} - 7.713\mathbf{j}$

10. $3\mathbf{u} - 5\mathbf{v} = 3(2\mathbf{i}-3\mathbf{j}) - 5(5\mathbf{i}+4\mathbf{j})$

$$= (6\mathbf{i}-9\mathbf{j}) - (25\mathbf{i}+20\mathbf{j})$$

$$= (6-25)\mathbf{i} + (-9-20)\mathbf{j}$$

$$= -19\mathbf{i} - 29\mathbf{j}$$

11. $\mathbf{u} \cdot \mathbf{v} = (-2\mathbf{i}+3\mathbf{j}) \cdot (5\mathbf{i}+3\mathbf{j})$

$$= -10 + 9$$

$$= -1$$

12. $\cos\theta = \dfrac{\mathbf{u} \cdot \mathbf{v}}{\|\mathbf{u}\| \, \|\mathbf{v}\|} = \dfrac{\langle 3,5 \rangle \cdot \langle -6,2 \rangle}{\sqrt{3^2+5^2}\sqrt{(-6)^2+2^2}}$

$\cos\theta = \dfrac{-18+10}{\sqrt{34}\sqrt{40}} = \dfrac{-8}{\sqrt{34}\sqrt{40}}$

$\theta \approx 103°$

13.

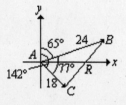

$A = 142° - 65° = 77°$

$R^2 = 24^2 + 18^2 - 2(24)(18)\cos 77°$

$R \approx 27$ mi

14.

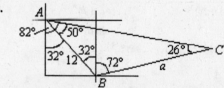

$A = 82° - 32° = 50°$

$C = 180° - 50° - 32° - 72° = 26°$

$$\frac{a}{\sin 50°} = \frac{12}{\sin 26°}$$

$$a = \frac{12\sin 50°}{\sin 26°}$$

$$a \approx 21 \text{ miles}$$

15.

$$S = \frac{1}{2}(112+165+140) = 208.5$$

$$K = \sqrt{208.5(208.5-112)(208.5-165)(208.5-140)}$$

$$K \approx 7743.0$$

$\text{cost} \approx 8.50(7743)$

$\text{cost} \approx \$65,800.$

SECTION 5.1, Page 287

1. $2+\sqrt{-9} = 2+i\sqrt{9} = 2+3i$

3. $4-\sqrt{-121} = 4-i\sqrt{121} = 4-11i$

5. $8+\sqrt{-3} = 8+i\sqrt{3}$

7. $\sqrt{-16}+7 = i\sqrt{16}+7 = 4i+7 = 7+4i$

9. $\sqrt{-81} = i\sqrt{81} = 9i$

11. $(2+5i)+(3+7i) = (2+3)+(5i+7i) = 5+12i$

13. $(-5-i)+(9-2i) = (-5+9)+(-i-2i) = 4-3i$

15. $(8-6i)-(10-i) = 8-6i-10+i = -2-5i$

17. $(7-3i)-(-5-i) = 7-3i+5+i = 12-2i$

19. $8i-(2-3i) = 8i-2+3i = -2+11i$

21. $3(2+7i)+5(2-i) = 6+21i+10-5i = 16+16i$

23. $(2+3i)(4-5i) = 8-10i+12i-15i^2 = 8+2i-15(-1) = 8+2i+15 = 23+2i$

25. $(5+7i)(5-7i) = 25-49i^2 = 25-49(-1) = 25+49 = 74$

27. $(8i+11)(-7+5i) = -56i+40i^2-77+55i = -56i+40(-1)-77+55i = -56i+55i-40-77 = -117-i$

29. $\dfrac{4+i}{3+5i} = \dfrac{4+i}{3+5i}\cdot\dfrac{3-5i}{3-5i} = \dfrac{12-20i+3i-5i^2}{9-25i^2} = \dfrac{12-17i-5(-1)}{9-25(-1)} = \dfrac{12-17i+5}{9+25} = \dfrac{17-17i}{34} = \dfrac{17}{34}-\dfrac{17}{34}i = \dfrac{1}{2}-\dfrac{1}{2}i$

31. $\dfrac{1}{7-3i} = \dfrac{1}{7-3i}\cdot\dfrac{7+3i}{7+3i} = \dfrac{7+3i}{49-9i^2} = \dfrac{7+3i}{49-9(-1)} = \dfrac{7+3i}{49+9} = \dfrac{7+3i}{58} = \dfrac{7}{58}+\dfrac{3}{58}i$

33. $\dfrac{3+2i}{3-2i} = \dfrac{3+2i}{3-2i}\cdot\dfrac{3+2i}{3+2i} = \dfrac{9+12i+4i^2}{9-4i^2} = \dfrac{9+12i+4(-1)}{9-4(-1)} = \dfrac{9+12i-4}{9+4} = \dfrac{5+12i}{13} = \dfrac{5}{13}+\dfrac{12}{13}i$

35. $\dfrac{2i}{11+i} = \dfrac{2i}{11+i}\cdot\dfrac{11-i}{11-i} = \dfrac{22i-2i^2}{121-i^2} = \dfrac{22i-2(-1)}{121-(-1)} = \dfrac{22i+2}{121+1} = \dfrac{2+22i}{122} = \dfrac{2}{122}+\dfrac{22i}{122} = \dfrac{1}{61}+\dfrac{11}{61}i$

37. $\dfrac{6+i}{i} = \dfrac{6+i}{i}\cdot\dfrac{i}{i} = \dfrac{6i+i^2}{i^2} = \dfrac{6i-1}{-1} = -6i+1 = 1-6i$

39. $(3-5i)^2 = 9-30i+25i^2 = 9-30i+25(-1) = 9-30i-25 = -16-30i$

41. $(1-i)-2(4+i)^2 = 1-i-2(16+8i+i^2) = 1-i-2(16+8i-1) = 1-i-2(16+8i-1) = 1-i-2(15+8i) = 1-i-30-16i = -29-17i$

43. $(1-i)^3 = (1-i)(1-i)^2 = (1-i)(1-2i+i^2) = (1-i)(1-2i-1) = (1-i)(-2i) = -2i+2i^2 = -2i+2(-1) = -2i-2 = -2-2i$

45. $(2i)(8i) = 16i^2 = 16(-1) = -16$ or $-16+0i$

47. $(5i)^2(-3i) = (25i^2)(-3i) = [25(-1)](-3i) = (-25)(-3i) = 75i$

49. $i^3 = i(i^2) = i(-1) = -i$

51. Remainder of $5\div 4$ is 1. Therefore, $i^5 = i^1 = i$.

53. Remainder of $10\div 4$ is 2. Therefore, $i^{10} = i^2 = -1$.

55. Remainder of $40\div 4$ is 0. Therefore, $-i^{40} = -i^0 = -1$.

57. Remainder of $223 \div 4$ is 3. Therefore $i^{223} = i^3 = -i$.

59. Remainder of $2001 \div 4$ is 1. Therefore $i^{2001} = i^1 = i$.

61. Remainder of $5042 \div 4$ is 2. Therefore $i^{5042} = i^2 = -1$.

63. $i^{-1} = i^{-1} \cdot 1 = i^{-1} \cdot i^4 = i^3 = i(i^2) = i(-1) = -i$

65. $\sqrt{-1}\sqrt{-4} = i\sqrt{1} \cdot i\sqrt{4} = i \cdot 2i = 2i^2 = 2(-1) = -2$

67. $\sqrt{-64}\sqrt{-5} = i\sqrt{64} \cdot i\sqrt{5} = 8i \cdot i\sqrt{5} = 8i^2\sqrt{5} = 8(-1)\sqrt{5} = -8\sqrt{5}$

69. $(3 + \sqrt{-2})(3 - \sqrt{-2}) = (3 + i\sqrt{2})(3 - i\sqrt{2}) = 3^2 - (i\sqrt{2})^2 = 9 - 2i^2 = 9 - 2(-1) = 9 + 2 = 11$

71. $(5 + \sqrt{-16})^2 = (5 + 4i)^2 = 25 + 40i + 16i^2 = 25 + 40i - 16 = 9 + 40i$

73. $\dfrac{-(-3) \pm \sqrt{(-3)^2 - 4(3)(3)}}{2(3)} = \dfrac{3 \pm \sqrt{9 - 36}}{6} = \dfrac{3 \pm \sqrt{-27}}{6} = \dfrac{3 \pm 3i\sqrt{3}}{6} = \dfrac{3}{6} \pm \dfrac{3i\sqrt{3}}{6} = \dfrac{1}{2} \pm \dfrac{\sqrt{3}}{2}i$

75. $\dfrac{-(4) \pm \sqrt{(4)^2 - 4(2)(4)}}{2(2)} = \dfrac{-4 \pm \sqrt{16 - 32}}{4} = \dfrac{-4 \pm \sqrt{-16}}{4} = \dfrac{-4 \pm 4i}{4} = \dfrac{-4}{4} \pm \dfrac{4i}{4} = -1 \pm i$

77. $\dfrac{-(6) \pm \sqrt{(6)^2 - 4(2)(6)}}{2(2)} = \dfrac{-6 \pm \sqrt{36 - 48}}{4} = \dfrac{-6 \pm \sqrt{-12}}{4} = \dfrac{-6 \pm 2i\sqrt{3}}{4} = \dfrac{-6}{4} \pm \dfrac{2i\sqrt{3}}{4} = -\dfrac{3}{2} \pm \dfrac{\sqrt{3}}{2}i$

79. $\dfrac{-(1) \pm \sqrt{(1)^2 - 4(2)(3)}}{2(2)} = \dfrac{-1 \pm \sqrt{1 - 24}}{4} = \dfrac{-1 \pm \sqrt{-23}}{4} = \dfrac{-1 \pm i\sqrt{23}}{4} = -\dfrac{1}{4} \pm \dfrac{\sqrt{23}}{4}i$

81. $|3 + 4i| = \sqrt{3^2 + 4^2} = \sqrt{9 + 16} = \sqrt{25} = 5$

83. $|2 - 5i| = \sqrt{2^2 + (-5)^2} = \sqrt{4 + 25} = \sqrt{29}$

85. $|7 - 4i| = \sqrt{7^2 + (-4)^2} = \sqrt{49 + 16} = \sqrt{65}$

87. $|-3i| = |0 - 3i| = \sqrt{0^2 + (-3)^2} = \sqrt{0 + 9} = \sqrt{9} = 3$

89. $|z| = |a + bi| = \sqrt{a^2 + b^2}$ and $|\bar{z}| = |a - bi| = \sqrt{a^2 + (-b)^2} = \sqrt{a^2 + b^2}$. Thus, the absolute value of a complex number and the absolute value of its conjugate are equal.

91. Let $z_1 = a + bi$ and $z_2 = c + di$ be two complex numbers. We need to prove that $\overline{z_1 + z_2} = \overline{z_1} + \overline{z_2}$.

$\overline{z_1 + z_2} = \overline{(a + bi) + (c + di)} = \overline{(a + c) + (b + d)i} = (a + c) - (b + d)i$

$\overline{z_1} + \overline{z_2} = (a - bi) + (c - di) = (a + c) + (-b - d)i = (a + c) - (b + d)i = \overline{z_1 + z_2}$

Thus, the conjugate of the sum of two complex numbers equals the sum of the conjugates of the two numbers.

93. If $x = 1 + i\sqrt{3}$, then $x^2 - 2x + 4 = (1 + i\sqrt{3})^2 - 2(1 + i\sqrt{3}) + 4 = 1 + 2i\sqrt{3} + 3i^2 - 2 - 2i\sqrt{3} + 4 = 1 + 2i\sqrt{3} - 3 - 2 - 2i\sqrt{3} + 4 = 0$.

SECTION 5.2, Page 294

1.

$|z| = \sqrt{(-2)^2 + (-2)^2}$
$= \sqrt{8} = 2\sqrt{2}$

3.

$|z| = \sqrt{(\sqrt{3})^2 + (-1)^2}$
$= \sqrt{3 + 1} = \sqrt{4}$
$= 2$

5.

$|z| = \sqrt{0^2 + (-2)^2}$
$= 2$

7.

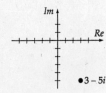

$|z| = \sqrt{3^2 + (-5)^2}$

$= \sqrt{34}$

9. $r = \sqrt{1^2 + (-1)^2}$

$r = \sqrt{2}$

$\alpha = \tan^{-1}\left|\frac{-1}{1}\right|$

$= \tan^{-1} 1$

$= 45°$

$\theta = 360° - 45° = 315°$

$z = \sqrt{2} \text{ cis } 315°$

11. $r = \sqrt{(\sqrt{3})^2 + (-1)^2}$

$r = 2$

$\alpha = \tan^{-1}\left|\frac{-1}{\sqrt{3}}\right|$

$= \tan^{-1} \frac{1}{\sqrt{3}} = 30°$

$\alpha = 360° - 30° = 330°$

$z = 2 \text{ cis } 330°$

13. $r = \sqrt{0^2 + 3^2}$

$r = 3$

$\theta = 90°$

$z = 3 \text{ cis } 90°$

15. $r = \sqrt{(-5)^2 + 0^2}$

$r = 5$

$\theta = 180°$

$z = 5 \text{ cis } 180°$

17. $z = 2(\cos 45° + i \sin 45°)$

$z = 2\left(\frac{\sqrt{2}}{2} + \frac{\sqrt{2}}{2}i\right)$

$z = \sqrt{2} + i\sqrt{2}$

19. $z = \cos 315° + i \sin 315°$

$z = \frac{\sqrt{2}}{2} - \frac{\sqrt{2}}{2}i$

21. $z = 6 \text{ cis } 135°$

$z = 6(\cos 135° + i \sin 135°)$

$z = 6\left(-\frac{\sqrt{2}}{2} + \frac{\sqrt{2}}{2}i\right)$

$z = -3\sqrt{2} + 3i\sqrt{2}$

23. $z = 8 \text{ cis } 0°$

$z = 8(\cos 0° + i \sin 0°)$

$z = 8(1 + 0i)$

$z = 8$

25. $z = 2\left(\cos\frac{5\pi}{6} + i \sin\frac{5\pi}{6}\right)$

$z = 2\left(-\frac{\sqrt{3}}{2} + \frac{1}{2}i\right)$

$z = -\sqrt{3} + i$

27. $z = 3\left(\cos\frac{3\pi}{2} + i \sin\frac{3\pi}{2}\right)$

$z = 3(0 - i)$

$z = -3i$

29. $z = 8 \text{ cis } \frac{3\pi}{4}$

$= 8\left(\cos\frac{3\pi}{4} + i \sin\frac{3\pi}{4}\right)$

$z = 8\left(-\frac{\sqrt{2}}{2} + \frac{i\sqrt{2}}{2}\right)$

$z = -4\sqrt{2} + 4i\sqrt{2}$

31. $z = 9 \text{ cis } \frac{11\pi}{6}$

$z = 9\left(\cos\frac{11\pi}{6} + i \sin\frac{11\pi}{6}\right)$

$z = 9\left(\frac{\sqrt{3}}{2} - \frac{1}{2}i\right)$

$z = \frac{9\sqrt{3}}{2} - \frac{9}{2}i$

33. $z = 2 \text{ cis } 2$

$z = 2(\cos 2 + i \sin 2)$

$z \approx 2(-0.4161 + 0.9093i)$

$z \approx -0.832 + 1.819i$

35. $z_1 z_2 = 2 \text{ cis } 30° \cdot 3 \text{ cis } 225°$

$z_1 z_2 = 6 \text{ cis}(30° + 225°)$

$z_1 z_2 = 6 \text{ cis } 255°$

37. $z_1 z_2 = 3(\cos 122° + i \sin 122°) \cdot 4(\cos 213° + i \sin 213°)$

$z_1 z_2 = 12[\cos(122° + 213°) + i \sin(122° + 213°)]$

$z_1 z_2 = 12(\cos 335° + i \sin 335°)$

$z_1 z_2 = 12 \text{ cis } 335°$

39. $z_1 z_2 = 5\left(\cos\frac{2\pi}{3} + i \sin\frac{2\pi}{3}\right) \cdot 2\left(\cos\frac{2\pi}{5} + i \sin\frac{2\pi}{5}\right)$

$z_1 z_2 = 10\left[\cos\left(\frac{2\pi}{3} + \frac{2\pi}{5}\right) + i \sin\left(\frac{2\pi}{3} + \frac{2\pi}{5}\right)\right]$

$z_1 z_2 = 10\left(\cos\frac{16\pi}{15} + i \sin\frac{16\pi}{15}\right)$

$z_1 z_2 = 10 \text{ cis } \frac{16\pi}{15}$

41. $z_1 z_2 = 4 \text{ cis } 2.4 \cdot 6 \text{ cis } 4.1$

$z_1 z_2 = 24 \text{ cis } (2.4 + 4.1)$

$z_1 z_2 = 24 \text{ cis } 6.5$

43. $\dfrac{z_1}{z_2} = \dfrac{32 \text{ cis } 30^\circ}{4 \text{ cis } 150^\circ}$

$\dfrac{z_1}{z_2} = 8 \text{ cis}(30^\circ - 150^\circ)$

$\dfrac{z_1}{z_2} = 8 \text{ cis}(-120^\circ)$

$\dfrac{z_1}{z_2} = 8\,(\cos 120^\circ - i \sin 120^\circ)$

$\dfrac{z_1}{z_2} = 8\left(-\dfrac{1}{2} - \dfrac{i\sqrt{3}}{2}\right) = -4 - 4i\sqrt{3}$

45. $\dfrac{z_1}{z_2} = \dfrac{27(\cos 315^\circ + i \sin 315^\circ)}{9(\cos 225^\circ + i \sin 225^\circ)}$

$\dfrac{z_1}{z_2} = 3\,[\cos(315^\circ - 225^\circ) + i \sin(315^\circ - 225^\circ)]$

$\dfrac{z_1}{z_2} = 3(\cos 90^\circ + i \sin 90^\circ) = 3(0 + i) = 3i$

47. $\dfrac{z_1}{z_2} = \dfrac{12\left(\cos \frac{2\pi}{3} + i \sin \frac{2\pi}{3}\right)}{4\left(\cos \frac{11\pi}{6} + i \sin \frac{11\pi}{6}\right)}$

$\dfrac{z_1}{z_2} = 3\left[\cos\left(\dfrac{2\pi}{3} - \dfrac{11\pi}{6}\right) + i \sin\left(\dfrac{2\pi}{3} - \dfrac{11\pi}{6}\right)\right]$

$\dfrac{z_1}{z_2} = 3\left(\cos \dfrac{7\pi}{6} - i \sin \dfrac{7\pi}{6}\right)$

$\dfrac{z_1}{z_2} = 3\left[-\dfrac{\sqrt{3}}{2} - \left(-\dfrac{1}{2}i\right)\right]$

$\dfrac{z_1}{z_2} = -\dfrac{3\sqrt{3}}{2} + \dfrac{3}{2}i$

49. $\dfrac{z_1}{z_2} = \dfrac{25 \text{ cis } 3.5}{5 \text{ cis } 1.5}$

$\dfrac{z_1}{z_2} = 5 \text{ cis } (3.5 - 1.5)$

$\dfrac{z_1}{z_2} = 5 \text{ cis } 2$

$\dfrac{z_1}{z_2} = 5\,(\cos 2 + i \sin 2)$

$\dfrac{z_1}{z_2} \approx 5\,(-0.4161 + 0.9093i)$

$\dfrac{z_1}{z_2} \approx -2.081 + 4.546i$

51. $z_1 = 1 - i\sqrt{3}$

$r_1 = \sqrt{1^2 + \left(\sqrt{3}\right)^2}$ $\qquad \alpha = \tan^{-1}\left|\dfrac{-\sqrt{3}}{1}\right| = 60^\circ$

$r_1 = 2$

$\qquad\qquad\qquad \theta_1 = 300^\circ$

$z_1 = 2(\cos 300^\circ + i \sin 300^\circ)$

$z_2 = 1 + i$

$r_2 = \sqrt{1^2 + 1^2}$ $\qquad \alpha = \tan^{-1}\left|\dfrac{1}{1}\right| = 45^\circ$

$r_2 = \sqrt{2}$

$\qquad\qquad\qquad \theta_2 = 45^\circ$

$z_2 = \sqrt{2}(\cos 45^\circ + i \sin 45^\circ)$

$z_1 z_2 = 2(\cos 300^\circ + i \sin 300^\circ) \cdot \sqrt{2}(\cos 45^\circ + i \sin 45^\circ)$

$z_1 z_2 = 2\sqrt{2}[\cos(300^\circ + 45^\circ) + i \sin(300^\circ + 45^\circ)]$

$z_1 z_2 = 2\sqrt{2}(\cos 345^\circ + i \sin 345^\circ)$

$z_1 z_2 \approx 2.732 - 0.732i$

53. $z_1 = 3 - 3i$

$r_1 = \sqrt{3^2 + (-3)^2}$ $\quad\quad \alpha = \tan^{-1}\left|\frac{-3}{3}\right| = 45°$

$r_1 = 3\sqrt{2}$ $\quad\quad\quad\quad\quad\quad\quad \theta_1 = 315°$

$z_2 = 1 + i$

$r_2 = \sqrt{1^2 + 1^2}$ $\quad\quad \alpha = \tan^{-1}\left|\frac{1}{1}\right| = 45°$

$r_2 = \sqrt{2}$ $\quad\quad\quad\quad\quad\quad \theta_2 = 45°$

$z_1 = 3\sqrt{2}(\cos 315° + i \sin 315°)$

$z_2 = \sqrt{2}(\cos 45° + i \sin 45°)$

$z_1 z_2 = 3\sqrt{2}(\cos 315° + i \sin 315°) \cdot \sqrt{2}(\cos 45° + i \sin 45°)$

$z_1 z_2 = 6[\cos(315° + 45°) + i \sin(315° + 45°)]$

$z_1 z_2 = 6(\cos 360° + i \sin 360°)$

$z_1 z_2 = 6 + 0i$

$z_1 z_2 = 6$

55. $z_1 = 1 + i\sqrt{3}$

$r_1 = \sqrt{1^2 + (\sqrt{3})^2}$ $\quad\quad \alpha_1 = \tan^{-1}\left|\frac{\sqrt{3}}{1}\right| = 60°$

$r_1 = 2$ $\quad\quad\quad\quad\quad\quad\quad\quad \theta_1 = 60°$

$z_1 = 2(\cos 60° + i \sin 60°)$

$z_2 = 1 - i\sqrt{3}$

$r_2 = \sqrt{1^2 + (\sqrt{3})^2}$ $\quad\quad \alpha = \tan^{-1}\left|\frac{-\sqrt{3}}{1}\right| = 60°$

$r_2 = 2$ $\quad\quad\quad\quad\quad\quad\quad\quad \theta_2 = 300°$

$z_2 = 2(\cos 300° + i \sin 300°)$

$\frac{z_1}{z_2} = \frac{2(\cos 60° + i \sin 60°)}{2(\cos 300° + i \sin 300°)}$

$\frac{z_1}{z_2} = \cos(60° - 300°) + i \sin(60° - 300°)$

$\frac{z_1}{z_2} = \cos 240° - i \sin 240° = -\frac{1}{2} + \frac{\sqrt{3}}{2}i$

57. $z_1 = \sqrt{2} - i\sqrt{2}$

$r_1 = \sqrt{(\sqrt{2})^2 + (\sqrt{2})^2}$ $\quad \alpha_1 = \tan^{-1}\left|\frac{-\sqrt{2}}{2}\right| = 45°$

$r_1 = 2$ $\quad\quad\quad\quad\quad\quad\quad\quad \theta_1 = 315°$

$z_1 = 2(\cos 315° + i \sin 315°)$

$z_2 = 1 + i$

$r_2 = \sqrt{1^2 + 1^2}$ $\quad\quad \alpha_2 = \tan^{-1}\left|\frac{1}{1}\right| = 45°$

$r_2 = \sqrt{2}$ $\quad\quad\quad\quad\quad\quad \theta_2 = 45°$

$z_2 = \sqrt{2}(\cos 45° + i \sin 45°)$

$\frac{z_1}{z_2} = \frac{2(\cos 315° + i \sin 315°)}{\sqrt{2}(\cos 45° + i \sin 45°)}$

$\frac{z_1}{z_2} = \sqrt{2}[\cos(315° - 45°) + i \sin(315° - 45°)]$

$\frac{z_1}{z_2} = \sqrt{2}(\cos 270° + i \sin 270°)$

$\frac{z_1}{z_2} = \sqrt{2}[0 + i(-1)] = \sqrt{2}(0 - 1i) = 0 - \sqrt{2}i = -\sqrt{2}i \text{ or } -i\sqrt{2}$

59.

$z_1 = \sqrt{3} - 1$

$r_1 = \sqrt{(\sqrt{3})^2 + (-1)^2}$

$r_1 = 2$

$\alpha_1 = \tan^{-1}\left|\dfrac{-1}{\sqrt{3}}\right| = 30°$

$\theta_1 = 330°$

$z_1 = 2(\cos 330° + i \sin 330°)$

$z_2 = 2 + 2i$

$r_2 = \sqrt{2^2 + 2^2}$

$r_2 = 2\sqrt{2}$

$\alpha_2 = \tan^{-1}\left|\dfrac{2}{2}\right| = 45°$

$\theta_2 = 45°$

$z_2 = 2\sqrt{2}(\cos 45° + i \sin 45°)$

$z_3 = 2 - 2i\sqrt{3}$

$r_3 = \sqrt{2^2 + (-2\sqrt{3})^2}$

$r_3 = 4$

$\alpha_3 = \tan^{-1}\left|\dfrac{-2\sqrt{3}}{2}\right| = 60°$

$\theta_3 = 300°$

$z_3 = 4(\cos 300° + i \sin 300°)$

$z_1 z_2 z_3 = 2(\cos 330° + i \sin 330°) \cdot 2\sqrt{2}(\cos 45° + i \sin 45°) \cdot 4(\cos 300° + i \sin 300°)$

$z_1 z_2 z_3 = 16\sqrt{2}[\cos(330° + 45° + 300°) + i \sin(330° + 45° + 300°)]$

$z_1 z_2 z_3 = 16\sqrt{2}(\cos 675° + i \sin 675°)$

$z_1 z_2 z_3 = 16\sqrt{2}(\cos 315° + i \sin 315°)$

$z_1 z_2 z_3 = 16\sqrt{2}\left(\dfrac{1}{\sqrt{2}} - \dfrac{1}{\sqrt{2}}i\right) = 16 - 16i$

61.

$z_1 = \sqrt{3} + i\sqrt{3}$

$r_1 = \sqrt{(\sqrt{3})^2 + (\sqrt{3})^2}$

$r_1 = \sqrt{6}$

$\alpha_1 = \tan^{-1}\left|\dfrac{\sqrt{3}}{\sqrt{3}}\right| = 45°$

$\theta_1 = 45°$

$z_1 = \sqrt{6}(\cos 45° + i \sin 45°)$

$z_2 = 1 - i\sqrt{3}$

$r_2 = \sqrt{1^2 + (-\sqrt{3})^2}$

$r_2 = 2$

$\alpha_2 = \tan^{-1}\left|\dfrac{-\sqrt{3}}{1}\right| = 60°$

$\theta_2 = 300°$

$z_2 = 2(\cos 300° + i \sin 300°)$

$z_3 = 2 - 2i$

$r_3 = \sqrt{2^2 + (-2)^2}$

$r_3 = 2\sqrt{2}$

$\alpha_3 = \tan^{-1}\left|\dfrac{-2}{2}\right| = 45°$

$\theta_3 = 315°$

$z_3 = 2\sqrt{2}(\cos 315° + i \sin 315°)$

$\dfrac{z_1}{z_2 z_3} = \dfrac{\sqrt{6}(\cos 45° + i \sin 45°)}{2(\cos 300° + i \sin 300°) \cdot 2\sqrt{2}(\cos 315° + i \sin 315°)}$

$\dfrac{z_1}{z_2 z_3} = \dfrac{\sqrt{6}(\cos 45° + i \sin 45°)}{4\sqrt{2}[\cos(300° + 315°) + i \sin(300° + 315°)]}$

$\dfrac{z_1}{z_2 z_3} = \dfrac{\sqrt{6}(\cos 45° + i \sin 45°)}{4\sqrt{2}(\cos 255° + i \sin 255°)}$

$\dfrac{z_1}{z_2 z_3} = \dfrac{\sqrt{3}}{4}[\cos(45° - 255°) + i \sin(45° - 255°)]$

$\dfrac{z_1}{z_2 z_3} = \dfrac{\sqrt{3}}{4}(\cos 210° - i \sin 210°) = \dfrac{\sqrt{3}}{4}\left(-\dfrac{\sqrt{3}}{2} + \dfrac{i}{2}\right) = -\dfrac{3}{8} + \dfrac{\sqrt{3}}{8}i$

63.

$z_1 = 1 - 3i$

$r_1 = \sqrt{1^2 + (-3)^2}$

$r_1 = \sqrt{10}$

$\alpha_1 = \tan^{-1}\left|\dfrac{-3}{1}\right| \approx 71.57^\circ$

$\theta_1 = 288.43^\circ$

$z_1 = \sqrt{10}(\cos 288.4^\circ + i\sin 288.4^\circ)$

$z_2 = 2 + 3i$

$r_2 = \sqrt{2^2 + 3^2}$

$r_2 = \sqrt{13}$

$\alpha_2 = \tan^{-1}\left|\dfrac{3}{2}\right| \approx 56.31^\circ$

$\theta_2 = 56.31^\circ$

$z_2 = \sqrt{13}(\cos 56.3^\circ + i\sin 56.3^\circ)$

$z_3 = 4 + 5i$

$r_3 = \sqrt{4^2 + 5^2}$

$r_3 = \sqrt{41}$

$\alpha_3 = \tan^{-1}\left|\dfrac{5}{4}\right| \approx 51.34^\circ$

$\theta_3 = 51.34^\circ$

$z_3 = \sqrt{41}(\cos 51.3^\circ + i\sin 51.3^\circ)$

$z_1 z_2 z_3 = \sqrt{10}(\cos 288.4^\circ + i\sin 288.4^\circ)\cdot\sqrt{13}(\cos 56.3^\circ + i\sin 56.3^\circ)\cdot\sqrt{41}(\cos 51.3^\circ + i\sin 51.3^\circ)$

$z_1 z_2 z_3 = \sqrt{10}\cdot\sqrt{13}\cdot\sqrt{41}[\cos(288.43^\circ + 56.31^\circ + 51.34^\circ) + i\sin(288.43^\circ + 56.31^\circ + 51.34^\circ)]$

$z_1 z_2 z_3 \approx 73.0(\cos 396.08^\circ + i\sin 396.08^\circ)$

$z_1 z_2 z_3 = 73.0(\cos 36.08^\circ + i\sin 36.08^\circ)$

$z_1 z_2 z_3 \approx 59.0 + 43.0i$

65. $z = r(\cos\theta + i\sin\theta) \qquad \overline{z} = r(\cos\theta - i\sin\theta)$

$z\cdot\overline{z} = r(\cos\theta + i\sin\theta)\cdot r(\cos\theta - i\sin\theta)$

$z\cdot\overline{z} = r(\cos\theta + i\sin\theta)\cdot r[\cos(-\theta) + i\sin(-\theta)]$

$z\cdot\overline{z} = r^2[\cos(\theta - \theta) + i\sin(\theta - \theta)]$

$z\cdot\overline{z} = r^2(\cos 0 + i\sin 0)$

$z\cdot\overline{z} = r^2$ or $a^2 + b^2$

SECTION 5.3, Page 299

1. $[2(\cos 30^\circ + i\sin 30^\circ)]^8 = 2^8[\cos(8\cdot 30^\circ) + i\sin(8\cdot 30^\circ)]$
$= 256(\cos 240^\circ + i\sin 240^\circ)$
$= -128 - 128i\sqrt{3}$

3. $[2(\cos 240^\circ + i\sin 240^\circ)]^5 = 2^5[\cos(5\cdot 240^\circ) + i\sin(5\cdot 240^\circ)]$
$= 32[\cos 1200^\circ + i\sin 1200^\circ]$
$= 32(\cos 120^\circ + i\sin 120^\circ)$
$= -16 + 16i\sqrt{3}$

5. $[2\operatorname{cis}(225^\circ)]^5 = 2^5\operatorname{cis}(5\cdot 225^\circ)$
$= 32(\cos 1125^\circ + i\sin 1125^\circ)$
$= 32(\cos 45^\circ + i\sin 45^\circ)$
$= 16\sqrt{2} + 16i\sqrt{2}$

7. $[2\operatorname{cis}(120^\circ)]^6 = 2^6\operatorname{cis}(6\cdot 2\pi/3)$
$= 64(\cos 720^\circ + i\sin 720^\circ)$
$= 64(\cos 0^\circ + i\sin 0^\circ)$
$= 64$

9. $z = 1 - i$

$r = \sqrt{1^2 + (-1)^2}$ $\qquad \alpha = \tan^{-1}\left|\dfrac{-1}{1}\right| = 45^\circ$

$r = \sqrt{2}$ $\qquad\qquad \theta = 315^\circ$

$z = \sqrt{2}(\cos 315^\circ + i\sin 315^\circ)$

$(1 - i)^{10} = [\sqrt{2}(\cos 315^\circ + i\sin 315^\circ)]^{10}$

$= (\sqrt{2})^{10}[\cos(10\cdot 315^\circ) + i\sin(10\cdot 315^\circ)]$

$= 32(\cos 3150^\circ + i\sin 3150^\circ)$

$= 32(\cos 270^\circ + i\sin 270^\circ)$

$= 0 - 32i = -32i$

11. $z = 2 + 2i$

$r = \sqrt{2^2 + 2^2}$ $\qquad \alpha = \tan^{-1}\left|\dfrac{2}{2}\right| = 45^\circ$

$r = 2\sqrt{2}$ $\qquad\qquad \theta = 45^\circ$

$z = 2\sqrt{2}(\cos 45^\circ + i\sin 45^\circ)$

$(2 + 2i)^7 = [2\sqrt{2}(\cos 45^\circ + i\sin 45^\circ)]^7$

$= 1024\sqrt{2}[\cos(7\cdot 45^\circ) + i\sin(7\cdot 45^\circ)]$

$= 1024\sqrt{2}(\cos 315^\circ + i\sin 315^\circ)$

$= 1024 - 1024i$

170   **Chapter 5/Complex Numbers**

13. $z = \frac{\sqrt{2}}{2} + i\frac{\sqrt{2}}{2}$

$r = \sqrt{(\sqrt{2}/2)^2 + (\sqrt{2}/2)^2}$ $\alpha = \tan^{-1}\left|\frac{\sqrt{2}/2}{\sqrt{2}/2}\right| = 45°$

$r = 1$

$\theta = 45°$

$z = \cos 45° + i\sin 45°$

$\left(\frac{\sqrt{2}}{2} - i\frac{\sqrt{2}}{2}\right)^6 = (\cos 45° + i\sin 45°)^6$

$= \cos(6\cdot 45°) + i\sin(6\cdot 45°)$

$= \cos 270° + i\sin 270°$

$= 0 - 1i = -i$

15. $9 = 9(\cos 0° + i\sin 0°)$

$w_k = 9^{1/2}\left(\cos\frac{0° + 360°k}{2} + i\sin\frac{0° + 360°k}{2}\right)$ $k = 0,1$

$w_0 = 3(\cos 0° + i\sin 0°)$

$w_0 = 3 + 0i = 3$

$w_1 = 3\left(\frac{\cos 0° + 360°}{2} + i\sin\frac{0° + 360°}{2}\right)$

$w_1 = 3(\cos 180° + i\sin 180°)$

$w_1 = -3 + 0i = -3$

17. $64 = 64(\cos 0° + i\sin 0°)$

$w_k = 64^{1/6}\left(\cos\frac{0° + 360°k}{6} + i\sin\frac{0° + 360°k}{6}\right)$ $k = 0, 1, 2, 3, 4, 5$

$w_0 = 2(\cos 0° + i\sin 0°)$
$w_0 = 2 + 0i = 2$

$w_1 = 2\left(\cos\frac{0 + 360°}{6} + i\sin\frac{0 + 360°}{6}\right)$
$w_1 = 2(\cos 60° + i\sin 60°)$
$w_1 = 2\left(\frac{1}{2} + i\frac{\sqrt{3}}{2}\right)$
$w_1 = 1 + i\sqrt{3}$

$w_2 = 2\left(\cos\frac{0° + 360°\cdot 2}{6} + i\sin\frac{0° + 360°\cdot 2}{6}\right)$
$w_2 = 2(\cos 120° + i\sin 120°)$
$w_2 = 2\left(-\frac{1}{2} + i\frac{\sqrt{3}}{2}\right)$
$w_2 = -1 + i\sqrt{3}$

$w_3 = 2\left(\cos\frac{0° + 360°\cdot 3}{6} + i\sin\frac{0° + 360°\cdot 3}{6}\right)$
$w_3 = 2(\cos 180° + i\sin 180°)$
$w_3 = 2(-1 + 0i)$
$w_3 = -2 + 0i = -2$

$w_4 = 2\left(\cos\frac{0° + 360°\cdot 4}{6} + i\sin\frac{0° + 360°\cdot 4}{6}\right)$
$w_4 = 2(\cos 240° + i\sin 240°)$
$w_4 = 2\left(-\frac{1}{2} - i\frac{\sqrt{3}}{2}\right)$
$w_4 = -1 - i\sqrt{3}$

$w_5 = 2\left(\cos\frac{0° + 360°\cdot 5}{6} + i\sin\frac{0° + 360°\cdot 5}{6}\right)$
$w_5 = 2(\cos 300° + i\sin 300°)$
$w_5 = 2\left(\frac{1}{2} - i\frac{\sqrt{3}}{2}\right)$
$w_5 = 1 - i\sqrt{3}$

19. $-1 = 1(\cos 180° + i\sin 180°)$

$$w_k = 1^{1/5}\left(\cos\frac{180° + 360°k}{5} + i\sin\frac{180° + 360°k}{5}\right) \quad k = 0, 1, 2, 3, 4$$

$w_0 = 1(\cos 36° + i\sin 36°)$
$w_0 \approx 0.809 + 0.588i$

$w_1 = \cos\dfrac{180° + 360°}{5} + i\sin\dfrac{180° + 360°}{5}$
$w_1 = \cos 108° + i\sin 108°$
$w_1 \approx -0.309 + 0.951i$

$w_2 = \cos\dfrac{180° + 360° \cdot 2}{5} + i\sin\dfrac{180° + 360° \cdot 2}{5}$
$w_2 = \cos 180° + i\sin 180°$
$w_2 = -1 + 0i = -1$

$w_3 = \cos\dfrac{180° + 360° \cdot 3}{5} + i\sin\dfrac{180° + 360° \cdot 3}{5}$
$w_3 = \cos 252° + i\sin 252°$
$w_3 \approx -0.309 - 0.951i$

$w_4 = \cos\dfrac{180° + 360° \cdot 4}{5} + i\sin\dfrac{180° + 360° \cdot 4}{5}$
$w_4 = \cos 324° + i\sin 324°$
$w_4 \approx 0.809 - 0.588i$

21. $1 = \cos 0° + i\sin 0°$

$$w_k = \cos\frac{0° + 360°k}{3} + i\sin\frac{0° + 360°k}{3} \quad k = 0, 1, 2$$

$w_0 = \cos\dfrac{0°}{3} + i\sin\dfrac{0°}{3}$
$w_0 = \cos 0° + i\sin 0°$
$w_0 = 1 + 0i = 1$

$w_1 = \cos\dfrac{0° + 360°}{3} + i\sin\dfrac{0° + 360°}{3}$
$w_1 = \cos 120° + i\sin 120°$
$w_1 = -\dfrac{1}{2} + \dfrac{\sqrt{3}}{2}i$

$w_2 = \cos\dfrac{0° + 360° \cdot 2}{3} + i\sin\dfrac{0° + 360° \cdot 2}{3}$
$w_2 = \cos 240° + i\sin 240°$
$w_2 = -\dfrac{1}{2} - \dfrac{\sqrt{3}}{2}i$

23. $1 + i = \sqrt{2}(\cos 45° + i\sin 45°)$

$$w_k = \left(\sqrt{2}\right)^{1/4}\left(\cos\frac{45° + 360°k}{4} + i\sin\frac{45° + 360°k}{4}\right) \quad k = 0, 1, 2, 3$$

$w_0 = 2^{1/8}\left(\cos\dfrac{45°}{4} + i\sin\dfrac{45°}{4}\right)$
$w_0 = 2^{1/8}(\cos 11.25° + i\sin 11.25°)$
$w_0 \approx 1.070 + 0.213i$

$w_1 = 2^{1/8}\left(\cos\dfrac{45° + 360°}{4} + i\sin\dfrac{45° + 360°}{4}\right)$
$w_1 = 2^{1/8}(\cos 101.25° = i\sin 101.25°)$
$w_1 \approx -0.213 - 1.070i$

$w_2 = 2^{1/8}\left(\cos\dfrac{45° + 360° \cdot 2}{4} + i\sin\dfrac{45° + 360° \cdot 2}{4}\right)$
$w_2 = 2^{1/8}(\cos 191.25° + i\sin 191.25°)$
$w_2 \approx -1.070 - 0.213i$

$w_3 = 2^{1/8}\left(\cos\dfrac{45° + 360° \cdot 3}{4} + i\sin\dfrac{45° + 360° \cdot 3}{4}\right)$
$w_3 = 2^{1/8}(\cos 281.25° + i\sin 281.25°)$
$w_3 \approx 0.213 - 1.070i$

172   **Chapter 5/Complex Numbers**

25. $2 - 2i\sqrt{3} = 4(\cos 300° + i\sin 300°)$ $k = 0, 1, 2$

$$w_k = 4^{1/3}\left(\cos\frac{300° + 360°k}{3} + i\sin\frac{300° + 360°k}{3}\right)$$

$$w_0 = 4^{1/3}\left(\cos\frac{300°}{3} + i\sin\frac{300°}{3}\right)$$

$$w_0 = 4^{1/3}(\cos 100° + i\sin 100°)$$

$$w_0 \approx -0.276 + 1.563i$$

$$w_1 = 4^{1/3}\left(\cos\frac{300° + 360°}{3} + i\sin\frac{300° + 360°}{3}\right)$$

$$w_1 = 4^{1/3}(\cos 220° + i\sin 220°)$$

$$w_1 \approx -1.216 - 1.020i$$

$$w_2 = 4^{1/3}\left(\cos\frac{300° + 360°\cdot 2}{3} + i\sin\frac{300° + 360°\cdot 2}{3}\right)$$

$$w_2 = 4^{1/3}(\cos 340° + i\sin 340°)$$

$$w_2 \approx 1.492 - 0.543i$$

27. $-16 + 16i\sqrt{3} = 32(\cos 120° + i\sin 120°)$

$$w_k = 32^{1/2}\left(\cos\frac{120° + 360°k}{2} + i\sin\frac{120° + 360°k}{2}\right)$$ $k = 0, 1$

$$w_0 = 4\sqrt{2}\left(\cos\frac{120°}{2} + i\sin\frac{120°}{2}\right)$$

$$w_0 = 4\sqrt{2}(\cos 60° + i\sin 60°)$$

$$w_0 \approx 2\sqrt{2} + 2i\sqrt{6}$$

$$w_1 = 4\sqrt{2}\left(\cos\frac{120° + 360°}{2} + i\sin\frac{120° + 360°}{2}\right)$$

$$w_1 = 4\sqrt{2}(\cos 240° + i\sin 240°)$$

$$w_1 \approx -2\sqrt{2} - 2i\sqrt{6}$$

29. $x^3 + 8 = 0$

$$x^3 = -8$$

Find the three cube roots of -8.

$$-8 = 8(\cos 180° + i\sin 180°)$$

$$x_k = 8^{1/3}\left(\cos\frac{180° + 360°k}{3} + i\sin\frac{180° + 360°k}{3}\right)$$ $k = 0,1,2$

$$w_0 = 2\left(\cos\frac{180°}{3} + i\sin\frac{180°}{3}\right)$$

$$w_0 = 2(\cos 60° + i\sin 60°)$$

$$w_0 = 2\text{ cis }60°$$

$$w_1 = 2\left(\cos\frac{180° + 360°}{3} + i\sin\frac{180° + 360°}{3}\right)$$

$$w_1 = 2(\cos 180° + i\sin 180°)$$

$$w_1 = 2\text{ cis }180°$$

$$w_2 = 2\left(\cos\frac{180° + 360°\cdot 2}{3} + i\sin\frac{180° + 360°\cdot 2}{3}\right)$$

$$w_2 = 2(\cos 300° + i\sin 300°)$$

$$w_2 = 2\text{ cis }300°$$

31. $x^4 + i = 0$

$\quad x^4 = -i$

Find the four fourth roots of $-i$.

$-i = (\cos 270^0 + i \sin 270^0)$

$w_k = \cos \dfrac{270^\circ + 360^\circ k}{4} + i \sin \dfrac{270^\circ + 360^\circ k}{4} \quad k = 0, 1, 2, 3$

$w_0 = \text{cis } \dfrac{270^\circ}{4} \qquad w_1 = \text{cis } \dfrac{270^\circ + 360^\circ}{4} \qquad w_2 = \text{cis } \dfrac{270^\circ + 360^\circ \cdot 2}{4} \qquad w_3 = \text{cis } \dfrac{270^\circ + 360^\circ \cdot 3}{4}$

$w_0 = \text{cis } 67.5^\circ \qquad w_1 = \text{cis } 157.5^\circ \qquad w_2 = \text{cis } 247.5^\circ \qquad w_3 = \text{cis } 337.5^\circ$

33. $x^3 - 27 = 0$

$\quad x^3 = 27$

Find the three cube roots of 27.

$27 = 27(\cos 0^\circ + i \sin 0^\circ)$

$w_k = 3\left(\cos \dfrac{0^\circ + 360^\circ k}{3} + i \sin \dfrac{0^\circ + 360^\circ k}{3} \right) \quad k = 0, 1, 2$

$w_0 = 3 \text{ cis } \dfrac{0^\circ}{3} \qquad\qquad w_1 = 3 \text{ cis } \dfrac{0^\circ + 360^\circ}{3} \qquad\qquad w_2 = 3 \text{ cis } \dfrac{0^\circ + 360^\circ \cdot 2}{3}$

$w_0 = 3 \text{ cis } 0^\circ \qquad\qquad w_1 = 3 \text{ cis } 120^\circ \qquad\qquad w_2 = 3 \text{ cis } 240^\circ$

35. $x^4 + 81 = 0$

$\quad x^4 = -81$

Find the four fourth roots of -81.

$-81 = 81(\cos 180^\circ + i \sin 180^\circ)$

$w_k = 81^{1/4}\left(\cos \dfrac{180^\circ + 360^\circ k}{4} + i \sin \dfrac{180^\circ + 360^\circ k}{4} \right) \quad k = 0, 1, 2, 3$

$w_0 = 3 \text{ cis } \dfrac{180^\circ}{4} \qquad w_1 = 3 \text{ cis } \dfrac{180^\circ + 360^\circ}{4} \qquad w_2 = 3 \text{ cis } \dfrac{0^\circ + 360^\circ \cdot 2}{5} \qquad w_3 = 3 \text{ cis } \dfrac{180^\circ + 360^\circ \cdot 3}{4}$

$w_0 = 3 \text{ cis } 45^\circ \qquad w_1 = 3 \text{ cis } 135^\circ \qquad w_2 = 3 \text{ cis } 225^\circ \qquad w_3 = 3 \text{ cis } 315^\circ$

37. $x^4 - (1 - i\sqrt{3}) = 0$

$\quad x^4 = 1 - i\sqrt{3}$

Find the four fourth roots of $1 - i\sqrt{3}$.

$1 - i\sqrt{3} = 2(\cos 300^\circ + i \sin 300^\circ)$

$w_k = 2^{1/4}\left(\cos \dfrac{300^\circ + 360^\circ k}{4} + i \sin \dfrac{300^\circ + 360^\circ k}{4} \right) \quad k = 0, 1, 2, 3$

$w_0 = \sqrt[4]{2} \text{ cis } \dfrac{300^\circ}{4} \qquad w_1 = \sqrt[4]{2} \text{ cis } \dfrac{300^\circ + 360^\circ}{4} \qquad w_2 = \sqrt[4]{2} \text{ cis } \dfrac{300^\circ + 360^\circ \cdot 2}{4} \qquad w_3 = \sqrt[4]{2} \text{ cis } \dfrac{300^\circ + 360^\circ \cdot 3}{4}$

$w_0 = \sqrt[4]{2} \text{ cis } 75^\circ \qquad w_1 = \sqrt[4]{2} \text{ cis } 165^\circ \qquad w_2 = \sqrt[4]{2} \text{ cis } 255^\circ \qquad w_3 = \sqrt[4]{2} \text{ cis } 345^\circ$

39. $x^3 + (1 + i\sqrt{3}) = 0$

$\qquad x^3 = -1 - i\sqrt{3}$

Find the three cube roots of $-1 - i\sqrt{3}$.

$-1 - i\sqrt{3} = 2(\cos 240° + i \sin 240°)$

$w_k = 2^{1/3}\left(\cos \dfrac{240° + 360°k}{3} + i \sin \dfrac{240° + 360°k}{3}\right) \qquad k = 0,\ 1,\ 2$

$w_0 = \sqrt[3]{2}\ \text{cis}\ \dfrac{240°}{3} \qquad\qquad w_1 = \sqrt[3]{2}\ \text{cis}\ \dfrac{240° + 360°}{3} \qquad\qquad w_2 = \sqrt[3]{2}\ \text{cis}\ \dfrac{240° + 360° \cdot 2}{3}$

$w_0 = \sqrt[3]{2}\ \text{cis}\ 80° \qquad\qquad w_1 = \sqrt[3]{2}\ \text{cis}\ 200° \qquad\qquad w_2 = \sqrt[3]{2}\ \text{cis}\ 320°$

41. Let $z = a + bi$. Then $\bar z = a - bi$ by definition.
Substitute $a = r \cos\theta$ and $b = r \sin\theta$.
Thus $\bar z = r\cos\theta - ri\sin\theta = r(\cos\theta - i\sin\theta)$

43. $z = r(\cos\theta + i\sin\theta)$

$z^2 = r^2(\cos 2\theta + i\sin 2\theta)$

$\dfrac{1}{z^2} = \dfrac{1}{r^2(\cos 2\theta + i\sin 2\theta)}$

$\qquad = \dfrac{\cos 2\theta - i\sin 2\theta}{r^2(\cos 2\theta + i\sin 2\theta)(\cos 2\theta - i\sin 2\theta)}$

$\qquad = \dfrac{\cos 2\theta - i\sin 2\theta}{r^2(\cos^2 2\theta - i^2 \sin^2 2\theta)} = \dfrac{\cos 2\theta - i\sin 2\theta}{r^2(\cos^2 2\theta + \sin^2 2\theta)}$

$z^{-2} = r^{-2}(\cos 2\theta - i\sin 2\theta)$

EXPLORING CONCEPTS WITH TECHNOLOGY, Page 300

1. The first application of step 2 yields: --- 0.34765625
The second application of step 2 yields   0.3708648682
The third application of step 2 yields: -- 0.3875407504

$\qquad\vdots$ ---------------------- $\vdots$

The 24th application of step 2 yields: --- 0.4679402539

a. The 25th application of step 2 yields: --- 0.4689680812
The 26th application of step 2 yields: --- 0.4699310612

$\qquad\vdots$ ---------------------- $\vdots$

The 49th application of step 2 yields: --- 0.4823939313

b. The 50th application of step 2 yields: --- 0.4827039050
The 51st application of step 2 yields:---- 0.4830030599

$\qquad\vdots$ ---------------------- $\vdots$

The 74th application of step 2 yields: --- 0.4878302056

c. The 75th application of step 2 yields: --- 0.4879783095
The 76th application of step 2 yields: --- 0.4881228306

$\qquad\qquad\vdots \qquad\qquad\vdots$

d. 0.5

CHAPTER 5 TRUE/FALSE EXERCISES, Page 302

1. False, because $\sqrt{(-1)^2} = \sqrt{1} = 1 \neq -i$.

3. True

5. False, $\dfrac{1+i}{1-i} = \dfrac{(1+i)(1+i)}{(1-i)(1+i)} = \dfrac{1 + 2i + i^2}{1+1} = \dfrac{1 + 2i - 1}{1+1} = \dfrac{2i}{2} = i$.

7. True

9. True

11. False, $\cos\pi + i\sin\pi = -1$

CHAPTER 5 REVIEW EXERCISES, Page 302

1. $3 - \sqrt{-64} = 3 - 8i$ and the conjugate of $3 - 8i$ is $3 + 8i$

3. $-2 + \sqrt{-5} = -2 + i\sqrt{5}$ and the conjugate of $-2 + i\sqrt{5}$ is $-2 - i\sqrt{5}$

5. $(\sqrt{-4})(\sqrt{-4}) = i\sqrt{4} \cdot i\sqrt{4} = 2i \cdot 2i = 4i^2 = 4(-1) = -4$

7. $(3 + 7i) + (2 - 5i) = (3 + 2) + (7 - 5)i = 5 + 2i$

9. $(6 - 8i) - (9 - 11i) = (6 - 9) + [-8 - (-11)]i = -3 + 3i$

11. $(5 + 3i)(2 - 5i) = 10 - 25i + 6i - 15i^2 = 25 - 19i$

13. $\dfrac{-2i}{3 - 4i} = \dfrac{-2i}{3 - 4i} \cdot \dfrac{3 + 4i}{3 + 4i} = \dfrac{-6i - 8i^2}{9 - 16i^2} = \dfrac{8 - 6i}{25} = \dfrac{8}{25} - \dfrac{6}{25}i$

15. $i(2i) - (1 + i)^2 = 2i^2 - (1 + 2i + i^2) = 2i^2 - 2i = -2 - 2i$

17. $(3 + \sqrt{-4}) - (-3 - \sqrt{-16}) = (3 + 2i) - (-3 - 4i) = [3 - (-3)] + [2 - (-4)]i = 6 + 6i$

19. $(2 - \sqrt{-3})(2 + \sqrt{-3}) = (2 - i\sqrt{3})(2 + i\sqrt{3}) = 4 - 3i^2 = 7$

21. $i^{27} = i^{24} \cdot i^3 = i^3 = -i$

23. $\dfrac{i}{i^{17}} = \dfrac{i}{i^{16} \cdot i} = \dfrac{i}{i} = 1$

25. $|-8i| = \sqrt{(-8)^2} = \sqrt{64} = 8$

27. $|-4 + 5i| = \sqrt{(-4)^2 + 5^2} = \sqrt{16 + 25} = \sqrt{41}$

29. $z = 2 - 2i$

$r = \sqrt{2^2 + (-2)^2}$ $\alpha = \tan^{-1}\left|\dfrac{-2}{2}\right| = 45°$

$r = \sqrt{8}$ $\theta = 360° - 45° = 315°$

$r = 2\sqrt{2}$

$z = 2\sqrt{2}\,(\cos 315° + i \sin 315°)$

31. $z = -3 + 2i$

$r = \sqrt{(-3)^2 + 2^2}$ $\alpha = \tan^{-1}\left|\dfrac{2}{-2}\right| \approx 33.7°$

$r = \sqrt{9 + 4}$ $\theta \approx 180° - 33.7°$

$r = \sqrt{13}$ $\theta \approx 146.3°$

$z \approx \sqrt{13} \text{ cis } 146.3°$

33. $z = 5(\cos 315° + i \sin 315°)$

$z = 5\left(\dfrac{\sqrt{2}}{2} - \dfrac{i\sqrt{2}}{2}\right)$

$z = \dfrac{5\sqrt{2}}{2} - \dfrac{5\sqrt{2}}{2}i$

$z \approx 3.536 - 3.536i$

35. $z = 2(\cos 2 + i \sin 2)$

$z \approx 2(-0.4161 + 0.9093i)$

$z \approx -0.832 + 1.819i$

37. $3(\cos 225° + i \sin 225°) \cdot 10(\cos 45° + i \sin 45°) = 30\,[\cos(225° + 45°) + i \sin(225° + 45°)]$

$= 30\,(\cos 270° + i \sin 270°)$

$= 30\,(0 - i)$

$= 0 - 30i$

$= -30i$

39. $3(\cos 12° + i \sin 12°) \cdot 4(\cos 126° + i \sin 126°) = 12\,[\cos(12° + 126°) + i \sin(12° + 126°)]$

$= 12\,(\cos 138° + i \sin 138°)$

$\approx -8.918 + 8.030i$

41. $3(\cos 1.8 + i \sin 1.8) \cdot 5(\cos 2.5 + i \sin 2.5) = 15\,[\cos(1.8 + 2.5) + i \sin(1.8 + 2.5)]$

$= 15\,(\cos 4.3 + i \sin 4.3)$

$\approx -6.012 - 13.742i$

43. $\dfrac{6(\cos 50° + i \sin 50°)}{2(\cos 150° + i \sin 150°)} = 3[\cos(50° - 150°) + \sin(50° - 150°)]$

$= 3[\cos(-100°) + i \sin(-100°)]$

$= 3 \text{ cis } (-100°)$

45. $\dfrac{40(\cos 66° + i\,\sin 66°)}{8(\cos\ 125° + i\,\sin 125°)} = 5[\cos(66° - 125°) + i\,\sin(66° - 125°)]$

$\qquad\qquad\qquad\qquad = 5[\cos(-59°) + i\,\sin(-59°)]$

$\qquad\qquad\qquad\qquad = 5\ \text{cis}\ (-59°)$

47. $\dfrac{10\left(\cos 3.7 + i\,\sin 3.7\right)}{6\left(\cos 1.8 + i\,\sin 1.8\right)} = \dfrac{5}{3}[\cos(3.7 - 1.8) + i\,\sin(3.7 - 1.8)]$

$\qquad\qquad\qquad\qquad = \dfrac{5}{3}(\cos 1.9 + i\,\sin 1.9)$

$\qquad\qquad\qquad\qquad = \dfrac{5}{3}\ \text{cis}\ 1.9$

49. $[3(\cos 45° + i\,\sin 45°)]^5 = 3^5[\text{cis}\ (5\cdot 45°)]$

$\qquad\qquad\qquad = 243(\cos 225° + i\,\sin 225°)$

$\qquad\qquad\qquad = 243\left(-\dfrac{\sqrt2}{2} - \dfrac{\sqrt2}{2}i\right)$

$\qquad\qquad\qquad = -\dfrac{243\sqrt2}{2} - \dfrac{243\sqrt2}{2}i$

$\qquad\qquad\qquad \approx -171.827 - 171.827i$

51. $(1 - i\sqrt3)^7 = (2\ \text{cis}\ 300°)^7$

$\qquad\qquad = 2^7[\text{cis}\ (7\cdot 300°)]$

$\qquad\qquad = 128\ \text{cis}\ 2100°$

$\qquad\qquad = 128\ \text{cis}\ 300°$

$\qquad\qquad = 128\left(\dfrac{1}{2} - \dfrac{\sqrt3}{2}i\right)$

$\qquad\qquad = 64 - 64i\sqrt3$

$\qquad\qquad \approx 64 - 110.851i$

53. $(\sqrt2 - i\sqrt2)^5 = [2(\cos 315° + i\,\sin 315°)]^5$

$\qquad\qquad = 32[\cos(5\cdot 315°) + i\,\sin(5\cdot 315°)]$

$\qquad\qquad = 32(\cos 1575° + i\,\sin 1575°)$

$\qquad\qquad = 32(\cos 135° + i\,\sin 135°)$

$\qquad\qquad = 32\left(-\dfrac{\sqrt2}{2} + \dfrac{\sqrt2}{2}i\right)$

$\qquad\qquad = -16\sqrt2 + 16i\sqrt2$

$\qquad\qquad \approx -22.627 + 22.627i$

55. $27i = 27(\cos 90° + i\,\sin 90°)$

$\qquad w_k = 27^{1/3}\left(\cos\dfrac{90° + 360°k}{3} + i\,\sin\dfrac{90° + 360°k}{3}\right) \qquad k = 0, 1, 2$

$k = 0 \Rightarrow w_0 = 3\left(\cos\dfrac{90°}{3} + i\,\sin\dfrac{90°}{3}\right)$

$\qquad\qquad w_0 = 3\left(\cos 30° + i\,\sin 30°\right)$

$\qquad\qquad w_0 = 3\left(\dfrac{\sqrt3}{2} + \dfrac{1}{2}i\right)$

$\qquad\qquad w_0 = \dfrac{3\sqrt3}{2} + \dfrac{3}{2}i$

$k = 1 \Rightarrow w_1 = 3(\cos 150° + i\,\sin 150°)$

$\qquad\qquad w_1 = 3\left(-\dfrac{\sqrt3}{2} + \dfrac{1}{2}i\right)$

$\qquad\qquad w_1 = -\dfrac{3\sqrt3}{2} + \dfrac{3}{2}i$

$k = 2 \Rightarrow w_2 = 3(\cos 270° + i\,\sin 270°)$

$\qquad\qquad w_2 = 3(0 - i)$

$\qquad\qquad w_2 = 0 - 3i$

$\qquad\qquad w_2 = -3i$

57. $w_k = 256^{1/4}\ \text{cis}\ \dfrac{0° + 360°k}{4} \quad k = 0, 1, 2, 3$

$k = 0 \quad w_0 = 4(\cos 0° + i\,\sin 0°) = 4 + 0i = 4$

$k = 1 \quad w_1 = 4(\cos 90° + i\,\sin 90°) = 0 + 4i = 4i$

$k = 2 \quad w_2 = 4(\cos 180° + i\,\sin 180°) = -4 + 0i = -4$

$k = 3 \quad w_3 = 4(\cos 270° + i\,\sin 270°) = 0 - 4i = -4i$

59. $81 = 81(\cos 0° + i\,\sin 0°)$

$\qquad w_k = 81^{1/4}\ \text{cis}\ \dfrac{0° + 360°k}{4} \quad k = 0, 1, 2, 3$

$k = 0 \quad w_0 = 3\left(\cos 0° + i\,\sin 0°\right) = 3 + 0i = 3$

$k = 1 \quad w_1 = 3(\cos 90° + i\,\sin 90°) = 0 + 3i = 3i$

$k = 2 \quad w_2 = 3(\cos 180° + i\,\sin 180°) = -3 + 0i = -3$

$k = 3 \quad w_3 = 3(\cos 270° + i\,\sin 270°) = 0 - 3i = -3i$

CHAPTER 5 TEST, Page 303

1. $6+\sqrt{-9}=6+3i$

2. $\sqrt{-18}=i\sqrt{18}$
   $$=3i\sqrt{2}$$

3. $(3+\sqrt{-4})+(7-\sqrt{-9})=(3+2i)+(7-3i)$
   $$=(3+7)+(2-3)i$$
   $$=10-i$$

4. $(-1+\sqrt{-25})-(8-\sqrt{-16})=(-1+5i)-(8-4i)$
   $$=(-1-8)+(5+4)i$$
   $$=-9+9i$$

5. $(\sqrt{-12})(\sqrt{-3})=(i\sqrt{12})(i\sqrt{3})$
   $$=i^2\sqrt{36}$$
   $$=-6$$

6. $i^{263}=i^3=-i$

7. $(3+7i)-(-2-9i)=5+16i$

8. $(-6-9i)(4+3i)=-24-18i-36i-27i^2$
   $$=-24-54i+27$$
   $$=3-54i$$

9. $(3-5i)(-3+5i)=-9+15i+15i-25i^2$
   $$=-9+30i+25$$
   $$=16+30i$$

10. $\dfrac{4-5i}{i}=\dfrac{(4-5i)}{i}\cdot\dfrac{i}{i}$
    $$=\dfrac{4i-5i^2}{i^2}$$
    $$=\dfrac{5+4i}{-1}$$
    $$=-5-4i$$

11. $\dfrac{2-7i}{4+3i}=\dfrac{(2-7i)}{(4+3i)}\cdot\dfrac{(4-3i)}{(4-3i)}$
    $$=\dfrac{8-6i-28i+21i^2}{16+9}$$
    $$=\dfrac{8-34i-21}{25}$$
    $$=\dfrac{-13-34i}{25}$$
    $$=-\dfrac{13}{25}-\dfrac{34}{25}i$$

12. $\dfrac{6+2i}{1-i}=\dfrac{(6+2i)}{(1-i)}\cdot\dfrac{(1+i)}{(1+i)}$
    $$=\dfrac{6+6i+2i+2i^2}{1+1}$$
    $$=\dfrac{6+8i-2}{2}$$
    $$=\dfrac{4+8i}{2}$$
    $$=2+4i$$

13. $|3-5i|=\sqrt{3^2+(-5)^2}$
    $$=\sqrt{9+25}$$
    $$=\sqrt{34}$$

14. $r=\sqrt{3^2+(-3)^2}=\sqrt{18}=3\sqrt{2}$
    $$a=\tan^{-1}\left|\dfrac{-3}{3}\right|=\tan^{-1}1=45°$$
    $$\theta=360°-45°=315°$$
    z is in the fourth quadrant
    $$z=3\sqrt{2}\text{ cis }315°$$

15. $-6i=6(\cos270°+i\,\sin270°)$
    $$=6\text{ cis }270°$$

16. $4(\cos120°+i\sin120°)=4\left(-\dfrac{1}{2}+i\dfrac{\sqrt{3}}{2}\right)$
    $$=-2+2i\sqrt{3}$$

17. $5(\cos225°+i\sin225°)=5\left(-\dfrac{\sqrt{2}}{2}-i\dfrac{\sqrt{2}}{2}\right)$
    $$=-\dfrac{5\sqrt{2}}{2}-\dfrac{5\sqrt{2}}{2}i$$

18. $3(\cos28°+i\sin\,28°)\cdot4(\cos17°+i\sin17°)=12\,(\cos45°+i\sin45°)$
    $$=12\left(\dfrac{\sqrt{2}}{2}+i\dfrac{\sqrt{2}}{2}\right)$$
    $$=6\sqrt{2}+6i\sqrt{2}$$

19. $5(\cos115°+i\sin115°)\cdot4(\cos10°+i\sin10°)=20(\cos125°+i\sin125°)$
    $$\approx-11.472+16.383i$$

178   **Chapter 5/Complex Numbers**

20. $\dfrac{24(\cos 258° + i\sin 258°)}{6(\cos 78° + i\sin 78°)} = \dfrac{24}{6}$ cis $(258° - 78°)$

$\qquad = 4(\cos 180° + i\sin 180°)$

$\qquad = 4(-1 + 0i)$

$\qquad = -4 + 0i$

$\qquad = -4$

21. $\dfrac{18(\cos 50° + i\sin 50°)}{3(\cos 140° + i\sin 140°)} = \dfrac{18}{3}$ cis $(50° - 140°)$

$\qquad = 6[\cos(-90°) + i\sin(-90°)]$

$\qquad = 6[0 + i(-1)]$

$\qquad = 0 - 6i$

$\qquad = -6i$

22. $2 - 2i\sqrt{3} = 4(\cos 300° + i\sin 300°)$

$[4(\cos 300° + i\sin 300°)]^{12} = 4^{12}(\cos 3600° + i\sin 3600°)$

$\qquad = 16,777,216(\cos 0° + i\sin 0°)$

$\qquad = 16,777,216(1 + 0i)$

$\qquad = 16,777,216 + 0i$

$\qquad = 16,777,216$

23. $64 = 64(\cos 0° + i\sin 0°)$

$w_k = 64^{1/6}\left(\cos\dfrac{0° + 360°k}{6} + i\sin\dfrac{0° + 360°k}{6}\right) \qquad k = 0, 1, 2, 3, 4, 5$

$w_0 = 64^{1/6}$ cis $\dfrac{0°}{6}$
$w_0 = 2$ cis $0°$

$w_1 = 64^{1/6}$ cis $\dfrac{360°}{6}$
$w_1 = 2$ cis $60°$

$w_2 = 64^{1/6}$ cis $\dfrac{2\cdot360°}{6}$
$w_2 = 2$ cis $120°$

$w_3 = 64^{1/6}$ cis $\dfrac{3\cdot360°}{6}$
$w_3 = 2$ cis $180°$

$w_4 = 64^{1/6}$ cis $\dfrac{4\cdot360°}{6}$
$w_4 = 2$ cis $240°$

$w_5 = 64^{1/6}$ cis $\dfrac{5\cdot360°}{6}$
$w_5 = 2$ cis $300°$

24. $-1 + i\sqrt{3} = 2(\cos 120° + i\sin 120°)$

$w_k = 2^{1/3}\left(\cos\dfrac{120° + 360°k}{3} + i\sin\dfrac{120° + 360°k}{3}\right) = \sqrt[3]{2}$ cis $\dfrac{120° + 360°k}{3} \qquad k = 0, 1, 2$

$w_0 = \sqrt[3]{2}$ cis $\dfrac{120°}{3}$
$w_0 = \sqrt[3]{2}$ cis $40°$

$w_1 = \sqrt[3]{2}$ cis $\dfrac{120° + 360°}{3}$
$w_1 = \sqrt[3]{2}$ cis $160°$

$w_2 = \sqrt[3]{2}$ cis $\dfrac{120° + 2\cdot360°}{3}$
$w_2 = \sqrt[3]{2}$ cis $280°$

25. $z^5 + 32 = 0$

$\qquad z^5 = -32$

Find the five fifth roots of -32.

$-32 = 32(\cos 180° + i\sin 180°)$

$w_k = 32^{1/5}\left(\cos\dfrac{180° + 360°k}{5} + i\sin\dfrac{180° + 360°k}{5}\right) \qquad k = 0, 1, 2, 3, 4$

$w_0 = 32^{1/5}$ cis $\dfrac{180°}{5}$
$w_0 = 2$ cis $36°$

$w_1 = 32^{1/5}$ cis $\dfrac{180° + 360°}{5}$
$w_1 = 2$ cis $108°$

$w_2 = 32^{1/5}$ cis $\dfrac{180° + 2\cdot360°}{5}$
$w_2 = 2$ cis $180°$

$w_3 = 32^{1/5}$ cis $\dfrac{180° + 3\cdot360°}{5}$
$w_3 = 2$ cis $252°$

$w_4 = 32^{1/5}$ cis $\dfrac{180° + 4\cdot360°}{5}$
$w_4 = 2$ cis $324°$

SECTION 6.1 Page 312

1. $x^2 = -4y$
$4p = -4$
$p = -1$
vertex $= (0, 0)$
focus $= (0, -1)$
directrix: $y = 1$

3. $y^2 = \frac{1}{3}x$
$4p = \frac{1}{3}$
$p = \frac{1}{12}$
vertex $= (0, 0)$
focus $= \left(\frac{1}{12}, 0\right)$
directrix: $x = -\frac{1}{12}$

5. $(x - 2)^2 = 8(y + 3)$
vertex $= (2, -3)$
$4p = 8 \quad p = 2$
$(h, k + p) = (2, -3 + 2) = (2, -1)$
focus $= (2, -1)$
$k - p = -3 - 2 = -5$
directrix: $y = -5$

7. $(y + 4)^2 = -4(x - 2)$
vertex $= (2, -4)$
$4p = -4 \quad p = -1$
$(h + p, k) = (2 - 1, -4) = (1, -4)$
focus $= (1, -4)$
$h - p = 2 + 1 = 3$
directrix: $x = 3$

9. $(y - 1)^2 = 2(x + 4)$
vertex $= (-4, 1)$
$4p = 2 \Rightarrow p = \frac{1}{2}$
$(h + p, k) = \left(-4 + \frac{1}{2}, 1\right) = \left(-\frac{7}{2}, 1\right)$
focus $= \left(-\frac{7}{2}, 1\right)$
$h - p = -4 - \frac{1}{2} = -\frac{9}{2}$
directrix: $x = -\frac{9}{2}$

11. $(x - 2)^2 = 2(y - 2)$
vertex $= (2, 2)$
$4p = 2 \quad p = \frac{1}{2}$
$(h, k + p) = \left(2, 2 + \frac{1}{2}\right) = \left(2, \frac{5}{2}\right)$
focus $= \left(2, \frac{5}{2}\right)$
$k - p = 2 - \frac{1}{2} = \frac{3}{2}$
directrix: $y = \frac{3}{2}$

13. $x^2 + 8x - y + 6 = 0$
$x^2 + 8x = y - 6$
$x^2 + 8x + 16 = y - 6 + 16$
$(x + 4)^2 = y + 10$
vertex $= (-4, -10)$
$4p = 1, \quad p = \frac{1}{4}$
focus $= \left(-4, -\frac{39}{4}\right)$
directrix: $y = -\frac{41}{4}$

15. $x + y^2 - 3y + 4 = 0$
$y^2 - 3y = -x - 4$
$y^2 - 3y + 9/4 = -x - 4 + 9/4$
$\left(y - \frac{3}{2}\right)^2 = -\left(x + \frac{7}{4}\right)$
vertex $= \left(-\frac{7}{4}, \frac{3}{2}\right)$
$4p = -1, \quad p = -\frac{1}{4}$
focus $= \left(-2, \frac{3}{2}\right)$
directrix: $x = -\frac{3}{2}$

17. $2x - y^2 - 6y + 1 = 0$
$-y^2 - 6y = -2x - 1$
$y^2 + 6y = 2x + 1$
$y^2 + 6y + 9 = 2x + 1 + 9$
$(y + 3)^2 = 2(x + 5)$
vertex $= (-5, -3)$
$4p = 2, \quad p = \frac{1}{2}$
focus $= \left(-\frac{9}{2}, -3\right)$
directrix: $x = -\frac{11}{2}$

19. $x^2 + 3x + 3y - 1 = 0$

$x^2 + 3x = -3y + 1$

$x^2 + 3x + 9/4 = -3y + 1 + 9/4$

$\left(x + \frac{3}{2}\right)^2 = -3\left(y - \frac{13}{12}\right)$

vertex $\left(-\frac{3}{2}, \frac{13}{12}\right)$

$4p = -3, \ p = -\frac{3}{4}$

focus $= \left(-\frac{3}{2}, \frac{1}{3}\right)$

directrix: $y = \frac{11}{6}$

21. $2x^2 - 8x - 4y + 3 = 0$

$2(x^2 - 4x) = 4y - 3$

$2(x^2 - 4x + 4) = 4y - 3 + 8$

$2(x - 2)^2 = 4y + 5$

$(x - 2)^2 = 2y + \frac{5}{2}$

$(x - 2)^2 = 2\left(y + \frac{5}{4}\right)$

vertex $= \left(2, -\frac{5}{4}\right)$

$4p = 2, \ p = \frac{1}{2}$

focus $= \left(2, -\frac{3}{4}\right)$

directrix $y = -\frac{7}{4}$

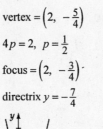

23. $2x + 4y^2 + 8y - 5 = 0$

$4y^2 + 8y = -2x + 5$

$4(y^2 + 2y) = -2x + 5$

$4(y^2 + 2y + 1) = -2x + 5 + 4$

$4(y + 1)^2 = -2x + 9$

$(y + 1)^2 = -\frac{1}{2}x + \frac{9}{4}$

$(y + 1)^2 = -\frac{1}{2}\left(x - \frac{9}{2}\right)$

vertex $= \left(\frac{9}{2}, -1\right)$

$4p = -\frac{1}{2}, \ p = -\frac{1}{8}$

focus $= \left(\frac{35}{8}, -1\right)$

directrix $x = \frac{37}{8}$

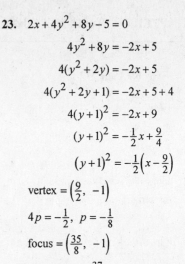
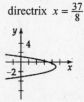

25. $(x - 1)^2 = 3\left(y - \frac{1}{9}\right)$

vertex $= \left(1, \frac{1}{9}\right)$

$4p = 3, \ p = \frac{3}{4}$

focus $= \left(1, \frac{31}{36}\right)$

directrix $y = -\frac{23}{36}$

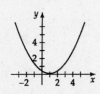

27. vertex: $(0, 0)$

focus: $(0, -4)$

$x^2 = 4py$

$p = -4$ since focus is $(0, p)$

$x^2 = 4(-4)y$

$x^2 = -16y$

29. vertex: $(-1, 2)$

focus: $(-1, 3)$

$(x - h)^2 = 4p(y - k)$

$h = -1, \ k = 2.$

The distance p from the vertex to the focus is 1.

$(x + 1)^2 = 4(1)(y - 2)$

$(x + 1)^2 = 4(y - 2)$

31. focus $(3, -3)$, directrix $y = -5$

The vertex is the midpoint of the line segment joining $(3, -3)$ and the point $(3, -5)$ on the directrix.

$(h, k) = \left(\frac{3 + 2}{2}, \frac{-3 + (-5)}{2}\right) = (3, -4)$

The distance p from the vertex to the focus is 1.

$4p = 4(1) = 4$

$(x - h)^2 = 4p(y - k)$

$(x - 3)^2 = 4(y + 4)$

33. vertex $= (-4, 1)$, point: $(-2, 2)$ on the parabola.

axis of symmetry: $x = -4$.

If $P_1 = (-2, 2)$, then $(x + 4)^2 = 4p(y - 1)$.

Since $(-2, 2)$ is on the curve, we get

$(-2 + 4)^2 = 4p(2 - 1)$

$4 = 4p \Rightarrow p = 1$

Thus, the equation in standard form is

$(x + 4)^2 = 4(y - 1)$

35. Place the satellite dish on an *xy*-coordinate system with its vertex at $(0, -1)$ as shown.

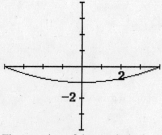

The equation of the parabola is

$$x^2 = 4p(y+1) \quad -1 \le y \le 0$$

Because $(4, 0)$ is a point on this graph, $(4, 0)$ must be a solution of the equation of the parabola. Thus,

$16 = 4p(0+1)$

$16 = 4p$

$4 = p$

Because p is the distance from the vertex to the focus, the focus is on the-axis 4 feet above the vertex.

37. The focus of the parabola is $(p, 0)$ where $y^2 = 4px$.
Half of 18.75 inches is 9.375 inches.
Therefore, the point $(3.66, 9.375)$ is on the parabola.

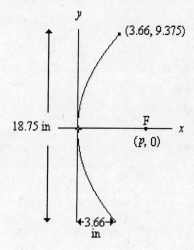

$$(9.375)^2 = 4p(3.66)$$

$$87.890625 = 14.64p$$

$$\frac{87.890625}{14.64} = p$$

$$p \approx 6.0 \text{ inches}$$

39. $S = \dfrac{\pi r}{6d^2}\left[\left(r^2 + 4d^2\right)^{3/2} - r^3\right]$

a. $r = 40.5$ feet
      $d = 16$ feet

$$S = \frac{\pi(40.5)}{6(16)^2}\left[\left([40.5]^2 + 4[16]^2\right)^{3/2} - (40.5)^3\right]$$

$$= \frac{40.5\pi}{1536}\left[(266.25)^{3/2} - 66430.125\right]$$

$$= \frac{40.5\pi}{1536}\left[137518.9228 - 66430.125\right]$$

$$= \frac{40.5\pi}{1536}\left[71088.79775\right]$$

$$\approx 5900 \text{ square feet}$$

b. $r = 125$ feet
      $d = 52$ feet

$$S = \frac{\pi(125)}{6(52)^2}\left[\left([125]^2 + 4[52]^2\right)^{3/2} - (125)^3\right]$$

$$= \frac{125\pi}{16224}\left[(26441)^{3/2} - 1953125\right]$$

$$= \frac{125\pi}{16224}\left[4299488.724 - 1953125\right]$$

$$= \frac{125\pi}{16224}\left[2346363.724\right]$$

$$\approx 56,800 \text{ square feet}$$

41. The equation of the mirror is given by

$$x^2 = 4py \quad -60 \le x \le 60$$

Because p is the distance from the vertex to the focus and the coordinates of the focus are $(0, 600)$, $p = 600$.
Therefore,

$$x^2 = 4(600)y$$

$$x^2 = 2400y$$

To determine a, substitute $(60, a)$ into the equation $x^2 = 2400y$ and solve for a.

$$x^2 = 2400y$$

$$60^2 = 2400a$$

$$3600 = 2400a$$

$$1.5 = a.$$

The concave depth of the mirror is 1.5 inches.

43. $(-0.3660, -0.3660)$, $(1.3660, 1.3660)$

45. $(-1.5616, 3.8769)$, $(2.5616, 12.1231)$

47. $x^2 = 4y$

$4p = 4$

$p = 1$

focus $= (0, 1)$

Substituting the vertical coordinate of the focus for y to obtain x-coordinates of endpoints (x_1, y_1) and (x_2, y_2), we have

$$x^2 = 4(1), \text{ or } x^2 = 4$$

$$x = \pm\sqrt{4}$$

$$x_1 = -2 \qquad x_2 = 2$$

Length of latus rectum $= |x_2 - x_1|$

$$= 2 - (-2) = 4.$$

49. If the parabola has a vertical axis of symmetry, then

$$(x - h)^2 = 4p(y - k)$$

focus $= (h, k + p)$

Substituting the vertical coordinate of the focus for y to obtain x-coordinates of endpoints (x_1, y_1), (x_2, y_2), we have

$$(x - h)^2 = 4p(k + p - k)$$

$$(x - h)^2 = 4p^2 \Rightarrow x - h = \pm 2p$$

$$x_1 = h - 2p \qquad x_2 = h + 2p$$

Solving for $|x_2 - x_1|$, we obtain

$$\Delta x = |x_2 - x_1| = |h + 2p - h + 2p| = 4|p|$$

If the parabola has a horizontal axis of symmetry, then

$$(y - k)^2 = 4p(x - h)$$

focus $= (h + p, k)$

Substituting the horizontal coordinate of the focus for x to obtain the y-coordinates of the endpoints (x_1, y_1), (x_2, y_2), we have

$$(y - k)^2 = 4p(h + p - h)$$

$$(y - k)^2 = 4p^2 \Rightarrow y - k = \pm 2p$$

$$y_1 = k - 2p \qquad y_2 = k + 2p$$

Solving for $|y_2 - y_1|$, we obtain

$$\Delta y = |y_2 - y_1| = |k + 2p - k + 2p| = 4|p|$$

Thus, the length of the latus rectum for any parabola is $4|p|$.

51.

$4p = -1$

$p = -\dfrac{1}{4}$

focus $\left(\dfrac{3}{4}, -4\right)$

one point: $\left(\dfrac{3}{4}, k + 2p\right) = \left(\dfrac{3}{4}, -\dfrac{9}{2}\right)$

one point: $\left(\dfrac{3}{4}, k - 2p\right) = \left(\dfrac{3}{4}, -\dfrac{7}{2}\right)$

53. Graph $y = \dfrac{7}{4} + \dfrac{1}{4}x|x|$.

55. By definition, any point on the curve (x, y) will be equidistant from both the focus $(1, 1)$ and the directrix, $(y_2 = -x_2 - 2)$.

If we let d_1 equal the distance from the focus to the point (x, y), we get $d_1 = \sqrt{(x - 1)^2 + (y - 1)^2}$.

To determine the distance d_2 from the point (x, y) to the line $y = -x - 2$, draw a line segment from (x, y) to the directrix so as to meet the directrix at a $90°$ angle.

Now drop a line segment parallel to the y-axis from (x, y) to the directrix. This segment will meet the directrix at a $45°$ angle, thus forming a right isosceles triangle with the directrix and the line segment perpendicular to the directrix from (x, y). The length of this segment, which is the hypotenuse of the triangle, is the difference between y and the y-value of the directrix at x, or $-x - 2$. Thus, the hypotenuse has a length of $y + x + 2$, and since the right triangle is also isosceles, each leg has a length of $\dfrac{y + x + 2}{\sqrt{2}}$.

But since d_2 is the length of the leg drawn from (x, y) to the directrix, $d_2 = \dfrac{y + x + 2}{\sqrt{2}}$.

Thus, $d_1 = \sqrt{(x - 1)^2 + (y - 1)^2}$ and $d_2 = \dfrac{x + y + 2}{\sqrt{2}}$. By definition, $d_1 = d_2$. So, by substitution,

$$\sqrt{(x - 1)^2 + (y - 1)^2} = \frac{x + y + 2}{\sqrt{2}}$$

$$\sqrt{2}\sqrt{(x - 1)^2 + (y - 1)^2} = x + y + 2$$

$$2[(x - 1)^2 + (y - 1)^2] = x^2 + y^2 + 4x + 4y + 2xy + 4$$

$$2(x^2 - 2x + 1 + y^2 - 2y + 1) = x^2 + y^2 + 4x + 4y + 2xy + 4$$

$$2x^2 - 4x + 2y^2 - 4y + 4 = x^2 + y^2 + 4x + 4y + 2xy + 4$$

$$x^2 + y^2 - 8x - 8y - 2xy = 0$$

SECTION 6.2, Page 324

1. $\frac{x^2}{16}+\frac{y^2}{25}=1$

$a^2=25 \to a=5$

$b^2=16 \to b=4$

$c=\sqrt{a^2-b^2}$
$=\sqrt{25-16}$
$=\sqrt{9}$
$=3$

center: (0, 0)
vertices: $(0,\ \pm 5)$
foci: $(0,\ \pm 3)$

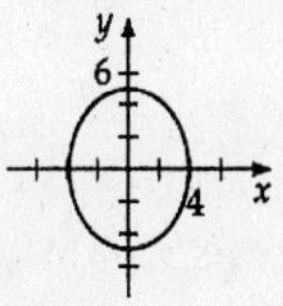

3. $\frac{x^2}{9}+\frac{y^2}{4}=1$

$a^2=9 \to a=3$

$b^2=4 \to b=2$

$c=\sqrt{a^2-b^2}$
$=\sqrt{9-4}$
$=\sqrt{5}$

center: (0, 0)
vertices: $(\pm 3,\ 0)$
foci: $(\pm\sqrt{5},\ 0)$

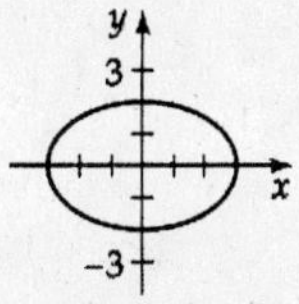

5. $\frac{x^2}{7}+\frac{y^2}{9}=1$

$a^2=9 \to a=3$

$b^2=7 \to b=\sqrt{7}$

$c=\sqrt{a^2-b^2}$
$=\sqrt{9-7}$
$=\sqrt{2}$

center: (0, 0)
vertices: $(0,\ \pm 3)$
foci: $(0,\ \pm\sqrt{2})$

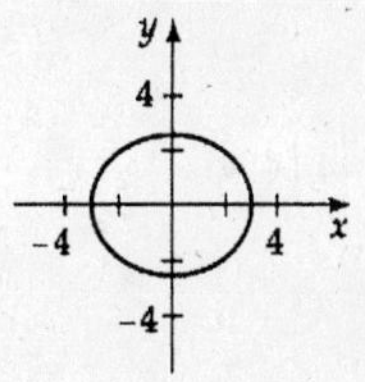

7. $\frac{4x^2}{9}+\frac{y^2}{16}=1$

Rewrite as

$\frac{x^2}{9/4}+\frac{y^2}{16}=1$

$a^2=16 \to a=4$

$b^2=9/4 \to b=3/2$

$c=\sqrt{a^2-b^2}$
$=\sqrt{16-9/4}$
$=\sqrt{55}/2$

center: (0, 0)
vertices: $(0,\ \pm 4)$
foci: $\left(0,\ \pm\frac{\sqrt{55}}{2}\right)$

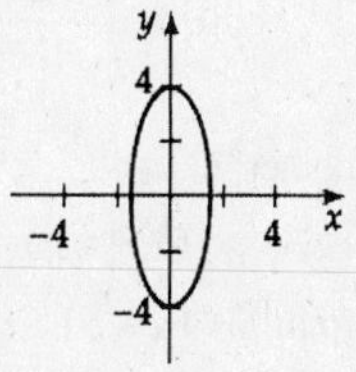

9. $\frac{(x-3)^2}{25}+\frac{(y+2)^2}{16}=1$
center: $(3,\ -2)$
vertices: $(3\pm 5,\ -2)=(8,\ -2),\ (-2,\ -2)$
foci: $(3\pm 3,\ -2)=(6,\ -2),\ (0,\ -2)$

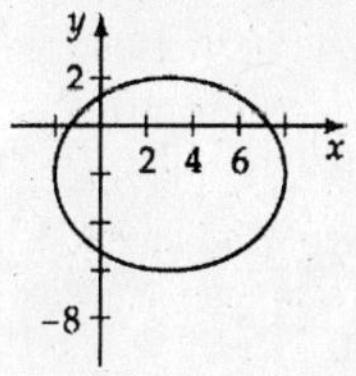

11. $\frac{(x+2)^2}{9}+\frac{y^2}{16}=1$
center: $(-2,\ 0)$
vertices: $(-2,\ 5),\ (-2,\ -5)$
foci: $(-2,\ 4),\ (-2,\ -4)$

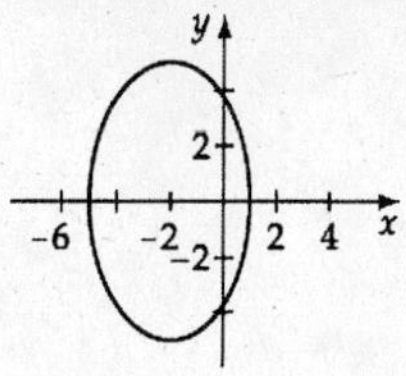

13. $\frac{(x-1)^2}{21}+\frac{(y-3)^2}{4}=1$
center: (1, 3)
vertices: $(1\pm\sqrt{21},\ 3)$
foci: $(1\pm\sqrt{17},\ 3)$

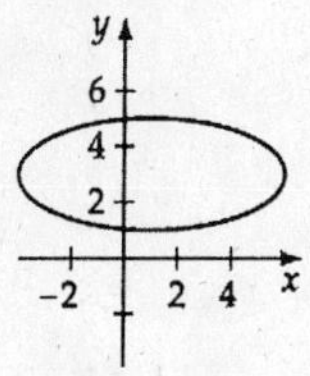

15. $\frac{9(x-1)^2}{16}+\frac{(y+1)^2}{9}=1$
center: $(1,\ -1)$
vertices: $(1,\ -1\pm 3)=(1,\ 2),\ (1,\ -4)$
foci: $(1,\ -1\pm\sqrt{65}/3)$

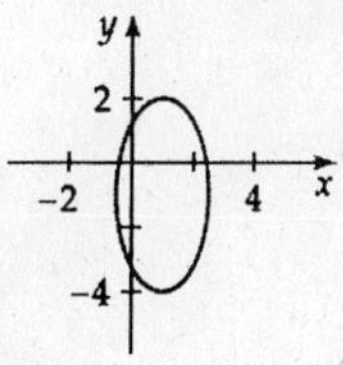

17. $3x^2 + 4y^2 = 12$

$$\frac{x^2}{4} + \frac{y^2}{3} = 1$$

center: $(0, 0)$

vertices: $(\pm 2, 0)$

foci: $(\pm 1, 0)$

19. $25x^2 + 16y^2 = 400$

$$\frac{x^2}{16} + \frac{y^2}{25} = 1$$

center: $(0, 0)$

vertices: $(0, \pm 5)$

foci: $(0, \pm 3)$

21. $64x^2 + 25y^2 = 400$

$$\frac{x^2}{\frac{25}{4}} + \frac{y^2}{16} = 1$$

center: $(0, 0)$

vertices: $(0, \pm 4)$

foci: $\left(0, \pm \frac{\sqrt{39}}{2}\right)$

23. $4x^2 + y^2 - 24x - 8y + 48 = 0$

$$4(x^2 - 6x) + (y^2 - 8y) = -48$$

$$4(x^2 - 6x + 9) + (y^2 - 8y + 16) = -48 + 36 + 16$$

$$4(x-3)^2 + (y-4)^2 = 4$$

$$\frac{(x-3)^2}{1} + \frac{(y-2)^2}{4} = 1$$

center: $(3, 4)$

vertices: $(3, 4 \pm 2) = (3, 6), (3, 2)$

foci: $(3, 4 \pm \sqrt{3})$

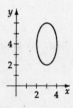
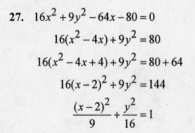

25. $5x^2 + 9y^2 - 20x + 54y + 56 = 0$

$$5(x^2 - 4x) + 9(y^2 + 6y) = -56$$

$$5(x^2 - 4x + 4) + 9(y^2 + 6y + 9) = -56 + 20 + 81$$

$$5(x-2)^2 + 9(y+3)^2 = 45$$

$$\frac{(x-2)^2}{9} + \frac{(y+3)^2}{5} = 1$$

center: $(2, -3)$

vertices: $(2 \pm 3, -3) = (-1, -3), (5, -3)$

foci: $(2 \pm 2, 3) = (0, -3), (4, -3)$

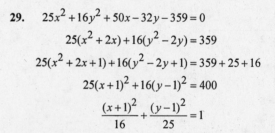

27. $16x^2 + 9y^2 - 64x - 80 = 0$

$$16(x^2 - 4x) + 9y^2 = 80$$

$$16(x^2 - 4x + 4) + 9y^2 = 80 + 64$$

$$16(x-2)^2 + 9y^2 = 144$$

$$\frac{(x-2)^2}{9} + \frac{y^2}{16} = 1$$

center: $(2, 0)$

vertices: $(2, \pm 4) = (2, 4), (2, -4)$

foci: $(2 \pm \sqrt{7})$

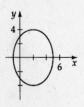

29. $25x^2 + 16y^2 + 50x - 32y - 359 = 0$

$$25(x^2 + 2x) + 16(y^2 - 2y) = 359$$

$$25(x^2 + 2x + 1) + 16(y^2 - 2y + 1) = 359 + 25 + 16$$

$$25(x+1)^2 + 16(y-1)^2 = 400$$

$$\frac{(x+1)^2}{16} + \frac{(y-1)^2}{25} = 1$$

center: $(-1, 1)$

vertices: $(-1, 1 \pm 5) = (-1, 6), (-1, -4)$

foci: $(-1, 1 \pm 3) = (-1, 4), (-1, -2)$

31.
$$8x^2 + 25y^2 - 48x + 50y + 47 = 0$$
$$8(x^2 - 6x) + 25(y^2 + 2y) = -47$$
$$8(x^2 - 6x + 9) + 25(y^2 + 2y + 1) = -47 + 72 + 25$$
$$8(x-3)^2 + 25(y+1)^2 = 50$$
$$\frac{(x-3)^2}{25/4} + \frac{(y+1)^2}{2} = 1$$

center: $(3, -1)$

vertices: $\left(3 \pm \dfrac{5}{2},\ -1\right) = \left(\dfrac{11}{2},\ -1\right), \left(\dfrac{1}{2},\ -1\right)$

foci: $\left(3 \pm \dfrac{\sqrt{17}}{2},\ -1\right)$

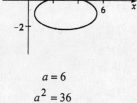

33.
$$2a = 10$$
$$a = 5$$
$$a^2 = 25$$
$$c = 4$$
$$c^2 = a^2 - b^2$$
$$16 = 25 - b^2$$
$$b^2 = 9$$
$$\frac{x^2}{25} + \frac{y^2}{9} = 1$$

35.
$$a = 6$$
$$a^2 = 36$$
$$b = 4$$
$$b^2 = 16$$
$$\frac{x^2}{36} + \frac{y^2}{16} = 1$$

37.
$$2a = 12$$
$$a = 6$$
$$a^2 = 36$$
$$\frac{x^2}{36} + \frac{y^2}{b^2} = 1$$
$$\frac{(2)^2}{36} + \frac{(-3)^2}{b^2} = 1$$
$$\frac{4}{36} + \frac{9}{b^2} = 1$$
$$\frac{9}{b^2} = \frac{8}{9}$$
$$8b^2 = 81$$
$$b^2 = \frac{81}{8}$$
$$\frac{x^2}{36} + \frac{y^2}{81/8} = 1$$

39.
$$c = 3$$
$$2a = 8$$
$$a = 4$$
$$a^2 = 16$$
$$c^2 = a^2 - b^2$$
$$9 = 16 - b^2$$
$$b^2 = 7$$
$$\frac{(x+2)^2}{16} + \frac{(y-4)^2}{7} = 1$$

41.
$$2a = 10$$
$$a = 5$$
$$a^2 = 25$$
Since the center of the ellipse is $(2, 4)$ and the point $(3, 3)$ is on the ellipse, we have
$$\frac{(x-2)^2}{b^2} + \frac{(y-4)^2}{a^2} = 1$$
$$\frac{(3-2)^2}{b^2} + \frac{(3-4)^2}{25} = 1$$
$$\frac{1}{b^2} = 1 - \frac{1}{25}$$
$$b^2 = \frac{25}{24}$$
$$\frac{(x-2)^2}{25/24} + \frac{(y-4)^2}{25} = 1$$

43. center $(5, 1)$
$$c = 3$$
$$2a = 10$$
$$a = 5$$
$$a^2 = 25$$
$$c^2 = a^2 - b^2$$
$$9 = 25 - b^2$$
$$b^2 = 16$$
$$\frac{(x-5)^2}{16} + \frac{(y-1)^2}{25} = 1$$

45. $2a = 10$

$a = 5$

$a^2 = 25$

$\dfrac{c}{a} = \dfrac{2}{5}$

$\dfrac{c}{5} = \dfrac{2}{5}$

$c = 2$

$c^2 = a^2 - b^2$

$4 = 25 - b^2$

$b^2 = 21$

$\dfrac{x^2}{25} + \dfrac{y^2}{21} = 1$

47. center: (0, 0)

$c = 4$

$\dfrac{c}{a} = \dfrac{2}{3}$

$\dfrac{4}{a} = \dfrac{2}{3}$

$a = 6$

$c^2 = a^2 - b^2$

$16 = 36 - b^2$

$b^2 = 20$

$\dfrac{x^2}{20} + \dfrac{y^2}{36} = 1$

49. center: (1, 3)

$c = 2$

$\dfrac{c}{a} = \dfrac{2}{5}$

$\dfrac{2}{a} = \dfrac{2}{5}$

$a = 5$

$c^2 = a^2 - b^2$

$4 = 25 - b^2$

$b^2 = 21$

$\dfrac{(x-1)^2}{25} + \dfrac{(y-3)^2}{21} = 1$

51. $2a = 24$

$a = 12$

$\dfrac{c}{a} = \dfrac{2}{3}$

$\dfrac{c}{12} = \dfrac{2}{3}$

$c = 8$

$c^2 = a^2 - b^2$

$64 = 144 - b^2$

$b^2 = 80$

$\dfrac{x^2}{80} + \dfrac{y^2}{144} = 1$

53. Aphelion $= 2a - $ perihelion

$934.34 = 2a - 835.14$

$a = 884.74$ million miles

Aphelion $= a + c = 934.34$

$884.74 + c = 934.34$

$c = 49.6$ million miles

$b = \sqrt{a^2 - c^2}$

$= \sqrt{884.74^2 - 49.6^2}$

≈ 883.35 million miles

An equation of the orbit of Saturn is

$\dfrac{x^2}{884.74^2} + \dfrac{y^2}{883.35^2} = 1$

55. $a = $ semimajor axis $= 50$ feet

$b = $ height $= 30$ feet

$c^2 = a^2 - b^2$

$c^2 = 50^2 - 30^2$

$c = \sqrt{1600} = 40$

The foci are located 40 feet to the right and to the left of center.

57. $2a = 36$ $2b = 9$

$a = 18$ $b = \dfrac{9}{2}$

$c^2 = a^2 - b^2$

$c^2 = 18^2 - \left(\dfrac{9}{2}\right)^2$

$c^2 = 324 - \dfrac{81}{4}$

$c^2 = \dfrac{1215}{4}$

$c = \dfrac{9\sqrt{15}}{2}$

Since one focus is at (0, 0), the center of the ellipse is at $(9\sqrt{15}/2, 0) \approx (17.43, 0)$. The equation of the path of Halley's Comet in astronomical units is

$\dfrac{\left(x - 9\sqrt{15}/2\right)^2}{324} + \dfrac{y^2}{81/4} = 1$

59. $2a = 16 \Rightarrow a = 8$

$2b = 10 \Rightarrow b = 5$

$p = \pi\sqrt{2\left(a^2 + b^2\right)}$

$= \pi\sqrt{2\left(8^2 + 5^2\right)}$

$= \pi\sqrt{2(64 + 25)}$

$= \pi\sqrt{2(89)}$

$= \pi\sqrt{178}$ inches

1 mile $= 5280$ feet $= 5280(12)$ inches $= 63360$ inches

$\dfrac{63360 \text{ inches}}{\pi\sqrt{178} \text{ inches per revolution}} \approx 1512$ revolutions

61. a. $2a = 615 \Rightarrow a = 307.5$

$2b = 510 \Rightarrow b = 255$

$\dfrac{x^2}{307.5^2} + \dfrac{y^2}{255^2} = 1$

b. $A = \pi ab$

$= \pi(307.5)(255)$

$= 78412.5\pi$

$\approx 246,300$ square feet

63. $9y^2 + 36y + 16x^2 - 108 = 0$

$$y = \frac{-36 \pm \sqrt{36^2 - 4(9)(16x^2 - 108)}}{2(9)}$$

$$= \frac{-36 \pm \sqrt{1296 - 36(16x^2 - 108)}}{18}$$

$$= \frac{-36 \pm \sqrt{1296 - 576x^2 + 3888}}{18}$$

$$= \frac{-36 \pm \sqrt{-576x^2 + 5184}}{18}$$

$$= \frac{-36 \pm \sqrt{576(-x^2 + 9)}}{18}$$

$$= \frac{-36 \pm 24\sqrt{(-x^2 + 9)}}{18}$$

$$= \frac{-6 \pm 4\sqrt{(-x^2 + 9)}}{3}$$

67. $9y^2 + 18y + 4x^2 + 24x + 44 = 0$

$$y = \frac{-18 \pm \sqrt{18^2 - 4(9)(4x^2 + 24x + 44)}}{2(9)}$$

$$= \frac{-18 \pm \sqrt{324 - 36(4x^2 + 24x + 44)}}{18}$$

$$= \frac{-18 \pm \sqrt{324 - 144x^2 - 864x - 1584}}{18}$$

$$= \frac{-18 \pm \sqrt{-144x^2 - 864x - 1260}}{18}$$

$$= \frac{-18 \pm \sqrt{36(-4x^2 - 24x - 35)}}{18}$$

$$= \frac{-18 \pm 6\sqrt{-4x^2 - 24x - 35}}{18}$$

$$= \frac{-3 \pm \sqrt{-4x^2 - 24x - 35}}{3}$$

65. $9y^2 - 54y + 16x^2 - 64x + 1 = 0$

$$y = \frac{-(-54) \pm \sqrt{(-54)^2 - 4(9)(16x^2 - 64x + 1)}}{2(9)}$$

$$= \frac{54 \pm \sqrt{2916 - 36(16x^2 - 64x + 1)}}{18}$$

$$= \frac{54 \pm \sqrt{2916 - 576x^2 + 2304x - 36}}{18}$$

$$= \frac{54 \pm \sqrt{-576x^2 + 2304x + 2880}}{18}$$

$$= \frac{54 \pm \sqrt{576(-x^2 + 4x + 5)}}{18}$$

$$= \frac{54 \pm 24\sqrt{-x^2 + 4x + 5}}{18}$$

$$= \frac{9 \pm 4\sqrt{-x^2 + 4x + 5}}{3}$$

69. The sum of the distances between the two foci and a point on the ellipse is $2a$.

$$2a = \sqrt{\left(\frac{9}{2} - 0\right)^2 + (3 - 3)^2} + \sqrt{\left(\frac{9}{2} - 0\right)^2 + (3 + 3)^2}$$

$$= \sqrt{\left(\frac{9}{2}\right)^2} + \sqrt{\frac{225}{4}}$$

$$= \frac{9}{2} + \frac{15}{2}$$

$$= 12$$

$$a = 6$$

$$c = 3$$

$$c^2 = a^2 - b^2$$

$$9 = 36 - b^2$$

$$b^2 = 27$$

$$\frac{x^2}{36} + \frac{y^2}{27} = 1$$

71. The sum of the distances between the two foci and a point on the ellipse is $2a$.

$$2a = \sqrt{(5-2)^2 + (3+1)^2} + \sqrt{(5-2)^2 + (3-3)^2}$$
$$= \sqrt{25} + \sqrt{3^2}$$
$$= 5 + 3$$
$$= 8$$

$$a = 4$$
$$c = 2$$

$$c^2 = a^2 - b^2$$
$$4 = 16 - b^2$$
$$b^2 = 12$$

$$\frac{(x-1)^2}{16} + \frac{(y-2)^2}{12} = 1$$

73. center: $(1, -1)$

$$c^2 = a^2 - b^2$$
$$c^2 = 16 - 9$$
$$c^2 = 7$$
$$c = \sqrt{7}$$

The latus rectum is on the graph of $y = -1 + \sqrt{7}$, or $y = -1 - \sqrt{7}$

$$\frac{(x-1)^2}{9} + \frac{(y+1)^2}{16} = 1$$

$$\frac{(x-1)^2}{9} + \frac{\left(-1+\sqrt{7}+1\right)^2}{16} = 1 \quad \text{or} \quad \frac{(x-1)^2}{9} + \frac{\left(-1-\sqrt{7}+1\right)^2}{16} = 1$$

$$\frac{(x-1)^2}{9} + \frac{7}{16} = 1$$

$$\frac{(x-1)^2}{9} = \frac{9}{16}$$

$$16(x-1)^2 = 81$$

$$(x-1)^2 = \frac{81}{16}$$

$$x - 1 = \pm\sqrt{\frac{81}{16}}$$

$$x - 1 = \pm\frac{9}{4}$$

$$x = \frac{13}{4} \text{ and } -\frac{5}{4}$$

The x-coordinates of the endpoints of the latus rectum are $\frac{13}{4}$ and $-\frac{5}{4}$.

$$\left| \frac{13}{4} - \left(-\frac{5}{4}\right) \right| = \frac{9}{2}$$

The length of the latus rectum is $\frac{9}{2}$.

75. Let us transform the general equation of an ellipse into an $x'y'$ - coordinate system where the center is at the origin by replacing $(x-h)$ by x' and $(y-k)$ by y'.

We have $\dfrac{x'^2}{a^2}+\dfrac{y'^2}{b^2}=1.$

Letting $x'=c$ and solving for y' yields

$$\dfrac{(c)^2}{a^2}+\dfrac{y'^2}{b^2}=1$$

$$b^2c^2+a^2y'^2=a^2b^2$$

$$a^2y'^2=a^2b^2-b^2c^2$$

$$a^2y'^2=b^2(a^2-c^2)$$

But since $c^2=a^2-b^2$, $b^2=a^2-c^2$, we can substitute to obtain

$$a^2y'^2=b^2(b^2)$$

$$y'^2=\dfrac{b^4}{a^2}$$

$$y'=\pm\sqrt{\dfrac{b^4}{a^2}}=\pm\dfrac{b^2}{a}$$

The endpoints of the latus rectum, then, are $\left(c,\ \dfrac{b^2}{a}\right)$ and $\left(c,\ -\dfrac{b^2}{a}\right)$.

The distance between these points is $\dfrac{2b^2}{a}$.

77. $a^2=9$

$b^2=4$

$c^2=a^2-b^2=9-4=5$

$c=\sqrt{5}$

$x=\pm\dfrac{a^2}{c}=\pm\dfrac{9}{\sqrt{5}}=\pm\dfrac{9\sqrt{5}}{5}$

The directrices are $x=\dfrac{9\sqrt{5}}{5}$ and $x=-\dfrac{9\sqrt{5}}{5}$.

79. The eccentricity is $e=\dfrac{c}{a}=\dfrac{2}{\sqrt{12}}=\dfrac{1}{\sqrt{3}}$. Let $P(x,y)$ be a point on the ellipse $\dfrac{x^2}{12}+\dfrac{y^2}{8}=1$ and $F(2,0)$ be a focus.

Then $d(P,F)=\sqrt{(x-2)^2+y^2}$. The distance from the line $x=6$ to $P(x,y)$ is $|x-6|$. Thus, solving the equation of the ellipse for y^2, we have

$$\dfrac{\sqrt{(x-2)^2+y^2}}{|x-6|}=\dfrac{\sqrt{(x-2)^2+8(1-x^2/12)}}{|x-6|}$$

$$=\dfrac{\sqrt{x^2-4x+4+8-\frac{2}{3}x^2}}{|x-6|}$$

$$=\dfrac{\sqrt{\frac{1}{3}x^2-4x+12}}{|x-6|}=\dfrac{\sqrt{(x^2-12x+36)/3}}{|x-6|}$$

$$=\dfrac{\sqrt{(x-6)^2/3}}{|x-6|}=\dfrac{|x-6|/\sqrt{3}}{|x-6|}=\dfrac{1}{\sqrt{3}}=e\ \text{(eccentricity)}$$

SECTION 6.3, Page 338

1. $\dfrac{x^2}{16} - \dfrac{y^2}{25} = 1$

Center $(0, 0)$

Vertices $(\pm 4, 0)$

Foci $\left(\pm\sqrt{41}, 0\right)$

Asymptotes $y = \pm\dfrac{5}{4}x$

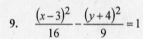

3. $\dfrac{y^2}{4} - \dfrac{x^2}{25} = 1$

Center $(0, 0)$

Vertices $(0, \pm 2)$

Foci $\left(0, \pm\sqrt{29}\right)$

Asymptotes $y = \pm\dfrac{2}{5}x$

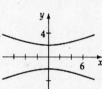

5. $\dfrac{x^2}{7} - \dfrac{y^2}{9} = 1$

Center $(0, 0)$

Vertices $\left(\pm\sqrt{7}, 0\right)$

Foci $(\pm 4, 0)$

Asymptotes $y = \pm\dfrac{3\sqrt{7}}{7}x$

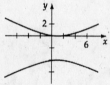

7. $\dfrac{4x^2}{9} - \dfrac{y^2}{16} = 1$

Center $(0, 0)$

Vertices $\left(\pm\dfrac{3}{2}, 0\right)$

Foci $\left(\pm\dfrac{\sqrt{73}}{2}, 0\right)$

Asymptotes $y = \pm\dfrac{8}{3}x$

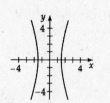

9. $\dfrac{(x-3)^2}{16} - \dfrac{(y+4)^2}{9} = 1$

Center $(3, -4)$

Vertices $(3 \pm 4, -4) = (7, -4), (-1, -4)$

Foci $(3 \pm 5, -4) = (8, -4), (-2, -4)$

Asymptotes $y + 4 = \pm\dfrac{3}{4}(x - 3)$

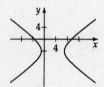

11. $\dfrac{(y+2)^2}{4} - \dfrac{(x-1)^2}{16} = 1$

Center $(1, -2)$

Vertices $(1, -2 \pm 2) = (1, 0), (1, -4)$

Foci $\left(1, -2 \pm 2\sqrt{5}\right) = \left(1, -2 + 2\sqrt{5}\right), \left(1, -2 - 2\sqrt{5}\right)$

Asymptotes $y + 2 = \pm\dfrac{1}{2}(x - 1)$

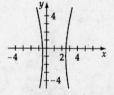

13. $\dfrac{(x+2)^2}{9} - \dfrac{y^2}{25} = 1$

Center $(-2, 0)$

Vertices $(-2 \pm 3, 0) = (1, 0), (-5, 0)$

Foci $\left(-2 \pm \sqrt{34}, 0\right)$

Asymptotes $y = \pm\dfrac{5}{3}(x + 2)$

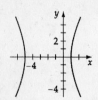

15. $\dfrac{9(x-1)^2}{16} - \dfrac{(y+1)^2}{9} = 1$

$\dfrac{(x-1)^2}{16/9} - \dfrac{(y+1)^2}{9}$

Center $(1, -1)$

Vertices $\left(1 \pm \dfrac{4}{3}, -1\right) = \left(\dfrac{7}{3}, -1\right), \left(-\dfrac{1}{3}, -1\right)$

Foci $\left(1 \pm \dfrac{\sqrt{97}}{3}, -1\right)$

Asymptotes $(y + 1) = \pm\dfrac{9}{4}(x - 1)$

17. $x^2 - y^2 = 9$

$$\frac{x^2}{9} - \frac{y^2}{9} = 1$$

Center $(0, 0)$

Vertices $(\pm 3, 0)$

Foci $\left(\pm 3\sqrt{2}, 0\right)$

Asymptotes $y = \pm x$

19. $16y^2 - 9x^2 = 144$

$$\frac{y^2}{9} - \frac{x^2}{16} = 1$$

Center $(0, 0)$

Vertices $(0, \pm 3)$

Foci (0 ± 5)

Asymptotes $y = \pm \frac{3}{4}x$

21. $9y^2 - 36x^2 = 4$

$$\frac{y^2}{4/9} - \frac{x^2}{1/9} = 1$$

Center $(0, 0)$

Vertices $\left(0, \pm \frac{2}{3}\right)$

Foci $\left(0, \pm \frac{\sqrt{5}}{3}\right)$

Asymptotes $y = \pm 2x$

23. $x^2 - y^2 - 6x + 8y = 3$

$$(x^2 - 6x) - (y^2 - 8y) = 3$$

$$(x^2 - 6x + 9) - (y^2 - 8y + 16) = 3 + 9 - 16$$

$$(x - 3)^2 - (y - 4)^2 = -4$$

$$\frac{(y - 4)^2}{4} - \frac{(x - 3)^2}{4} = 1$$

Center $(3, 4)$

Vertices $(3, 4 \pm 2) = (3, 6), (3, 2)$

Foci $(3, 4 \pm 2\sqrt{2}) = (3, 4 + 2\sqrt{2}), (3, 4 - 2\sqrt{2})$

Asymptotes $y - 4 = \pm(x - 3)$

25. $9x^2 - 4y^2 + 36x - 8y + 68 = 0$

$$9x^2 + 36x - 4y^2 - 8y = -68$$

$$9(x^2 + 4x) - 4(y^2 + 2y) = -68$$

$$9(x^2 + 4x + 4) - 4(y^2 + 2y + 1) = -68 + 36 - 4$$

$$9(x + 2)^2 - 4(y + 1)^2 = -36$$

$$\frac{(y + 1)^2}{9} - \frac{(x + 2)^2}{4} = 1$$

Center $(-2, -1)$

Vertices $(-2, -1 \pm 3) = (-2, 2), (-2, -4)$

Foci $(-2, -1 \pm \sqrt{13}) = (-2, -1 + \sqrt{13}), (-2, -1 - \sqrt{13})$

Asymptotes $y + 1 = \pm \frac{3}{2}(x + 2)$

27. $y = \dfrac{-6 \pm \sqrt{6^2 - 4(-1)(4x^2 + 32x + 39)}}{2(-1)}$

$= \dfrac{-6 \pm \sqrt{36 + 4(4x^2 + 32x + 39)}}{-2}$

$= \dfrac{-6 \pm \sqrt{16x^2 + 128x + 192}}{-2}$

$= \dfrac{-6 \pm \sqrt{16(x^2 + 8x + 12)}}{-2}$

$= \dfrac{-6 \pm 4\sqrt{x^2 + 8x + 12}}{-2}$

$= 3 \pm 2\sqrt{x^2 + 8x + 12}$

29. $y = \dfrac{64 \pm \sqrt{(-64)^2 - 4(-16)(9x^2 - 36x + 116)}}{2(-16)}$

$= \dfrac{64 \pm \sqrt{4096 + 64(9x^2 - 36x + 116)}}{-32}$

$= \dfrac{64 \pm \sqrt{64(9x^2 - 36x + 116 + 64)}}{-32}$

$= \dfrac{64 \pm 8\sqrt{(9x^2 - 36x + 180)}}{-32}$

$= \dfrac{64 \pm 8\sqrt{9(x^2 - 4x + 20)}}{-32}$

$= \dfrac{64 \pm 24\sqrt{x^2 - 4x + 20}}{-32}$

$= \dfrac{-8 \pm 3\sqrt{x^2 - 4x + 20}}{4}$

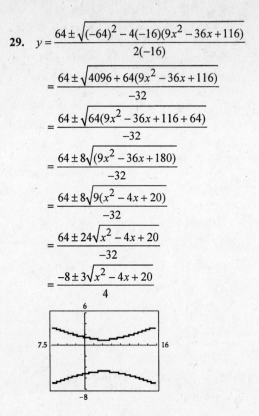

31. $y = \dfrac{18 \pm \sqrt{(-18)^2 - 4(-9)(4x^2 + 8x - 6)}}{2(-9)}$

$= \dfrac{18 \pm \sqrt{324 + 36(4x^2 + 8x - 6)}}{-18}$

$= \dfrac{18 \pm \sqrt{36(4x^2 + 8x - 6 + 9)}}{-18}$

$= \dfrac{18 \pm 6\sqrt{(4x^2 + 8x + 3)}}{-18}$

$= \dfrac{-3 \pm \sqrt{4x^2 + 8x + 3}}{3}$

33. vertices $(3, 0)$ and $(-3, 0)$, foci $(4, 0)$ and $(-4, 0)$

Traverse axis is on x-axis. For a standard hyperbola, the vertices are at $(h + a, k)$ and $(h - a, k)$, $h + a = 3$, $h - a = -3$, and $k = 0$..

If $h + a = 3$ and $h - a = -3$, then $h = 0$ and $a = 3$.

The foci are located at $(4, 0)$ and $(-4, 0)$. Thus, $h = 0$ and $c = 4$.

Since $c^2 = a^2 + b^2$, $b^2 = c^2 - a^2$

$b^2 = (4)^2 - (3)^2 = 16 - 9 = 7$

$\dfrac{(x - h)^2}{a^2} - \dfrac{(y - k)^2}{b^2} = 1$

$\dfrac{(x - 0)^2}{(3)^2} - \dfrac{(y - 0)^2}{7} = 1$

$\dfrac{x^2}{9} - \dfrac{y^2}{7} = 1$

35. foci $(0, 5)$ and $(0, -5)$, asymptotes $y = 2x$ and $y = -2x$

Transverse axis is on y-axis. Since foci are at $(h, k+c)$ and $(h, k-c)$, $k+c = 5$, $k-c = -5$, and $h = 0$.

Therefore, $k = 0$ and $c = 5$.

Since one of the asymptotes is $y = \dfrac{a}{b}x$, $\dfrac{a}{b} = 2$ and $a = 2b$.

$a^2 + b^2 = c^2$; then substituting $a = 2b$ and $c = 5$ yields $(2b)^2 + b^2 = (5)^2$, or $5b^2 = 25$.

Therefore, $b^2 = 5$ and $b = \sqrt{5}$.
Since $a = 2b$, $a = 2(\sqrt{5}) = 2\sqrt{5}$.

$$\frac{(y-k)^2}{a^2} - \frac{(x-h)^2}{b^2} = 1$$
$$\frac{y^2}{(2\sqrt{5})^2} - \frac{x^2}{5} = 1$$
$$\frac{y^2}{20} - \frac{x^2}{5} = 1$$

39. vertices $(0, 4)$ and $(0, -4)$, asymptotes $y = \frac{1}{2}x$ and $y = -\frac{1}{2}x$.

The length of the transverse axis, or the distance between the vertices, is equal to $2a$.

$2a = 4 - (-4) = 8$, or $a = 4$

The center of the hyperbola, or the midpoint of the line segment joining the vertices, is

$\left(\dfrac{0+0}{2}, \dfrac{4+(-4)}{2}\right)$, or $(0, 0)$

Since both vertices lie on the y-axis, the transverse axis must lie on the y-axis. Therefore, the asymptotes are given by $y = \frac{a}{b}x$ and $y = -\frac{a}{b}x$. One asymptote is $y = \frac{1}{2}x$.

Thus $\frac{a}{b} = \frac{1}{2}$ or $b = 2a$.

Since $b = 2a$ and $a = 4$, $b = 2(4) = 8$.

Thus, the equation is
$$\frac{y^2}{4^2} - \frac{x^2}{8^2} = 1 \text{ or } \frac{y^2}{16} - \frac{x^2}{64} = 1$$

37. vertices $(0, 3)$ and $(0, -3)$, point $(2, 4)$

The distance between the two vertices is the length of the transverse axis, which is $2a$.

$$2a = |3 - (-3)| = 6 \text{ or } a = 3.$$

Since the midpoint of the transverse axis is the center of the hyperbola, the center is given by

$\left(\dfrac{0+0}{2}, \dfrac{3+(-3)}{2}\right)$, or $(0, 0)$

Since both vertices lie on the y-axis, the transverse axis must be on the y-axis.

Taking the standard form of the hyperbola, we have
$$\frac{y^2}{a^2} - \frac{x^2}{b^2} = 1$$

Substituting the point $(2, 4)$ for x and y, and 3 for a, we have
$$\frac{16}{9} - \frac{4}{b^2} = 1$$

Solving for b^2 yields $b^2 = \dfrac{36}{7}$.

Therefore, the equation is
$$\frac{y^2}{9} - \frac{x^2}{36/7} = 1$$

41. vertices $(6, 3)$ and $(2, 3)$, foci $(7, 3)$ and $(1, 3)$

Length of transverse axis = distance between vertices
$2a = |6 - 2|$
$a = 2$

The center of the hyperbola (h, k) is the midpoint of the line segment joining the vertices, or the point $\left(\dfrac{6+2}{2}, \dfrac{3+3}{2}\right)$.

Thus, $h = \dfrac{6+2}{2}$, or 4, and $k = \dfrac{3+3}{3}$, or 3.

Since both vertices lie on the horizontal line $y = 3$, the transverse axis is parallel to the x-axis. The location of the foci is given by $(h + c, k)$ and $(h - c, k)$, or specifically $(7, 3)$ and $(1, 3)$. Thus $h + c = 7$, $h - c = 1$, and $k = 3$. Solving for h and c simultaneously yields $h = 4$ and $c = 3$.

Since $c^2 = a^2 + b^2$, $b^2 = c^2 - a^2$.

Substituting, we have $b^2 = 3^2 - 2^2 = 9 - 4 = 5$.

Substituting $a = 2$, $b^2 = 5$, $h = 4$, and $k = 3$ in the standard equation
$$\frac{(x-h)^2}{a^2} - \frac{(y-k)^2}{b^2} = 1$$

yields $\dfrac{(x-4)^2}{4} - \dfrac{(y-3)^2}{5} = 1$.

43. foci $(1, -2)$ and $(7, -2)$, slope of an asymptote $= \frac{5}{4}$

Both foci lie on the horizontal line $y = -2$; therefore, the transverse axis is parallel to the x-axis.

The foci are given by $(h + c, k)$ and $(h - c, k)$.

Thus, $h - c = 1, h + c = 7$, and $k = -2$. Solving simultaneously for h and c yields $h = 4$ and $c = 3$.

Since $y - k = \frac{b}{a}(x - h)$ is the equation for an asymptote, and the slope of an asymptote is given as $\frac{5}{4}$, $\frac{b}{a} = \frac{5}{4}$, $b = \frac{5a}{4}$, and $b^2 = \frac{25a^2}{16}$.

Because $a^2 + b^2 = c^2$, substituting $c = 3$ and $b^2 = \frac{25a^2}{16}$ yields $a^2 = \frac{144}{41}$.

Therefore, $b^2 = \frac{3600}{656} = \frac{225}{41}$.

Substituting in the standard equation for a hyperbola yields

$$\frac{(x-4)^2}{144/41} - \frac{(y+2)^2}{225/41} = 1$$

45. Because the transverse axis is parallel to the y-axis and the center is $(7, 2)$, the equation of the hyperbola is

$$\frac{(y-2)^2}{a^2} - \frac{(x-7)^2}{b^2} = 1$$

Because $(9, 4)$ is a point on the hyperbola,

$$\frac{(4-2)^2}{a^2} - \frac{(9-7)^2}{b^2} = 1$$

The slope of the asymptote is $\frac{1}{2}$. Therefore $\frac{1}{2} = \frac{a}{b}$ or $b = 2a$.

Substituting, we have

$$\frac{4}{a^2} - \frac{4}{4a^2} = 1$$

$$\frac{4}{a^2} - \frac{1}{a^2} = 1, \text{ or } a^2 = 3$$

Since b $= 2a$, $b^2 = 4a^2$, or $b^2 = 12$. The equation is $\dfrac{(y-2)^2}{3} - \dfrac{(x-7)^2}{12} = 1$.

47. vertices $(1, 6)$ and $(1, 8)$, eccentricity $= 2$

Length of transverse axis = distance between vertices

$$2a = |6 - 8| = 2$$
$$a = 1 \text{ and } a^2 = 1$$

Center (midpoint of transverse axis) is $\left(\frac{1+1}{2}, \frac{6+8}{2}\right)$, or $(1, 7)$.

Therefore, $h = 1$ and $k = 7$.

Since both vertices lie on the vertical line $x = 1$, the transverse axis is parallel to the y-axis.

Since $e = \frac{c}{a}$, $c = ae = (1)(2) = 2$.

Because $b^2 = c^2 - a^2$, $b^2 = (2)^2 - (1)^2 = 4 - 1 = 3$.

Substituting h, k, a^2, and b^2 into the standard equation yields

$$\frac{(y-7)^2}{1} - \frac{(x-1)^2}{3} = 1$$

49. foci $(4, 0)$ and $(-4, 0)$, eccentricity $= 2$

Center (midpoint of line segment joining foci) is $\left(\dfrac{4+(-4)}{2}, \dfrac{0+0}{2}\right)$, or $(0, 0)$

Thus, $h = 0$ and $k = 0$.

Since both foci lie on the horizontal line $y = 0$, the transverse axis is parallel to the x-axis. The locations of the foci are given by $(h + c, k)$ and $(h - c, k)$, or specifically $(4, 0)$ and $(-4, 0)$

Since $h = 0$, $c = 4$.

Because $e = \dfrac{c}{a}$, $a = \dfrac{c}{e} = \dfrac{4}{2} = 2$ and $a^2 = 4$.

Because $b^2 = c^2 - a^2$, $b^2 = 4^2 - 2^2 = 16 - 4 = 12$.

Substituting h, k, a^2 and b^2 into the standard formula for a hyperbola yields

$\dfrac{x^2}{4} - \dfrac{y^2}{12} = 1$

51. conjugate axis length $= 4$, center $(4, 1)$, eccentricity $= \dfrac{4}{3}$

$2b = $ conjugate axis length $= 4$

$b = 2$ and $b^2 = 4$

Since

$e = \dfrac{c}{a} = \dfrac{4}{3}$, $c = \dfrac{4a}{3}$ and $c^2 = \dfrac{16a^2}{9}$. Since $a^2 + b^2 = c^2$, substituting $b^2 = 4$ and $c^2 = \dfrac{16a^2}{9}$ and solving for a^2 yields $a^2 = \dfrac{36}{7}$.

Substituting into the two standard equations of a hyperbola yields

$\dfrac{(x-4)^2}{36/7} - \dfrac{(y-1)^2}{4} = 1$ and $\dfrac{(y-1)^2}{36/7} - \dfrac{(x-4)^2}{4} = 1$

53. a. Because the transmitters are 250 miles apart,

$2c = 250$ and $c = 125$.

$2a = $ rate $\times$ time

$2a = 0.186 \times 500 = 93$

Thus, $a = 46.5$ miles.

$b = \sqrt{c^2 - a^2} = \sqrt{125^2 - 46.5^2} = \sqrt{13,462.75}$ miles

The ship is located on the hyperbola given by

$\dfrac{x^2}{2,162.25} - \dfrac{y^2}{13,462.75} = 1$

b. $x = 100$

$\dfrac{10,000}{2,162.25} - \dfrac{y^2}{13,462.75} = 1$

$\dfrac{-y^2}{13,462.75} \approx -3.6248121$

$y^2 \approx 48,799.939$

$y \approx 221$

The ship is 221 miles from the coastline.

55. $4x^2 + 9y^2 - 16x - 36y + 16 = 0$

$4(x^2 - 4x) + 9(y^2 - 4y) = -16$

$4(x^2 - 4x + 4) + 9(y^2 - 4y + 4) = -16 + 16 + 36$

$4(x - 2)^2 + 9(y - 2)^2 = 36$

$\dfrac{(x-2)^2}{9} + \dfrac{(y-2)^2}{4} = 1$

ellipse

center $(2, 2)$

vertices $(2 \pm 3, 2) = (5, 2), (-1, 2)$

foci $(2 \pm \sqrt{5}, 2) = (2 + \sqrt{5}, 2), (2 - \sqrt{5}, 2)$

57. $5x - 4y^2 + 24y - 11 = 0$

$$-4(y^2 - 6y) = -5x + 11$$

$$-4(y^2 - 6y + 9) = -5x + 11 - 36$$

$$-4(y - 3)^2 = -5x - 25$$

$$-4(y - 3)^2 = -5(x + 5)$$

$$(y - 3)^2 = \frac{5}{4}(x + 5)$$

parabola

vertex $(-5, 3)$

focus $\left(-5 + \frac{5}{16}, 3\right) = \left(-\frac{75}{16}, 3\right)$

directrix $x = -5 - \frac{5}{16}$, or $x = \frac{-85}{16}$

59. $x^2 + 2y - 8x = 0$

$$x^2 - 8x = -2y$$

$$x^2 - 8x + 16 = -2y + 16$$

$$(x - 4)^2 = -2(y - 8)$$

parabola

vertex $(4, 8)$

foci $\left(4, 8 - \frac{1}{2}\right) = \left(4, \frac{15}{2}\right)$

directrix $y = 8 + \frac{1}{2}$, or $y = \frac{17}{2}$

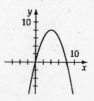

61. $25x^2 + 9y^2 - 50x - 72y - 56 = 0$

$$25(x^2 - 2x) + 9(y^2 - 8y) = 56$$

$$25(x^2 - 2x + 1) + 9(y^2 - 8y + 16) = 56 + 25 + 144$$

$$25(x - 1)^2 + 9(y - 4)^2 = 225$$

$$\frac{(x - 1)^2}{9} + \frac{(y - 4)^2}{25} = 1$$

ellipse

center $(1, 4)$

vertices $(1, 4 \pm 5) = (1, 9), (1, -1)$

foci $(1, 4 \pm 4) = (1, 8), (1, 0)$

63. foci $F_1(2, 0)$, $F_2(-2, 0)$ passing through $P_1(2, 3)$

$$d(P_1, F_2) - d(P_1, F_1) = \sqrt{(2 + 2)^2 + 3^2} - \sqrt{(2 - 2)^2 + 3^2} = 5 - 3 =$$

Let $P(x, y)$ be any point on the hyperbola. Since the difference between $F_1 P$ and $F_2 P$ is the same as the difference between $F_1 P_1$ and $F_2 P_1$, we have

$$2 = \sqrt{(x - 2)^2 + y^2} - \sqrt{(x + 2)^2 + y^2}$$

$$\sqrt{(x - 2)^2 + y^2} = 2 + \sqrt{(x + 2)^2 + y^2}$$

$$x^2 - 4x + 4 + y^2 = 4 + 4\sqrt{(x + 2)^2 + y^2} + x^2 + 4x + 4 + y^2$$

$$-8x - 4 = 4\sqrt{(x + 2)^2 + y^2}$$

$$-2x - 1 = \sqrt{(x + 2)^2 + y^2}$$

$$4x^2 + 4x + 1 = x^2 + 4x + 4 + y^2$$

$$3x^2 - y^2 = 3$$

$$\frac{x^2}{1} - \frac{y^2}{3} = 1$$

65. foci $(0, 4)$ and $(0, -4)$, point $\left(\dfrac{7}{3}, 4\right)$

Difference in distances from (x, y) to foci = difference of distances from $\left(\dfrac{7}{3}, 4\right)$ to foci

$$\sqrt{(x-0)^2 + (y-4)^2} - \sqrt{(x-0)^2 + (y+4)^2} = \sqrt{\left(\dfrac{7}{3} - 0\right)^2 + (4-4)^2} - \sqrt{\left(\dfrac{7}{3} - 0\right)^2 + (4+4)^2}$$

$$\sqrt{x^2 + y^2 - 8y + 16} - \sqrt{x^2 + y^2 + 8y + 16} = \dfrac{7}{3} - \dfrac{25}{3} = -6$$

$$\sqrt{x^2 + y^2 - 8y + 16} = \sqrt{x^2 + y^2 + 8y + 16} - 6$$

$$x^2 + y^2 - 8y + 16 = x^2 + y^2 + 8y + 16 - 12\sqrt{x^2 + y^2 + 8y + 16} + 36$$

$$-16y - 36 = -12\sqrt{x^2 + y^2 + 8y + 16}$$

$$4y + 9 = 3\sqrt{x^2 + y^2 + 8y + 16}$$

$$16y^2 + 72y + 81 = 9x^2 + 9y^2 + 72y + 144$$

$$7y^2 - 9x^2 = 63$$

$$\dfrac{y^2}{9} - \dfrac{x^2}{7} = 1$$

67. $\dfrac{x^2}{16} - \dfrac{y^2}{25} = 1$

$$a^2 = 16$$

$$b^2 = 25$$

$$c^2 = a^2 + b^2 = 16 + 25 = 41$$

$$c = \sqrt{41}$$

$$\dfrac{a^2}{c} = \dfrac{16}{\sqrt{41}} = \dfrac{16\sqrt{41}}{41}.$$

Thus, the directrices are $x = \pm\dfrac{16\sqrt{41}}{41}$.

69. $\frac{x^2}{9} - \frac{y^2}{16} = 1$, focus (5, 0), directrix $x = \frac{9}{5}$

Solving for y^2 gives us

$16x^2 - 9y^2 = 144$

$9y^2 = 16x^2 - 144$

$y^2 = \frac{16}{9}x^2 - 16$

Let $k = \dfrac{\text{distance from } P(x, y) \text{ to focus (5, 0)}}{\text{distance from } P(x, y) \text{ to directrix } (x = 9/5)} = \dfrac{\sqrt{(x-5)^2 + (y-0)^2}}{|x - 9/5|} = \dfrac{\sqrt{x^2 - 10x + 25 + y^2}}{|x - 9/5|}$

But since $P(x, y)$ lies on the curve, $y^2 = \frac{16}{9}x^2 - 16$.

Substituting gives us

$k = \dfrac{\sqrt{x^2 - 10x + 25 + 16x^2/9 - 16}}{|x - 9/5|} = \dfrac{\sqrt{\frac{25x^2}{9} - \frac{90x}{9} + \frac{81}{9}}}{|x - 9/5|} = \dfrac{\sqrt{25x^2 - 90x + 81}}{3|x - 9/5|}$.

Solving for k^2 yields

$k^2 = \dfrac{25\left(x^2 - \frac{18}{5} + \frac{81}{25}\right)}{9\left(x^2 - \frac{18}{5} + \frac{81}{25}\right)} = \dfrac{25}{9}$

$k = \pm\sqrt{\dfrac{25}{9}} = \pm\dfrac{5}{3}$

But since $k = \dfrac{\text{distance from } P \text{ to focus}}{\text{distance from } P \text{ to directrix}}$, and the ratio of two distances must be positive, $k = \dfrac{5}{3}$.

$a^2 = 9$

$b^2 = 16$

$c^2 = a^2 + b^2 = 9 + 16 = 25$

$c = 5$ and $a = 3$

Since $e = \frac{c}{a} = \frac{5}{3}$ and $k = \frac{5}{3}$, $e = k$.

71.

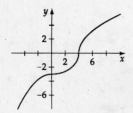

SECTION 6.4, Page 346

1. $xy = 3$

$A = 0, B = 1, C = 0$

$\cot 2\alpha = \dfrac{A - C}{B}$

$\cot 2\alpha = \dfrac{0 - 0}{1}$

$\cot 2\alpha = 0$

$2\alpha = 90°$

$\alpha = 45°$

3. $9x^2 - 24xy + 16y^2 - 320x - 240y = 0$

$A = 9, B = -24, C = 16$

$\cot 2\alpha = \dfrac{A - C}{B}$

$\cot 2\alpha = \dfrac{9 - 16}{-24}$

$\cot 2\alpha = \dfrac{-7}{-24}$

$\cot 2\alpha = \dfrac{7}{24}$

$2\alpha \approx 73.74°$

$\alpha \approx 36.9°$

5. $5x^2 - 6\sqrt{3}xy - 11y^2 + 4x - 3y + 2 = 0$

$A = 5, B = -6\sqrt{3}, C = -11$

$\cot 2a = \dfrac{A-C}{B} = \dfrac{5-(-11)}{-6\sqrt{3}} = \dfrac{5+11}{-6\sqrt{3}} = \dfrac{16}{-6\sqrt{3}} = -\dfrac{8}{3\sqrt{3}} = -\dfrac{8\sqrt{3}}{9}$

$\quad 2\alpha \approx 147°$

$\quad \alpha \approx 73.5°$

7. $\quad xy = 4$

$xy - 4 = 0$

$A = 0, B = 1, C = 0, F = -4$

$\cot 2\alpha = \dfrac{A-C}{B} = \dfrac{0-0}{1} = 0$

$\csc^2 2\alpha = \cot^2 2\alpha + 1$

$\csc^2 2\alpha = 0^2 + 1 = 1$

$\quad \csc 2\alpha = +1 \quad (2\alpha \text{ is in the first quadrant.})$

$\sin 2\alpha = \dfrac{1}{\csc 2\alpha} = \dfrac{1}{1} = 1$

$\sin^2 2\alpha + \cos^2 2\alpha = 1$

$\quad\quad \cos^2 2\alpha = 1 - \sin^2 2\alpha$

$\quad\quad \cos^2 2\alpha = 1 - (1)^2$

$\quad\quad \cos^2 2\alpha = 0$

$\quad\quad \cos 2\alpha = 0$

$\sin \alpha = \sqrt{\dfrac{1-(0)}{2}} = \dfrac{\sqrt{2}}{2} \quad\quad \cos \alpha = \sqrt{\dfrac{1+(0)}{2}} = \dfrac{\sqrt{2}}{2}$

$\quad \alpha = 45°$

$A' = A\cos^2 \alpha + B\cos \alpha \sin \alpha + C\sin^2 \alpha = 0\left(\dfrac{\sqrt{2}}{2}\right)^2 + 1\left(\dfrac{\sqrt{2}}{2}\right)\left(\dfrac{\sqrt{2}}{2}\right) + 0\left(\dfrac{\sqrt{2}}{2}\right)^2 = \dfrac{1}{2}$

$C' = A\sin^2 \alpha - B\cos \alpha \sin \alpha + C\cos^2 \alpha = 0\left(\dfrac{\sqrt{2}}{2}\right)^2 - 1\left(\dfrac{\sqrt{2}}{2}\right)\left(\dfrac{\sqrt{2}}{2}\right) + 0\left(\dfrac{\sqrt{2}}{2}\right)^2 = -\dfrac{1}{2}$

$F' = F = -4$

$\dfrac{1}{2}(x')^2 - \dfrac{1}{2}(y')^2 - 4 = 0 \quad \text{or} \quad \dfrac{(x')^2}{8} - \dfrac{(y')^2}{8} = 1$

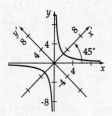

9. $6x^2 - 6xy + 14y^2 - 45 = 0$

$A = 6, B = -6, C = 14, F = -45$

$$\cot 2\alpha = \frac{A-C}{B} = \frac{6-14}{-6} = \frac{4}{3}$$

$$\csc^2 2\alpha = \cot^2 2\alpha + 1$$

$$\csc^2 2\alpha = \left(\frac{4}{3}\right)^2 + 1 = \frac{25}{9}$$

$$\csc 2\alpha = +\sqrt{\frac{25}{9}} = \frac{5}{3} \qquad (2\alpha \text{ is in the first quadrant.})$$

$$\sin 2\alpha = \frac{1}{\csc 2\alpha} = \frac{3}{5}$$

$$\sin^2 \alpha + \cos^2 2\alpha = 1$$

$$\cos^2 2\alpha = 1 - \sin^2 2\alpha$$

$$\cos^2 2\alpha = 1 - \left(\frac{3}{5}\right)^2 = \frac{16}{25}$$

$$\cos 2\alpha = +\sqrt{\frac{16}{25}} = \frac{4}{5} \qquad (2\alpha \text{ is in the first quadrant.})$$

$$\sin \alpha = \sqrt{\frac{1-\left(\frac{4}{5}\right)}{2}} = \frac{\sqrt{10}}{10} \qquad \cos \alpha = \sqrt{\frac{1+\left(\frac{4}{5}\right)}{2}} = \frac{3\sqrt{10}}{10}$$

$$\alpha = 18.4°$$

$$A' = A\cos^2 \alpha + B\cos \alpha \sin \alpha + C\sin^2 \alpha = 6\left(\frac{3\sqrt{10}}{10}\right)^2 - 6\left(\frac{3\sqrt{10}}{10}\right)\left(\frac{\sqrt{10}}{10}\right) + 14\left(\frac{\sqrt{10}}{10}\right)^2 = 5$$

$$C' = A\sin^2 \alpha - B\cos \alpha \sin \alpha + C\cos^2 \alpha = 6\left(\frac{\sqrt{10}}{10}\right)^2 + 6\left(\frac{3\sqrt{10}}{10}\right)\left(\frac{\sqrt{10}}{10}\right) + 14\left(\frac{3\sqrt{10}}{10}\right)^2 = 15$$

$$F' = F = -45$$

$$5(x')^2 + 15(y')^2 - 45 = 0 \text{ or } \frac{(x')^2}{9} + \frac{(y')^2}{3} = 1$$

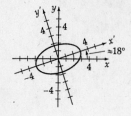

11. $x^2 - 4xy + 2y^2 - 1 = 0$

$A = 1, B = 4, C = -2, F = -1$

$$\cot 2\alpha = \frac{A-C}{B} = \frac{1-(-2)}{4} = \frac{3}{4}$$
$$\csc^2 2\alpha = \cot^2 2\alpha + 1$$
$$\csc^2 2\alpha = \left(\frac{3}{4}\right)^2 + 1 = \frac{25}{16}$$
$$\csc 2\alpha = +\sqrt{\frac{25}{16}} = \frac{5}{4} \qquad (2\alpha \text{ is in the first quadrant.})$$

$$\sin 2\alpha = \frac{1}{\csc 2\alpha} = \frac{4}{5}$$

$$\sin^2\alpha + \cos^2 2\alpha = 1$$
$$\cos^2 2\alpha = 1 - \sin^2\alpha$$
$$\cos^2 2\alpha = 1 - \left(\frac{4}{5}\right)^2 = \frac{9}{25}$$
$$\cos 2\alpha = +\sqrt{\frac{9}{25}} = \frac{3}{5} \qquad \left(2\alpha \text{ is in first quadrant}\right)$$

$$\sin\alpha = \sqrt{\frac{1-\left(\frac{3}{5}\right)}{2}} = \frac{\sqrt{5}}{5} \qquad\qquad \cos\alpha = \sqrt{\frac{1+\left(\frac{3}{5}\right)}{2}} = \frac{2\sqrt{5}}{5}$$

$\alpha \approx 26.6°$

$$A' = A\cos^2\alpha + B\cos\alpha\sin\alpha + C\sin^2\alpha = 1\left(\frac{2\sqrt{5}}{5}\right)^2 + 4\left(\frac{2\sqrt{5}}{5}\right)\left(\frac{\sqrt{5}}{5}\right) - 2\left(\frac{\sqrt{5}}{5}\right)^2 = 2$$

$$C' = A\sin^2\alpha - B\cos\alpha\sin\alpha + C\cos^2\alpha = 1\left(\frac{\sqrt{5}}{5}\right)^2 - 4\left(\frac{2\sqrt{5}}{5}\right)\left(\frac{\sqrt{5}}{5}\right) - 2\left(\frac{2\sqrt{5}}{5}\right)^2 = -3$$

$F' = F = -1$

$2(x')^2 - 3(y')^2 = 1$

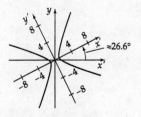

13. $3x^2 + 2\sqrt{3}xy + y^2 + 2x - 2\sqrt{3}y + 16 = 0$

$A = 3, B = 2\sqrt{3}, C = 1, D = 2, E = -2\sqrt{3}, F = 16$

$\cot 2\alpha = \dfrac{A-C}{B} = \dfrac{3-1}{2\sqrt{3}} = \dfrac{\sqrt{3}}{3}$

$\csc^2 2\alpha = \cot^2 2\alpha + 1$

$\csc^2 2\alpha = \left(\dfrac{\sqrt{3}}{3}\right)^2 + 1 = \dfrac{4}{3}$

$\csc 2\alpha = +\sqrt{\dfrac{4}{3}} = \dfrac{2\sqrt{3}}{3}$ (2α is in the first quadrant.)

$\sin 2\alpha = \dfrac{1}{\csc 2\alpha} = \dfrac{\sqrt{3}}{2}$

$\sin^2 \alpha + \cos^2 2\alpha = 1$

$\quad \cos^2 2\alpha = 1 - \sin^2 2\alpha$

$\quad \cos^2 2\alpha = 1 - \left(\dfrac{\sqrt{3}}{2}\right)^2 = \dfrac{1}{4}$

$\quad\quad \cos 2\alpha = +\sqrt{\dfrac{1}{4}} = \dfrac{1}{2}$ (2α is in the first quadrant.)

$\sin \alpha = \sqrt{\dfrac{1-\left(\frac{1}{2}\right)}{2}} = \dfrac{1}{2}$ $\cos \alpha = \sqrt{\dfrac{1+\left(\frac{1}{2}\right)}{2}} = \dfrac{\sqrt{3}}{2}$

$\alpha = 30°$

$A' = A\cos^2 \alpha + B\cos \alpha \sin \alpha + C\sin^2 \alpha = 3\left(\dfrac{\sqrt{3}}{2}\right)^2 + 2\sqrt{3}\left(\dfrac{\sqrt{3}}{2}\right)\left(\dfrac{1}{2}\right) + 1\left(\dfrac{1}{2}\right)^2 = 4$

$C' = A\sin^2 \alpha - B\cos \alpha \sin \alpha + C\cos^2 \alpha = 3\left(\dfrac{1}{2}\right)^2 - 2\sqrt{3}\left(\dfrac{\sqrt{3}}{2}\right)\left(\dfrac{1}{2}\right) + 1\left(\dfrac{\sqrt{3}}{2}\right)^2 = 0$

$D' = D\cos \alpha + E\sin \alpha = 2\left(\dfrac{\sqrt{3}}{2}\right) - 2\sqrt{3}\left(\dfrac{1}{2}\right) = 0$

$E' = -D\sin \alpha + E\cos \alpha = -2\left(\dfrac{1}{2}\right) - 2\sqrt{3}\left(\dfrac{\sqrt{3}}{2}\right) = -4$

$F' = F = 16$

$4(x')^2 - 4y' + 16 = 0$ or $y' = (x')^2 + 4$

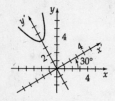

15. $9x^2 + 24xy + 16y^2 - 40x - 30y + 100 = 0$

$A = 9, B = -24, C = 16, D = -40, E = -30\ F = 100$

$$\cot 2\alpha = \frac{A - C}{B} = \frac{9 - 16}{-24} = \frac{7}{24}$$

$$\csc^2 2\alpha = \cot^2 2\alpha + 1$$

$$\csc^2 2\alpha = \left(\frac{7}{24}\right)^2 + 1 = \frac{625}{576}$$

$$\csc 2\alpha = +\sqrt{\frac{625}{576}} = \frac{25}{24} \qquad (2\alpha \text{ is in the first quadrant.})$$

$$\sin 2\alpha = \frac{1}{\csc 2\alpha} = \frac{24}{25}$$

$$\sin^2 \alpha + \cos^2 2\alpha = 1$$

$$\cos^2 2\alpha = 1 - \sin^2 2\alpha$$

$$\cos^2 2\alpha = 1 - \left(\frac{24}{25}\right)^2 = \frac{49}{625}$$

$$\cos 2\alpha = +\sqrt{\frac{49}{625}} = \frac{7}{25} \qquad (2\alpha \text{ is in first quadrant.})$$

$$\sin \alpha = \sqrt{\frac{1 - \left(\frac{7}{25}\right)}{2}} = \frac{3}{5} \qquad \cos \alpha = \sqrt{\frac{1 + \left(\frac{7}{25}\right)}{2}} = \frac{4}{5}$$

$\alpha \approx 36.9°$

$$A' = A\cos^2 \alpha + B\cos \alpha \sin \alpha + C\sin^2 \alpha = 9\left(\frac{4}{5}\right)^2 - 24\left(\frac{4}{5}\right)\left(\frac{3}{5}\right) + 16\left(\frac{3}{5}\right)^2 = 0$$

$$C' = A\sin^2 \alpha - B\cos \alpha \sin \alpha + C\cos^2 \alpha = 9\left(\frac{3}{5}\right)^2 + 24\left(\frac{4}{5}\right)\left(\frac{3}{5}\right) + 16\left(\frac{4}{5}\right)^2 = 25$$

$$D' = D\cos \alpha + E\sin \alpha = -40\left(\frac{4}{5}\right) - 30\left(\frac{3}{5}\right) = -50$$

$$E' = -D\sin \alpha + E\cos \alpha = 40\left(\frac{3}{5}\right) - 30\left(\frac{4}{5}\right) = 0$$

$$F' = F = 100$$

$$25(y')^2 - 50x' + 100 = 0 \text{ or } (y')^2 = 2(x - 2)$$

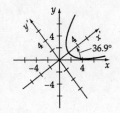

17. $6x^2 + 24xy - y^2 - 12x + 26y + 11 = 0$

$A = 6, B = 24, C = -1, D = -12, E = 26\ F = 11$

$$\cot 2\alpha = \frac{A - C}{B} = \frac{6 - (-1)}{24} = \frac{7}{24}$$

$$\csc^2 2\alpha = \cot^2 2\alpha + 1$$

$$\csc^2 2\alpha = \left(\frac{7}{24}\right)^2 + 1 = \frac{625}{576}$$

$$\csc 2\alpha = +\sqrt{\frac{625}{576}} = \frac{25}{24} \qquad (2\alpha \text{ is in the first quadrant.})$$

$$\sin 2\alpha = \frac{1}{\csc 2\alpha} = \frac{24}{25}$$

$$\sin^2 2\alpha + \cos^2 2\alpha = 1$$

$$\cos^2 2\alpha = 1 - \sin^2 2\alpha$$

$$\cos^2 2\alpha = 1 - \left(\frac{24}{25}\right)^2 = \frac{49}{625}$$

$$\cos 2\alpha = +\sqrt{\frac{49}{625}} = \frac{7}{25} \qquad (2\alpha \text{ is in the first quadrant.})$$

$$\sin \alpha = \sqrt{\frac{1 - \left(\frac{7}{25}\right)}{2}} = \frac{3}{5} \qquad\qquad \cos \alpha = \sqrt{\frac{1 + \left(\frac{7}{25}\right)}{2}} = \frac{4}{5}$$

$$\alpha \approx 36.9°$$

$$A' = A\cos^2 \alpha + B\cos \alpha \sin \alpha + C\sin^2 \alpha = 6\left(\frac{4}{5}\right)^2 + 24\left(\frac{4}{5}\right)\left(\frac{3}{5}\right) - 1\left(\frac{3}{5}\right)^2 = 15$$

$$C' = A\sin^2 \alpha - B\cos \alpha \sin \alpha + C\cos^2 \alpha = 6\left(\frac{3}{5}\right)^2 - 24\left(\frac{4}{5}\right)\left(\frac{3}{5}\right) - 1\left(\frac{4}{5}\right)^2 = -10$$

$$D' = D\cos \alpha + E\sin \alpha = -12\left(\frac{4}{5}\right) + 26\left(\frac{3}{5}\right) = 6$$

$$E' = -D\sin \alpha + E\cos \alpha = 12\left(\frac{3}{5}\right) + 26\left(\frac{4}{5}\right) = 28$$

$$F' = F = 11$$

$$15(x')^2 - 10(y')^2 + 6x' + 28y' + 11 = 0$$

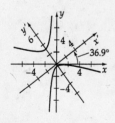

19. $A = 6, B = -1, C = 2, D = 4, E = -12, F = 7$

Graph $y = \dfrac{-(-x-12) \pm \sqrt{(-x-12)^2 - 8(6x^2 + 4x + 7)}}{4}$

The graph will appear disconnected at the endpoints of the minor axes on a graphing utility.

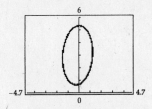

21. $A = 1, B = -6, C = 1, D = -2, E = -5, F = 4$

Graph $y = \dfrac{-(-6x-5) \pm \sqrt{(-6x-5)^2 - 4(x^2 - 2x + 4)}}{2}$

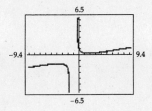

23. $A = 3, B = -6, C = 3, D = 10, E = -8, F = -2$

Graph $y = \dfrac{-(-6x-8) \pm \sqrt{(-6x-8)^2 - 12(3x^2 + 10x - 2)}}{6}$

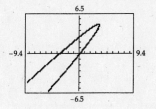

25. $\dfrac{2(x')^2}{1} - \dfrac{3(y')^2}{1} = 1 \quad \sin\alpha = \dfrac{\sqrt{5}}{5} \quad \cos\alpha = \dfrac{2\sqrt{5}}{5}$

$a^2 = \dfrac{1}{2}; \quad a = \dfrac{\sqrt{2}}{2}$

$b^2 = \dfrac{1}{3}; \quad b = \dfrac{\sqrt{3}}{3}$

Asymptotes: $y' = \pm\dfrac{b}{a}x'$ or $y' = \pm\dfrac{\sqrt{6}}{3}x'$

Using the transformation formulas for x' and y' yields

$$y\cos\alpha - x\sin\alpha = \pm\dfrac{\sqrt{6}}{3}(x\cos\alpha + y\sin\alpha)$$

$$\dfrac{2\sqrt{5}}{5}y - \dfrac{\sqrt{5}}{5}x = \pm\dfrac{\sqrt{6}}{3}\left(\dfrac{2\sqrt{5}}{5}x + \dfrac{\sqrt{5}}{5}y\right)$$

$$\dfrac{2\sqrt{5}}{5}y - \dfrac{\sqrt{5}}{5}x = \pm\left(\dfrac{2\sqrt{30}}{15} + \dfrac{\sqrt{30}}{15}y\right)$$

Multiplying both sides of the equation by $15/\sqrt{5}$ yields

$$6y - 3x = \pm\left(2\sqrt{6}x + \sqrt{6}y\right)$$

$6y - 3x = 2\sqrt{6}x + \sqrt{6}y \quad$ and $\quad 6y - 3x = -(2\sqrt{6}x + \sqrt{6}y)$

$6y - \sqrt{6}y = 3x + 2\sqrt{6}x \qquad\qquad 6y + \sqrt{6}y = 3x - 2\sqrt{6}x$

$\qquad y = \dfrac{3 + 2\sqrt{6}}{6 - \sqrt{6}}x \qquad\qquad\qquad y = \dfrac{3 - 2\sqrt{6}}{6 + \sqrt{6}}x$

Rationalizing the denominators, we obtain

$$y = \dfrac{2 + \sqrt{6}}{2}x \qquad\text{and}\qquad y = \dfrac{2 - \sqrt{6}}{2}x$$

27. From Exercise 9, $\dfrac{(x')^2}{9} + \dfrac{(y')^2}{3} = 1$, $\sin\alpha = \dfrac{\sqrt{10}}{10}$, $\cos\alpha = \dfrac{3\sqrt{10}}{10}$, $a^2 = 9, b^2 = 3, c^2 = 9 - 3 = 6$.

Thus $c = \sqrt{6}$.

Foci in $x'y'$-coordinates are $(\pm\sqrt{6}, 0)$. Thus $x' = \pm\sqrt{6}, y' = 0$.

$x = x'\cos\alpha - y'\sin\alpha \qquad\qquad y = y'\cos\alpha + x'\sin\alpha$

$= \pm\sqrt{6}\left(\dfrac{3\sqrt{10}}{10}\right) - 0\cdot\dfrac{\sqrt{10}}{10} \qquad\qquad = 0\cdot\cos\alpha \pm \sqrt{6}\left(\dfrac{\sqrt{10}}{10}\right)$

$= \pm\dfrac{3\sqrt{15}}{5} \qquad\qquad\qquad\qquad\qquad = \pm\dfrac{\sqrt{15}}{5}$

Foci in the xy-coordinate sysetm are $\left(\dfrac{3\sqrt{15}}{5}, \dfrac{\sqrt{15}}{5}\right)$ and $\left(-\dfrac{3\sqrt{15}}{5}, -\dfrac{\sqrt{15}}{5}\right)$.

29. $x^2 + xy - y^2 - 40 = 0$

$A = 1, B = 1, C = -1$

Since $B^2 - 4AC = 1^2 - 4(1)(-1) = 5 > 0$, the graph is a hyperbola.

31. $3x^2 + 2\sqrt{3}xy + y^2 - 3x + 2y + 20 = 0$

$A = 3, B = 2\sqrt{3}, C = 1$

Since $B^2 - 4AC = (2\sqrt{3})^2 - 4(3)(1) = 0$, the graph is a parabola.

33. $4x^2 - 4xy + y^2 - 12y - 20 = 0$

    $A = 4, B = -4, C = 1$

    Since $B^2 - 4AC = (-4)^2 - 4(4)(1) = 0$, the graph is a parabola.

35. $5x^2 - 6\sqrt{3}xy - 11y^2 + 4x - 3y + 2 = 0$

    $A = 5, B = -6\sqrt{3}, C = -11$

    Since $B^2 - 4AC = (-6\sqrt{3})^2 - 4(5)(-11) = 328 > 0$, the graph is a hyperbola.

37. $x^2 + y^2 = r^2$

    Substitute $x = x'\cos\alpha - y'\sin\alpha$ and $y = y'\cos\alpha + x'\sin\alpha$.

$$(x'\cos\alpha - y'\sin\alpha)^2 + (y'\cos\alpha + x'\sin\alpha)^2 = r^2$$
$$(x')^2\cos^2\alpha - 2x'y'\cos\alpha\sin\alpha + (y')^2\sin^2\alpha + (y')^2\cos^2\alpha + 2x'y'\cos\alpha\sin\alpha + x'^2\sin^2\alpha = r^2$$
$$(x')^2(\cos^2\alpha + \sin^2\alpha) + x'y'(2\cos\alpha\sin\alpha - 2\cos\alpha\sin\alpha) + (y')^2(\sin^2\alpha + \cos^2\alpha) = r^2$$
$$(x')^2(1) + x'y'(0) + (y')^2(1) = r^2$$
$$(x')^2 + (y')^2 = r^2$$

39. Vertices $(2, 4)$ and $(-2, -4)$ imply that the major axis is on the line $y = 2x$. Consider an $x'y'$-coordinate system rotated an angle α, where $\tan\alpha = 2$. From this equation, using identities, $\cos\alpha = \dfrac{\sqrt{5}}{5}$ and $\sin\alpha = \dfrac{2\sqrt{5}}{5}$.

    The equation of the ellipse in the $x'y'$-coordinate system is

$$\frac{x'^2}{20} + \frac{y'^2}{10} = 1 \qquad (1)$$

   This follows from the fact that $(2, 4)$ in xy-coordinates is $\left(\sqrt{20},\ 0\right)$ in $x'y'$-coordinates.

   Therefore, $\alpha = \sqrt{20}$. Also, $\left(\sqrt{2},\ 2\sqrt{2}\right)$ in xy-coordinates is $\left(\sqrt{10},\ 0\right)$ in $x'y'$-coordinates.

   Therefore, $c = \sqrt{10}$. Thus $b = \sqrt{10}$.

   Now let $x' = \dfrac{\sqrt{5}}{5}x + \dfrac{2\sqrt{5}}{5}y$ and $y' = \dfrac{\sqrt{5}}{5}y - \dfrac{2\sqrt{5}}{5}x$ and substitute into Equation (1):

$$\frac{\left(\frac{\sqrt{5}}{5}x + \frac{2\sqrt{5}}{5}y\right)^2}{20} + \frac{\left(\frac{\sqrt{5}}{5}y - \frac{2\sqrt{5}}{5}x\right)^2}{10} = 1$$

   Simplifying, we have

$$\frac{\left(\frac{1}{5}x^2 + \frac{4}{5}xy + \frac{4}{5}y^2\right)}{20} + \frac{\left(\frac{4}{5}x^2 - \frac{4}{5}xy + \frac{1}{5}y^2\right)}{10} = 1$$

$$\frac{\frac{9}{5}x^2 - \frac{4}{5}xy + \frac{6}{5}y^2}{20} = 1$$

$$\frac{9x^2 - 4xy + 6y^2}{100} = 1$$

$$\text{or } 9x^2 - 4xy + 6y^2 = 100$$

41. $A' + C' = A\cos^2\alpha + B\cos\alpha\sin\alpha + C\sin^2\alpha + A\sin^2\alpha - B\cos\alpha\sin\alpha + C\cos^2\alpha$

    $= A(\cos^2\alpha + \sin^2\alpha) + B(\cos\alpha\sin\alpha - \cos\alpha\sin\alpha) + C(\sin^2\alpha + \cos^2\alpha)$

    $= A + C$

43. Ellipse with major axis parallel to x-axis:

$$\frac{(x-h)^2}{a^2} + \frac{(y-k)^2}{b^2} = 1$$

$$b^2(x-h)^2 + a^2(y-k)^2 = a^2b^2$$
$$b^2(x^2 - 2hx + h^2) + a^2(y^2 - 2ky + k^2) = a^2b^2$$
$$b^2x^2 - 2b^2hx + b^2h^2 + a^2y^2 - 2a^2ky + a^2k^2 = a^2b^2$$
$$b^2x^2 + a^2y^2 - 2b^2hx - 2a^2ky + b^2h^2 + a^2k^2 - a^2b^2 = 0$$

$$A = b^2, B = 0, C = a^2$$
$$B^2 - 4AC = 0^2 - 4b^2a^2 = 4a^2b^2 < 0$$

$B^2 - 4AC < 0$ for an ellipse whose major axis is parallel to the x-axis.

Ellipse with major axis parallel to y-axis:

$$\frac{(y-k)^2}{a^2} + \frac{(x-h)^2}{b^2} = 1$$

$$b^2(y-k)^2 + a^2(x-h)^2 = a^2b^2$$
$$b^2(y^2 - 2ky + k^2) + a^2(x^2 - 2hx + h^2) = a^2b^2$$
$$b^2y^2 - 2b^2ky + b^2k^2 + a^2x^2 - 2a^2hx + a^2h^2 = a^2b^2$$
$$b^2y^2 + a^2x^2 - 2b^2ky - 2a^2hx + b^2k^2 + a^2h^2 - a^2b^2 = 0$$

$$A = b^2, B = 0, C = a^2$$
$$B^2 - 4AC = 0^2 - 4b^2a^2 = -4a^2b^2 < 0$$

$B^2 - 4AC < 0$ for an ellipse whose major axis is parallel to the y-axis.

Parabola with axis of symmetry parallel to y-axis:

$$(x-h)^2 = 4p(y-k)$$
$$x^2 - 2hx + h^2 = 4py - 4pk$$
$$x^2 - 2hx - 4py + h^2 + 4pk = 0$$
$$A = 1, B = 0, C = 0$$
$$B^2 - 4AC = 0^2 - 4(1)(0) = 0$$

$B^2 - 4AC = 0$ for a parabola with axis of symmetry parallel to the y-axis.

Parabola with axis of symmetry parallel to x-axis:

$$(y-k)^2 = 4p(x-h)$$
$$y^2 - 2ky + k^2 = 4px - 4ph$$
$$y^2 - 2ky - 4px + k^2 + 4ph = 0$$
$$A = 1, B = 0, C = 0$$
$$B^2 - 4AC = 0^2 - 4(1)(0) = 0$$

$B^2 - 4AC = 0$ for a parabola with axis of symmetry parallel to the x-axis.

Hyperbola with the transverse axis parallel to x-axis:

$$\frac{(x-h)^2}{a^2} - \frac{(y-k)^2}{b^2} = 1$$

$$b^2(x-h)^2 - a^2(y-k)^2 = a^2b^2$$
$$b^2(x^2 - 2hx + h^2) - a^2(y^2 - 2ky + k^2) = a^2b^2$$
$$b^2x^2 - 2b^2hx + b^2h^2 - a^2y^2 + 2a^2ky - a^2k^2 = a^2b^2$$
$$b^2x^2 - a^2y^2 - 2b^2hx + 2a^2ky + b^2h^2 - a^2k^2 - a^2b^2 = 0$$

$A = b^2, B = 0, C = -a^2$

$B^2 - 4AC = 0^2 - 4b^2(-a^2) = 4a^2b^2 > 0$

$B^2 - 4AC > 0$ for a hyperbola whose transverse axis is parallel to the x-axis.

Hyperbola with transverse axis parallel to y-axis:

$$\frac{(y-k)^2}{a^2} - \frac{(x-h)^2}{b^2} = 1$$

$$b^2(y-k)^2 - a^2(x-h)^2 = a^2b^2$$
$$b^2(y^2 - 2ky + k^2) - a^2(x^2 - 2hx + h^2) = a^2b^2$$
$$b^2y^2 - 2b^2ky + b^2k^2 - a^2x^2 + 2a^2hx - a^2h^2 = a^2b^2$$
$$b^2y^2 - a^2x^2 - 2b^2ky + 2a^2hx + b^2k^2 - a^2h^2 - a^2b^2 = 0$$

$A = -a^2, B = 0, C = b^2$

$B^2 - 4AC = 0^2 - 4b^2(-a^2) = 4a^2b^2 > 0$

$B^2 - 4AC > 0$ for a hyperbola whose transverse axis is parallel to the y-axis.

SECTION 6.5, Page 358

1.

3.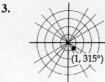

5.

7.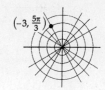

9. $0 \le \theta \le 2\pi$

11.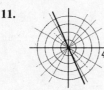

13. $0 \le \theta \le \pi$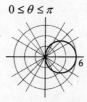

15. $0 \le \theta \le 2\pi$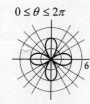

17. $0 \le \theta \le \pi$

19. $0 \le \theta \le 2\pi$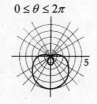

21. $0 \le \theta \le 2\pi$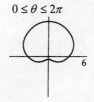

23. $0 \le \theta \le 2\pi$

25. Graph for $0 \le \theta \le 2\pi$.

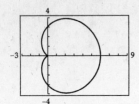

27. Graph for $0 \le \theta \le \pi$.

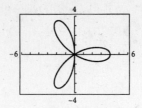

29. Graph for $0 \le \theta \le \pi$.

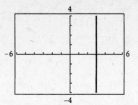

31. Graph for $0 \le \theta \le \pi$.

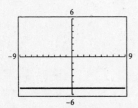

33. Graph for $0 \le \theta \le 4\pi$.

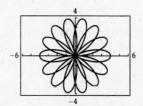

35. Graph for $0 \le \theta \le 6\pi$.

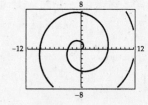

37. Graph for $0 \le \theta \le 2\pi$.

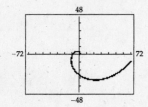

39. Graph for $0 \le \theta \le 2\pi$ with $\theta_{\text{step}} = \pi/200$.

(Some graphing utilities may draw a false asymptote in "connected" mode.)

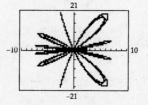

41.
$$r = \sqrt{x^2 + y^2} \qquad\qquad \theta = \tan^{-1}\frac{y}{x}$$
$$= \sqrt{1^2 + (-\sqrt{3})^2} \qquad = \tan^{-1}\left(\frac{-\sqrt{3}}{1}\right)$$
$$= \sqrt{1 + 3}$$
$$= \sqrt{4} \qquad\qquad\qquad = \tan^{-1}\left(-\frac{\sqrt{3}}{1}\right)$$
$$= 2 \qquad\qquad\qquad\qquad = -60^{\circ}$$

The polar coordinates of the point are $(2, -60^{\circ})$.

43.
$$x = r\cos\theta \qquad\qquad\qquad y = r\sin\theta$$
$$= (-3)\left(\cos\frac{2\pi}{3}\right) \qquad = (-3)\left(\sin\frac{2\pi}{3}\right)$$
$$= (-3)\left(-\frac{1}{2}\right) \qquad\qquad = (-3)\left(\frac{\sqrt{3}}{2}\right)$$
$$= \frac{3}{2} \qquad\qquad\qquad\qquad = -\frac{3\sqrt{3}}{2}$$

The rectangular coordinates of the point are $\left(\frac{3}{2}, -\frac{3\sqrt{3}}{2}\right)$.

45.
$$x = r\cos\theta \qquad\qquad y = r\sin\theta$$
$$= 0\cos\left(-\frac{\pi}{2}\right) \qquad = 0\sin\left(-\frac{\pi}{2}\right)$$
$$= 0 \qquad\qquad\qquad = 0$$

The rectangular coordinates of the point are $(0, 0)$

47.
$$r = \sqrt{x^2 + y^2} \qquad\qquad \theta = \tan^{-1}\frac{y}{x}$$
$$= \sqrt{(3)^2 + (4)^2} \qquad = \tan^{-1}\frac{4}{3}$$
$$= \sqrt{9 + 16} \qquad\qquad \approx 53.1^{\circ}$$
$$= \sqrt{25}$$
$$= 5$$

The approximate polar coordinates of the point are $(5, 53.1^{\circ})$.

49.
$$r = 3\cos\theta$$
$$r - 3\cos\theta = 0$$
$$r^2 - 3r\cos\theta = 0$$
$$x^2 + y^2 - 3x = 0$$

51.
$$r = 3\sec\theta$$
$$r = \frac{3}{\cos\theta}$$
$$r\cos\theta = 3$$
$$x = 3$$

53.
$$r = 4$$
$$\sqrt{x^2 + y^2} = 4$$
$$x^2 + y^2 = 16$$

55.
$$r = \tan\theta$$
$$r = \frac{\sin\theta}{\cos\theta}$$
$$r\cos\theta = \sin\theta$$
$$r\cos\theta - \sin\theta = 0$$
$$r^2\cos\theta - r\sin\theta = 0$$
$$\sqrt{x^2 + y^2}\,(x) - y = 0$$
$$\sqrt{x^2 + y^2} = \frac{y}{x}$$
$$x^2 + y^2 = \frac{y^2}{x^2}$$
$$x^4 - y^2 + x^2 y^2 = 0$$

57.
$$r = \frac{2}{1 + \cos\theta}$$
$$r + r\cos\theta = 2$$
$$\sqrt{x^2 + y^2} + x = 2$$
$$\sqrt{x^2 + y^2} = 2 - x$$
$$x^2 + y^2 = 4 - 4x + x^2$$
$$y^2 + 4x - 4 = 0$$

59.
$$r(\sin\theta - 2\cos\theta) = 6$$
$$r\sin\theta - 2r\cos\theta = 6$$
$$y - 2x = 6$$
$$y = 2x + 6$$

61.
$$y = 2$$
$$r\sin\theta = 2$$
$$r = 2\csc\theta$$

63.
$$x^2 + y^2 = 4$$
$$r^2 = 4$$
$$r = 2$$

65.
$$x^2 = 8y$$
$$r^2\cos^2\theta = 8r\sin\theta$$
$$r\cos^2\theta = 8\sin\theta$$

67.
$$x^2 - y^2 = 25$$
$$r^2\cos^2\theta - r^2\sin^2\theta = 25$$
$$r^2(\cos^2\theta - \sin^2\theta) = 25$$
$$r^2(\cos 2\theta) = 25$$

69.
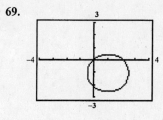

71.
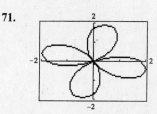

73.
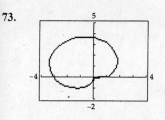

75.
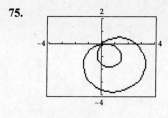

77. $\cos\theta = \pm\sqrt{\cos^2\theta}$ is *not* an identity.

79. Enter as $r = \sqrt{4\cos 2\theta}$ and $r = -\sqrt{4\cos 2\theta}$ for $0 \le \theta \le 4\pi$.

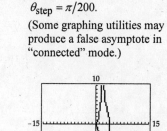

81. Graph for $0 \le \theta \le 2\pi$ with $\theta_{\text{step}} = \pi/200$.

(Some graphing utilities may produce a false asymptote in "connected" mode.)

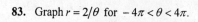

83. Graph $r = 2/\theta$ for $-4\pi < \theta < 4\pi$.

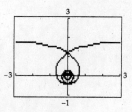

85. Graph for $-30 \le \theta \le 30$.

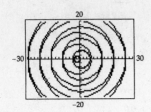

87. a. $0 \le \theta \le 5\pi$

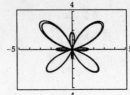

b. $0 \le \theta \le 20\pi$

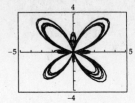

SECTION 6.6, Page 364

1. $r = \dfrac{12}{3 - 6\cos\theta}$

$= \dfrac{4}{1 - 2\cos\theta}$

$e = 2$ The graph is a hyperbola.
The transverse axis is on the polar axis.

Let $\theta = 0$.

$r = \dfrac{12}{3 - 6\cos 0} = \dfrac{12}{3 - 6} = -4$

Let $\theta = \pi$.

$r = \dfrac{12}{3 - 6\cos\pi} = \dfrac{12}{3 + 6} = \dfrac{4}{3}$

The vertices are at

$(-4, \ 0)$ and $\left(\dfrac{4}{3}, \ \pi \right)$.

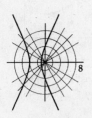

3. $r = \dfrac{8}{4 + 3\sin\theta}$

$= \dfrac{2}{1 + \frac{3}{4}\sin\theta}$

$e = \dfrac{3}{4}$ The graph is an ellipse.
The major axis is on the line

$\theta = \dfrac{\pi}{2}$.

Let $\theta = \dfrac{\pi}{2}$.

$r = \dfrac{8}{4 + 3\sin\frac{\pi}{2}} = \dfrac{8}{4 + 3} = \dfrac{8}{7}$

Let $\theta = \dfrac{3\pi}{2}$.

$r = \dfrac{8}{4 + 3\sin\frac{3\pi}{2}} = \dfrac{8}{4 - 3} = 8$

Vertices on major axis are at

$\left(\dfrac{8}{7}, \ \dfrac{\pi}{2} \right)$ and $\left(8, \ \dfrac{3\pi}{2} \right)$.

Let $\theta = 0$.

$r = \dfrac{8}{4 + 3\sin 0} = \dfrac{8}{4 + 0} = 2$

Let $\theta = \pi$.

$r = \dfrac{8}{4 + 3\sin\pi} = \dfrac{8}{4 + 0} = 2$

The curve also goes through $(2, 0)$
and $(2, \pi)$.

5. $r = \dfrac{9}{3 - 3\sin\theta}$

$= \dfrac{3}{1 - \sin\theta}$

$e = 1$ The graph is a parabola.
The axis of symmetry is $\theta = \dfrac{\pi}{2}$.

When $\theta = \dfrac{\pi}{2}$, r is undefined.

Let $\theta = \dfrac{3\pi}{2}$.

$r = \dfrac{9}{3 - 3\sin\frac{3\pi}{2}} = \dfrac{9}{3 + 3} = \dfrac{3}{2}$

Vertex is at $\left(\dfrac{3}{2}, \dfrac{3\pi}{2} \right)$.

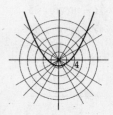

7. $r = \dfrac{10}{5 + 6\cos\theta}$

$= \dfrac{2}{1 + \frac{6}{5}\cos\theta}$

$e = \frac{6}{5}$ The graph is a hyperbola.

The transverse axis is on the polar axis.

Let $\theta = 0$.

$r = \dfrac{10}{5 + 6\cos 0} = \dfrac{10}{5 + 6} = \dfrac{10}{11}$

Let $\theta = \pi$.

$r = \dfrac{10}{5 + 6\cos\pi} = \dfrac{10}{5 - 6} = -10$

The vertices are at $\left(\frac{10}{11}, 0\right)$ and $(-10, \pi)$.

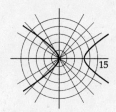

9. $r = \dfrac{4\sec\theta}{2\sec\theta - 1}$

$= \dfrac{\dfrac{4}{\cos\theta}}{\dfrac{2}{\cos\theta} - 1} = \dfrac{4}{2 - \cos\theta}$

$= \dfrac{2}{1 - \frac{1}{2}\cos\theta}$

$e = \frac{1}{2}$ The graph is an ellipse.

The major axis is on the polar axis.

However, the original equation is undefined at $\frac{\pi}{2}$ and at $\frac{3\pi}{2}$.

Thus, the ellipse will have holes at those angles.

Let $\theta = 0$.

$r = \dfrac{4}{2 - \cos 0} = \dfrac{4}{2 - 1} = 4$

Let $\theta = \pi$.

$r = \dfrac{4}{2 - \cos\pi} = \dfrac{4}{2 + 1} = \dfrac{4}{3}$

Vertices on major axis are at $(4, 0)$ and $\left(\frac{4}{3}, \pi\right)$.

$\theta = \dfrac{\pi}{2}$.

$r = \dfrac{4}{2 - \cos\frac{\pi}{2}} = \dfrac{4}{2 - 0} = 2$

Let $\theta = \dfrac{3\pi}{2}$.

$r = \dfrac{4}{2 - \cos\frac{3\pi}{2}} = \dfrac{4}{2 - 0} = 2$

Vertices on minor axis of $\dfrac{2}{1 - \frac{1}{2}\cos\theta}$ are at

$\left(2, \frac{\pi}{2}\right)$ and $\left(2, \frac{3\pi}{2}\right)$.

Thus, the equation $r = \dfrac{4\sec\theta}{2\sec\theta - 1}$ will have holes at

$\left(2, \frac{\pi}{2}\right)$ and $\left(2, \frac{3\pi}{2}\right)$

214   **Chapter 6/Topics in Analytic Geometry**

11. $r = \dfrac{12\csc\theta}{6\csc\theta - 2}$

$= \dfrac{12/\sin\theta}{6/\sin\theta - 2} = \dfrac{12}{6 - 2\sin\theta} = \dfrac{2}{1 - \frac{1}{3}\sin\theta}$

$e = \frac{1}{3}$ The graph is an ellipse.

The major axis is on $\theta = \frac{\pi}{2}$.

Let $\theta = \frac{\pi}{2}$.

$r = \dfrac{12}{6 - 2\sin\frac{\pi}{2}} = \dfrac{12}{6 - 2} = 3$

Let $\theta = \frac{3\pi}{2}$.

$r = \dfrac{12}{6 - 2\sin\frac{3\pi}{2}} = \dfrac{12}{6 + 2} = \dfrac{3}{2}$

Vertices on major axis are at $\left(3, \frac{\pi}{2}\right)$ and $\left(\frac{3}{2}, \frac{3\pi}{2}\right)$.

The equation $r = \dfrac{12\csc\theta}{6\csc\theta - 2}$ has holes at $(2, 0)$ and $(2, \pi)$.

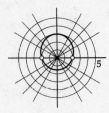

13. $r = \dfrac{3}{\cos\theta - 1}$

$= \dfrac{-3}{1 - \cos\theta}$

$e = 1$ The graph is a parabola.
The axis of symmetry is the polar axis.

Let $\theta = \pi$.

$r = \dfrac{-3}{1 - \cos\pi} = \dfrac{-3}{1 - (-1)} = \dfrac{-3}{1 + 1} = -\dfrac{3}{2}$

Vertex is at $\left(-\frac{3}{2}, \pi\right)$.

15.
$$r = \dfrac{12}{3 - 6\cos\theta}$$
$$r(3 - 6\cos\theta) = 12$$
$$3r - 6r\cos\theta = 12$$
$$3\sqrt{x^2 + y^2} - 6x = 12$$
$$3\sqrt{x^2 + y^2} = 6x + 12$$
$$\sqrt{x^2 + y^2} = 2x + 4$$
$$x^2 + y^2 = 4x^2 + 16x + 16$$
$$3x^2 - y^2 + 16x + 16 = 0$$

17.
$$r = \dfrac{8}{4 + 3\sin\theta}$$
$$r(4 + 3\sin\theta) = 8$$
$$4r + 3r\sin\theta = 8$$
$$4\sqrt{x^2 + y^2} + 3y = 8$$
$$4\sqrt{x^2 + y^2} = -3y + 8$$
$$16x^2 + 16y^2 = 9y^2 - 48y + 64$$
$$16x^2 + 7y^2 + 48y - 64 = 0$$

19.
$$r = \dfrac{9}{3 - 3\sin\theta}$$
$$r(3 - 3\sin\theta) = 9$$
$$3r - 3r\sin\theta = 9$$
$$3\sqrt{x^2 + y^2} - 3y = 9$$
$$3\sqrt{x^2 + y^2} = 3y + 9$$
$$3\sqrt{x^2 + y^2} = 3(y + 3)$$
$$\sqrt{x^2 + y^2} = y + 3$$
$$x^2 + y^2 = y^2 + 6y + 9$$
$$x^2 - 6y - 9 = 0$$

21. $e = 2$, $r\cos\theta = -1$,

$d = |-1| = 1$

$r = \dfrac{ed}{1 - e\cos\theta}$

$= \dfrac{(2)(1)}{1 - (2)\cos\theta}$

$= \dfrac{2}{1 - 2\cos\theta}$

23. $e = 1$, $r\sin\theta = 2$, $d = |2| = 2$

$r = \dfrac{ed}{1 + e\sin\theta}$

$= \dfrac{(1)(2)}{1 + (1)\sin\theta}$

$= \dfrac{2}{1 + \sin\theta}$

25. $e = \dfrac{2}{3},\ r\sin\theta = -4,$

$d = |-4| = 4$

$r = \dfrac{ed}{1 - e\sin\theta}$

$= \dfrac{\left(\frac{2}{3}\right)(4)}{1 - \left(\frac{2}{3}\right)\sin\theta}$

$= \dfrac{\frac{8}{3}}{1 - \frac{2}{3}\sin\theta}$

$= \dfrac{8}{3 - 2\sin\theta}$

27. $e = \dfrac{3}{2},\ r = 2\sec\theta$

$r = \dfrac{2}{\cos\theta}$

$r\cos\theta = 2, \quad d = |2| = 2$

$r = \dfrac{ed}{1 + e\cos\theta}$

$= \dfrac{\left(\frac{3}{2}\right)(2)}{1 + \left(\frac{3}{2}\right)\cos\theta}$

$= \dfrac{3}{1 + \frac{3}{2}\cos\theta}$

$= \dfrac{6}{2 + 3\cos\theta}$

29. vertex: $(2,\ \pi)$, curve: parabola

$r = \dfrac{ed}{1 - e\cos\theta}$ $e = 1$ (by definition of a parabola)

When $\theta = \pi, r = 2$. Substituting into

$r = \dfrac{ed}{1 - e\cos\theta}$,

we have

$2 = \dfrac{1 \cdot d}{1 - 1 \cdot \cos(\pi)} = \dfrac{d}{2}$

Therefore, $d = 4$. Substituting $e = 1$ and $d = 4$ yields

$r = \dfrac{(1)\,(4)}{1 - (1)\cos\theta}$ or $r = \dfrac{4}{1 - \cos\theta}$

31. vertex: $(1,\ 3\pi/2),\ e = 2$

$r = \dfrac{ed}{1 - e\sin\theta}$

When $\theta = \dfrac{3\pi}{2}, r = 1.$
Substituting into

$r = \dfrac{ed}{1 - e\sin\theta}$, we have

$1 = \dfrac{2d}{1 - 2\sin\left(\frac{3\pi}{2}\right)} = \dfrac{2d}{3}$

Therefore $d = \dfrac{3}{2}$.

Substituting $e = 2$ and $d = \dfrac{3}{2}$ yields

$r = \dfrac{(2)\left(\frac{2}{3}\right)}{1 - (2)\sin\theta}$

$= \dfrac{3}{1 - 2\sin\theta}$

33.
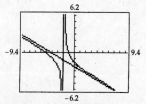

Rotate the graph of Exercise 1 $\dfrac{\pi}{6}$ radians counterclockwise about the pole.

35.
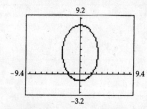

Rotate the graph of Exercise 3 π radians counterclockwise about the pole.

37.
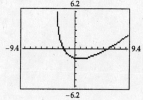

Rotate the graph of Exercise 5 $\dfrac{\pi}{6}$ radians clockwise about the pole.

39.
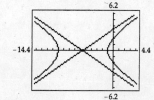

Rotate the graph of Exercise 7 π radians clockwise about the pole.

216   **Chapter 6/Topics in Analytic Geometry**

41.

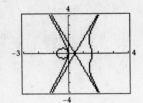

43.

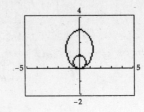

45. $0 \le \theta \le 12\pi$

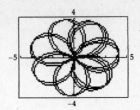

47. Convert the equations for the conic and the directrix to rectangular form.

$$r = \frac{ed}{1 - e\cos\theta} \qquad\qquad d = -r\cos\theta \ \ \text{(by definition)}$$

$$r(1 - e\cos\theta) = ed \qquad\qquad\quad = -x$$

$$r - er\cos\theta = ed \qquad\qquad\quad x = -d$$

$$\sqrt{x^2 + y^2} - ex = ed$$

$$\sqrt{x^2 + y^2} = ex + ed = e(x + d)$$

$$x^2 + y^2 = e^2(x^2 + 2dx + d^2)$$

$$x^2 + y^2 - e^2 x^2 - 2e^2 dx - e^2 d^2 = 0$$

$$(1 - e^2)x^2 + y^2 - (2e^2 d)x - (e^2 d^2) = 0$$

Solving for y^2 yields $y^2 = (e^2 - 1)x^2 + (2ed)x + (e^2 d^2)$.

Now, with $y^2 = (e^2 - 1)x^2 + (2ed)x + (e^2 d^2)$ and a directrix of $x = -d$, let $k = \dfrac{d(P,F)}{d(P,D)}$, where the focus is at the origin (by definition).

$$k = \frac{d(P,F)}{d(P,D)} = \frac{\sqrt{x^2 + y^2}}{|x + d|}$$

We can substitute $y^2 = (e^2 - 1)x^2 + (2ed)x + (e^2 d^2)$ to obtain

$$k = \frac{\sqrt{x^2 + (e^2 - 1)x^2 + (2ed)x + (e^2 d^2)}}{|x + d|} = \frac{\sqrt{e^2 x^2 + 2e^2 dx + e^2 d^2}}{|x + d|}$$

Solving for k^2 gives us $k^2 = \dfrac{e^2 x^2 + 2e^2 dx + e^2 d^2}{x^2 + 2dx + d^2} = \dfrac{e^2(x^2 + 2dx + d^2)}{x^2 + 2dx + d^2} = e^2$.

Since $k^2 = e^2$, $k = \pm\sqrt{e^2} = \pm e$. But, since $k = \dfrac{d(P,F)}{d(P,D)}$, and the ratio of two distances must be positive, k cannot be negative.

Therefore, $k = e$, or $\dfrac{d(P,F)}{d(P,D)} = e$.

SECTION 6.7, Page 370

1. A table of five arbitrarily chosen values of t and the corresponding values of x and y are shown in the table below.

t	$x = 2t$	$y = -t$	(x, y)
-2	-4	2	$(-4, 2)$
-1	-2	1	$(-2, 1)$
0	0	0	$(0, 0)$
1	2	-1	$(2, -1)$
2	4	-2	$(4, -2)$

Plotting points for several values of t yields the following graph.

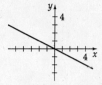

3. A table of five arbitrarily chosen values of t and the corresponding values of x and y are shown in the table below.

t	$x = -t$	$y = t^2 - 1$	(x, y)
-2	2	3	$(2, 3)$
-1	1	0	$(1, 0)$
0	0	-1	$(0, -1)$
1	-1	0	$(-1, 0)$
2	-2	3	$(-2, 3)$

Plotting points for several values of t yields the following graph.

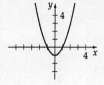

5. A table of five arbitrarily chosen values of t and the corresponding values of x and y are shown the table below.

t	$x = t^2$	$y = t^3$	(x, y)
-2	4	-8	$(4, -8)$
-1	1	-1	$(1, -1)$
0	0	0	$(0, 0)$
1	1	1	$(1, 1)$
2	4	8	$(4, 8)$

Plotting points for several values of t yields the following graph.

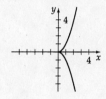

7. A table of eight values of t in the specified interval and the corresponding values of x and y are shown the table below.

t	$x = 2\cos t$	$y = 3\sin t$	(x, y)
0	2	0	$(2, 0)$
$\pi/4$	$\sqrt{2}$	$3\sqrt{2}/2$	$(\sqrt{2}, 3\sqrt{2}/2)$
$\pi/2$	0	3	$(0, 3)$
$3\pi/4$	$-\sqrt{2}$	$3\sqrt{2}/2$	$(-\sqrt{2}, 3\sqrt{2}/2)$
π	-2	0	$(-2, 0)$
$5\pi/4$	$-\sqrt{2}$	$-3\sqrt{2}/2$	$(-\sqrt{2}, -3\sqrt{2}/2)$
$3\pi/2$	0	-3	$(0, -3)$
$7\pi/4$	$\sqrt{2}$	$-3\sqrt{2}/2$	$(\sqrt{2}, -3\sqrt{2}/2)$

Plotting points for several values of t yields the following graph.

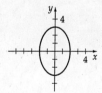

9. A table of five arbitrarily chosen values of t and the corresponding values of x and y are shown the table below.

t	$x = 2^t$	$y = 2^{t+1}$	(x, y)
-2	$1/4$	$1/2$	$(1/4, 1/2)$
-1	$1/2$	1	$(1/2, 1)$
0	1	2	$(1, 2)$
1	2	4	$(2, 4)$
2	4	8	$(4, 8)$

Plotting points for several values of t yields the following graph.

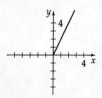

11. $x = \sec t \qquad -\dfrac{\pi}{2} < t < \dfrac{\pi}{2}$

$y = \tan t$

$\tan^2 t + 1 = \sec^2 t$

$y^2 + 1 = x^2$

$x^2 - y^2 - 1 = 0 \qquad x \geq 1, \; y \in R$

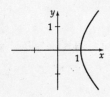

13. $x = 2 - t^2 \qquad t \in R$

$y = 3 + 2t^2$

$x = 2 - t^2 \rightarrow t^2 = 2 - x$

$\qquad y = 3 + 2(2 - x)$

$\qquad y = -2x + 7$

Because $x = 2 - t^2$ and $t^2 \geq 0$ for all real numbers t, $x \leq 2$ for all t. Similarly, $y \geq 3$ for all t.

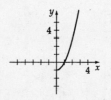

15. $x = \cos^3 t \qquad 0 \leq t < 2\pi$

$y = \sin^3 t$

$\cos^2 t = x^{2/3}$

$\sin^2 t = y^{2/3}$

$\cos^2 t + \sin^2 t = 1 \qquad -1 \leq x \leq 1$

$x^{2/3} + y^{2/3} = 1 \qquad -1 \leq y \leq 1$

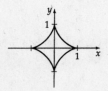

17. $x = \sqrt{t + 1} \qquad t \geq -1$

$y = t$

$x = \sqrt{y + 1} \qquad x \geq 0$

$y = x^2 - 1 \qquad y \geq -1$

19. $x = t^3 \qquad t > 0, \; x > 0$

$y = 3 \ln t \qquad y \in R$

$x = t^3 \rightarrow t = x^{1/3}$

$y = 3 \ln x^{1/3}$

$y = \ln x$ for $x > 0$ and $y \in R$

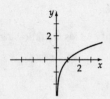

21. $C_1 : x = 2 + t^2$

$\qquad y = 1 - 2t^2$

$x = 2 + t^2 \rightarrow t^2 = x - 2$

$\qquad\qquad y = 1 - 2(x - 2)$

$\qquad\qquad y = -2x + 5 \qquad x \geq 2, \; y \leq 1$

$C_2 : x = 2 + t$

$\qquad y = 1 - 2t$

$x = 2 + t \rightarrow t = x - 2$

$\qquad\qquad y = 1 - 2(x - 2)$

$\qquad\qquad y = -2x + 5 \qquad x \in R, \; y \in R$

The graph of C_1 is a ray beginning at $(2, 1)$ with slope -2. The graph of C_2 is a line passing through $(2, 1)$ with slope -2.

23. $x = \sin t$
$y = \csc t$

$$\csc t = \frac{1}{\sin t}$$

$$y = \frac{1}{x}$$

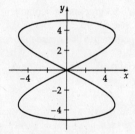

Range for graph 1: $0 < t \le \dfrac{\pi}{2}$

$$0 < x \le 1$$
$$y \ge 1$$

Range for graph 2: $\pi \le t \le \dfrac{3\pi}{2}$

$$-1 \le x \le 0$$
$$y \le -1$$

25.

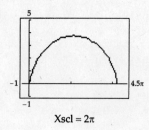

Xscl $= 2\pi$

27.

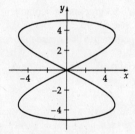

29.

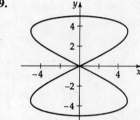

31.

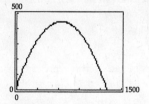

Maximum height (to the nearest foot) of 462 feet is attained when $t \approx 5.38$ seconds.

The projectile has a range (to the nearest foot) of 1295 feet and hits the ground in about 10.75 seconds.

33.

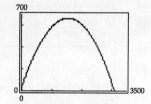

Maximum height (to the nearest foot) of 694 feet is attained when $t \approx 6.59$ seconds.

The projectile has a range (to the nearest foot) of 3084 feet and hits the ground in about 13.17 seconds.

35. Let $P_1(x_1, y_1)$ and $P_2(x_2, y_2)$ be two distinct points on a line.

If $P(x, y)$ is any other point on the line, then $\dfrac{y - y_1}{x - x_1} = \dfrac{y_2 - y_1}{x_2 - x_1}$.   (Slope is constant along entire line.)

This equation can be rewritten as $\dfrac{y - y_1}{y_2 - y_1} = \dfrac{x - x_1}{x_2 - x_1}$.   Let this value equal t.

Thus, $\dfrac{x - x_1}{x_2 - x_1} = t$ and $\dfrac{y - y_1}{y_2 - y_1} = t$.

Solving for x and y, respectively, we have

$$x = x_1 + t(x_2 - x_1) \quad \text{and} \quad y = y_1 + t(y_2 - y_1)$$

37. radius $= a$, $\theta = \angle TOR$

The x-coordinate of $P(x, y)$ is given by $x = OR + QP$.
The y-coordinate is given by $y = TR - TQ$.

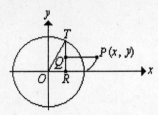

From the figure,

$OR = a\cos\theta$ and $QP = a\theta\sin\theta$. Thus,

$\quad x = a\cos\theta + a\theta\sin\theta$

$TR = a\sin\theta$ and $TQ = a\theta\cos\theta$. Thus,

$\quad y = a\sin\theta - a\theta\cos\theta$

The parametric equations are

$x = a\cos\theta + a\theta\sin\theta$

$y = a\sin\theta - a\theta\cos\theta$

39. Because the circle moves without slipping, $b\theta = a\alpha$.

Therefore, $\alpha = \dfrac{b\theta}{a}$. Let $P(x, y)$ be the coordinates of the moving point.

Angle $\phi = \dfrac{\pi}{2} - \left(\dfrac{b-a}{a}\right)\theta$

Thus, $x = (b-a)\cos\theta + a\sin\left[\dfrac{\pi}{2} - \left(\dfrac{b-a}{a}\right)\theta\right]$

$\qquad y = (b-a)\sin\theta - a\cos\left[\dfrac{\pi}{2} - \left(\dfrac{b-a}{a}\right)\theta\right]$

Simplifying, we have

$x = (b-a)\cos\theta + a\cos\left(\dfrac{b-a}{a}\theta\right)$

$y = (b-a)\sin\theta - a\sin\left(\dfrac{b-a}{a}\theta\right)$

EXPLORING CONCEPTS WITH TECHNOLOGY, Page 372

1. The procedure for a TI-83 calculator is illustrated below.

$z = -27 = 27(\cos 180° + i\sin 180°)$, thus, $r = 27$, $\theta = 180°$.

cube roots $\rightarrow n - 3$

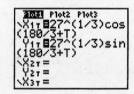

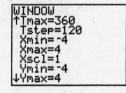
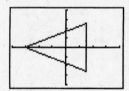

Use the TRACE feature and the arrow keys to display and move to each of the vertices of the polygon.

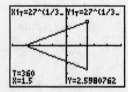
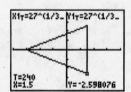

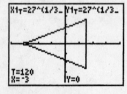

Thus, the three cube roots of -27 are

$1.5 + 2.598076i$, $1.5 - 2.598076i$, and -3.

3. Here is the procedure for a TI-83 graphing calculator.

Be sure the calculator is in parametric and degree mode.

In the WINDOW menu, set Tmin=0, Tmax=360, and, since $360/4 = 90$, set Tstep=90.

Set Xmin, Xmax, Ymin, and Ymax to appropriate values that will allow the roots to be seen.

Since $z = \sqrt{8} + \sqrt{8}i = 4(\cos 45° + i\sin 45°)$, in the Y- menu, enter X₁ᴛ=4^(1/4)cos(45/4+T) and Y₁ᴛ=4^(1/4)sin(45/4+T).

Press GRAPH to display a polygon.
The x- and y-coordinates of each vertex represent a root of z in the rectangular form $x + yi$.

Use the TRACE feature and the arrow keys to display and move to each of the vertices of the polygon.

The fourth roots of $\sqrt{8} + \sqrt{8}i$ are $1.38704 + 0.2758994i$, $-0.2758994 + 1.38704i$, $-1.38704 - 0.2758994i$, and $0.2758994 - 1.38704i$.

CHAPTER 6 TRUE/FALSE EXERCISES, Page 374

1. False, a parabola has no asymptotes.

3. False, $x^2 - y^2 = 1$ has a transverse axis of 2 and a conjugate axis of 2. By keeping the foci fixed and varying the asymptotes, we can make the conjugate axis any size needed.

5. False, parabolas have no asymptotes.

7. False, a parabola can be a function. The graph of the function $f(x) = x^2$ is a parabola.

9. True

CHAPTER 6 REVIEW EXERCISES, Page 374

1. $\qquad x^2 - y^2 = 4$

$$\frac{x^2}{4} - \frac{y^2}{4} = 1$$

hyperbola
center: (0, 0)
vertices: (±2, 0)
foci: $(\pm 2\sqrt{2},\ 0)$
asymptotes: $y = \pm x$

3. $\qquad x^2 + 4y^2 - 6x + 8y - 3 = 0$

$$x^2 - 6x + 4(y^2 + 2y) = 3$$

$$(x^2 - 6x + 9) + 4(y^2 + 2y + 1) = 3 + 9 + 4$$

$$(x-3)^2 + 4(y+1)^2 = 16$$

$$\frac{(x-3)^2}{16} + \frac{(y+1)^2}{4} = 1$$

ellipse
center: (3, −1)
vertices: $(3 \pm 4, -1) = (7, -1), (-1, -1)$
foci: $(3 \pm 2\sqrt{3},\ -1) = (3 + 2\sqrt{3},\ -1),\ (3 - 2\sqrt{3},\ -1)$

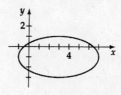

5. $\qquad 3x - 4y^2 + 8y + 2 = 0$

$$-4(y^2 - 2y) = -3x - 2$$

$$-4(y^2 - 2y + 1) = -3x - 2 - 4$$

$$-4(y - 1)^2 = -3(x + 2)$$

$$(y - 1)^2 = \frac{3}{4}(x + 2)$$

parabola
vertex: (−2, 1)
focus: $\left(-2 + \dfrac{3}{16}, 1\right) = \left(-\dfrac{29}{16}, 1\right)$

directrix: $x = -2 - \dfrac{3}{16},$ or $x = -\dfrac{35}{16}$

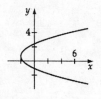

7. $\qquad 9x^2 + 4y^2 + 36x - 8y + 4 = 0$

$$9(x^2 + 4x) + 4(y^2 - 2y) = -4$$

$$9(x^2 + 4x + 4) + 4(y^2 - 2y + 1) = -4 + 36 + 4$$

$$9(x + 2)^2 + 4(y - 1)^2 = 36$$

$$\frac{(x+2)^2}{4} + \frac{(y-1)^2}{9} = 1$$

ellipse
center: (−2, 1)
vertices: $(-2,\ 1 \pm 3) = (-2, 4),\ (-2,\ -2)$
foci: $(-2,\ 1 \pm \sqrt{5}) = (-2,\ 1 + \sqrt{5}),\ (-2,\ 1 - \sqrt{5})$

9.
$$4x^2 - 9y^2 - 8x + 12y - 144 = 0$$

$$4(x^2 - 2x) - 9\left(y^2 - \frac{4}{3}y\right) = 144$$

$$4(x^2 - 2x + 1) - \left(9y^2 - \frac{4}{3}y + \frac{4}{9}\right) = 144 + 4 - 4$$

$$4(x-1)^2 - 9\left(y - \frac{2}{3}\right)^2 = 144$$

$$\frac{(x-1)^2}{36} - \frac{\left(y - \frac{2}{3}\right)^2}{16} = 1$$

hyperbola

center: $\left(1, \dfrac{2}{3}\right)$

vertices: $\left(1 \pm 6, \dfrac{2}{3}\right) = \left(7, \dfrac{2}{3}\right), \left(-5, \dfrac{2}{3}\right)$

foci: $\left(1 \pm 2\sqrt{13}, \dfrac{2}{3}\right) = \left(1 + 2\sqrt{13}, \dfrac{2}{3}\right), \left(1 - 2\sqrt{13}, \dfrac{2}{3}\right)$

asymptotes: $y - \dfrac{2}{3} = \pm \dfrac{2}{3}(x - 1)$

11.
$$4x^2 + 28x + 32y + 81 = 0$$

$$4(x^2 + 7x) = -32y - 81$$

$$4\left(x^2 + 7x + \frac{49}{4}\right) = -32y - 81 + 49$$

$$4\left(x + \frac{7}{2}\right)^2 = -32(y + 1)$$

$$\left(x + \frac{7}{2}\right)^2 = -8(y + 1)$$

parabola

$4p = -8 \Rightarrow p = -2$

vertex: $\left(-\dfrac{7}{2}, -1\right)$

focus: $\left(-\dfrac{7}{2}, -1 - 2\right) = \left(-\dfrac{7}{2}, -3\right)$

directrix: $y = 1$

13.
$$2a = |7 - (-3)| = 10$$
$$a = 5$$
$$a^2 = 25$$
$$2b = 8$$
$$b = 4$$
$$b^2 = 16$$

center $(2, 3)$

$$\frac{(x-2)^2}{25} + \frac{(y-3)^2}{16} = 1$$

15. center $(-2, 2)$, $c = 3$
$$2a = 4$$
$$a = 2$$
$$a^2 = 4$$
$$c^2 = a^2 + b^2$$
$$9 = 4 + b^2$$
$$b^2 = 5$$

$$\frac{(x+2)^2}{4} - \frac{(y-2)^2}{5} = 1$$

17. $(x - h)^2 = 4p(y - k)$ or $(y - k)^2 = 4p(x - h)$

$(3 - 0)^2 = 4p(4 + 2)$ $\qquad$ $(4 + 2)^2 = 4p(3 - 0)$

$\quad 9 = 4p(6)$ $\qquad\qquad\quad$ $36 = 4p(3)$

$\quad p = \dfrac{3}{8}$ $\qquad\qquad\qquad\quad$ $p = 3$

Thus, there are two parabolas that satisfy the given conditions:

$$x^2 = \frac{3}{2}(y + 2) \quad \text{or} \quad (y + 2)^2 = 12x$$

19. $a = 6$ and the transverse axis is on the x-axis.

$$\pm \frac{b}{a} = \pm \frac{1}{9}$$

$$\frac{b}{6} = \frac{1}{9}$$

$$b = \frac{2}{3}$$

$$\frac{x^2}{36} - \frac{y^2}{4/9} = 1$$

21. $A = 11, \ B = -6, \ C = 19, \ D = 0, \ E = 0, \ F = -40$

$\cot 2\alpha = \dfrac{11-19}{-6} = \dfrac{-8}{-6} = \dfrac{4}{3}$ (2α is in quadrant I.)

$\csc^2 2\alpha = 1 + \cot^2 2\alpha = 1 + \dfrac{16}{9} = \dfrac{25}{9}$

$\csc 2\alpha = \dfrac{5}{3}$

Thus, $\sin 2\alpha = \dfrac{3}{5}$ and $\cos 2\alpha = \dfrac{4}{5}$.

$\sin \alpha = \sqrt{\dfrac{1-\frac{4}{5}}{2}} = \dfrac{\sqrt{10}}{10}$ $\cos \alpha = \dfrac{3\sqrt{10}}{10}$

$A' = 11\left(\dfrac{3\sqrt{10}}{10}\right)^2 - 6\left(\dfrac{\sqrt{10}}{10}\right)\left(\dfrac{3\sqrt{10}}{10}\right) + 19\left(\dfrac{\sqrt{10}}{10}\right)^2$

$= \dfrac{99}{10} - \dfrac{18}{10} + \dfrac{19}{10} = \dfrac{100}{10} = 10$

$B' = 0$

$C' = 11\left(\dfrac{\sqrt{10}}{10}\right)^2 + 6\left(\dfrac{\sqrt{10}}{10}\right)\left(\dfrac{3\sqrt{10}}{10}\right) + 19\left(\dfrac{3\sqrt{10}}{10}\right)^2$

$= \dfrac{11}{10} + \dfrac{18}{10} + \dfrac{171}{10} = \dfrac{200}{10} = 20$

$F' = F$

$10(x')^2 + 20(y')^2 - 40 = 0$ or $(x')^2 + 2(y')^2 - 4 = 0$

The graph is an ellipse.

23. $A = 1, B = 2\sqrt{3}, C = 3, D = 8\sqrt{3}, E = -8, F = 32$

$\cot 2\alpha = \dfrac{1-3}{2\sqrt{3}} = -\dfrac{1}{\sqrt{3}}$. Thus $90° < 2\alpha < 180°$.

$\cot^2 2\alpha + 1 = \csc^2 2\alpha$

$\dfrac{1}{3} + 1 = \csc^2 2\alpha$

$\dfrac{4}{3} = \csc^2 2\alpha$, or $\csc 2\alpha = \dfrac{2}{\sqrt{3}}$

Therefore, $\sin 2\alpha = \dfrac{\sqrt{3}}{2}$ and $\cos 2\alpha = -\dfrac{1}{2}$.

Since 2α is in quadrant II,

$\cos \alpha = \sqrt{\dfrac{1+(-1/2)}{2}} = \dfrac{1}{2}$ $\sin \alpha = \sqrt{\dfrac{1-(-1/2)}{2}} = \dfrac{\sqrt{3}}{2}$

$A' = 1\left(\dfrac{1}{2}\right)^2 + 2\sqrt{3}\left(\dfrac{1}{2}\right)\left(\dfrac{\sqrt{3}}{2}\right) + 3\left(\dfrac{\sqrt{3}}{2}\right)^2 = \dfrac{1}{4} + \dfrac{6}{4} + \dfrac{9}{4} = 4$

$B' = 0$

$C' = 1\left(\dfrac{\sqrt{3}}{2}\right)^2 - 2\sqrt{3}\left(\dfrac{1}{2}\right)\left(\dfrac{\sqrt{3}}{2}\right) + 3\left(\dfrac{1}{2}\right)^2 = \dfrac{3}{4} - \dfrac{6}{4} + \dfrac{3}{4} = 0$

$D' = 8\sqrt{3}\left(\dfrac{1}{2}\right) - 8\left(\dfrac{\sqrt{3}}{2}\right) = 0$

$E' = 8\sqrt{3}\left(\dfrac{\sqrt{3}}{2}\right) - 8\left(\dfrac{1}{2}\right) = -12 - 4 = -16$

$F' = 32$

$4(x')^2 - 16y' + 32 = 0$ or $(x')^2 - 4y' + 8 = 0$

The graph is a parabola.

25.

27.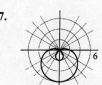

29.

31. horizontal line through $(0, 4)$

33.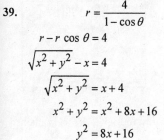

35. $y^2 = 16x$

$(r \sin \theta)^2 = 16(r \cos \theta)$

$r^2 \sin^2 \theta = 16r \cos \theta$

$r \sin^2 \theta = 16 \cos \theta$

37. $3x - 2y = 6$

$3r \cos \theta - 2r \sin \theta = 6$

39. $r = \dfrac{4}{1 - \cos \theta}$

$r - r \cos \theta = 4$

$\sqrt{x^2 + y^2} - x = 4$

$\sqrt{x^2 + y^2} = x + 4$

$x^2 + y^2 = x^2 + 8x + 16$

$y^2 = 8x + 16$

41. $r^2 = \cos 2\theta$

$r^2 = \cos^2\theta - \sin^2\theta$

$r^4 = r^2\cos^2\theta - r^2\sin^2\theta$

$(r^2)^2 = r^2\cos^2\theta - r^2\sin^2\theta$

$(x^2 + y^2)^2 = x^2 - y^2$

$x^4 + 2x^2y^2 + y^4 = x^2 - y^2$

$x^4 + y^4 + 2x^2y^2 - x^2 + y^2 = 0$

43. $r = \dfrac{4}{3 - 6\sin\theta}$

45. $r = \dfrac{2}{2 - \cos\theta}$

47. $x = 4t - 2,\ y = 3t + 1,\ t \in R$

$4t = x + 2$

$t = \dfrac{x+2}{4}$

$y = 3t + 1$

$y = 3\left(\dfrac{x+2}{4}\right) + 1$

$y = \dfrac{3}{4}x + \dfrac{5}{2}$

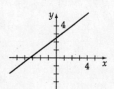

49. $x = 4\sin t$ $y = 3\cos t$ $0 \le t < 2\pi$

$\dfrac{1}{4}x = \sin t$ $\dfrac{1}{3}y = \cos t$

$\dfrac{1}{16}x^2 = \sin^2 t$ $\dfrac{1}{9}y^2 = \cos t^2$

Using the trignometric identity $\sin^2 t + \cos^2 t = 1$, we have

$\dfrac{1}{16}x^2 + \dfrac{1}{9}y^2 = 1$

$\dfrac{x^2}{19} + \dfrac{y^2}{9} = 1$

51. $x = \dfrac{1}{t}$ $y = -\dfrac{2}{t}$ $t > 0$

$y = -2\left(\dfrac{1}{t}\right)$

$y = -2x,\ x > 0$

53. $x = \sqrt{t},\ y = 2^{-t}, t \ge 0$

$t = x^2$

$y = 2^{-x^2},\ x \ge 0$

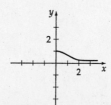

55. Graph $y = \dfrac{-(4x+5) \pm \sqrt{(4x+5)^2 - 8(x^2 - 2x + 1)}}{4}$.

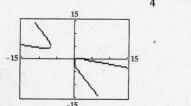

57.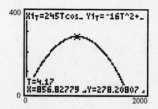

Graph in parametric mode. Use the TRACE feature to determine that the maximum height (to the nearest foot) of 278 feet is attained when $t \approx 4.17$ seconds.

CHAPTER 6 TEST, Page 375

1.
$$y = \frac{1}{8}x^2$$
$$x^2 = 8y$$
$$4p = 8$$
$$p = 2$$

vertex: $(0, 0)$
focus: $(0, 2)$
directrix: $y = -2$

2.

3.
$$25x^2 - 150x + 9y^2 + 18y + 9 = 0$$
$$25(x^2 - 6x + 9) + 9(y^2 + 2y + 1) = -9 + 255 + 9$$
$$25(x - 3)^2 + 9(y + 1)^2 = 225$$
$$\frac{(x - 3)^2}{9} + \frac{(y + 1)^2}{25} = 1$$
$$a = 5 \quad b = 3 \quad c = 4$$

vertices: $(3, 4), (3, -6)$
foci: $(3, 3), (3, -5)$

4.
$$2b = 6 \qquad c = 6$$
$$b = 3$$
$$a^2 = 9 + 36 = 45$$
$$\text{center} = (0, -3)$$
$$\frac{x^2}{45} + \frac{(y + 3)^2}{9} = 1$$

5.

6.
$$\frac{x^2}{36} - \frac{y^2}{64} = 1$$

vertices: $(6, 0), (-6, 0)$

foci: $(10, 0), (-10, 0)$

asymptotes: $y = \pm\frac{4}{3}x$

7.
$$\frac{(y + 1)^2}{4} - \frac{(x + 3)^2}{16} = 1$$

8.
$$x^2 - 4xy - 5y^2 + 3x - 5y - 20 = 0$$
$$A = 1 \quad B = -4 \quad C = -5 \quad D = 3 \quad E = -5 \quad F = -20$$
$$\cot 2\alpha = \frac{A - C}{B} = \frac{1 - (-5)}{-4} = -\frac{3}{2} \qquad 2\alpha \text{ is in quadrant II.}$$
$$\tan 2\alpha = -\frac{2}{3}$$
$$2\alpha = \tan^{-1}\left(-\frac{2}{3}\right)$$
$$2\alpha \approx (-33.69° + 180°)$$
$$\alpha \approx 73.15°$$

9.
$$A = 8 \quad B = 5 \quad C = 2 \quad D = -10 \quad E = 5 \quad F = 4$$
Since $B^2 - 4AC = (5)^2 - 4(8)(2) = -39 < 0$, the graph is an ellipse.

10. $P(1, -\sqrt{3})$

$$r = \sqrt{x^2 + y^2}$$
$$r = \sqrt{1^2 + (-\sqrt{3})^2}$$
$$r = 2$$

$$r\cos\theta = x$$
$$2\cos\theta = 1$$
$$\cos\theta = \frac{1}{2}$$

$$r\sin\theta = y$$
$$2\sin\theta = -\sqrt{3}$$
$$\sin\theta = -\frac{\sqrt{3}}{2}$$

θ is in quadrant IV.
$\theta = 300°$

$$P(1, -\sqrt{3}) = P(2, 300°)$$

11. $r = 4\cos\theta$

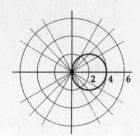

12. $r = 3(1-\sin\theta)$

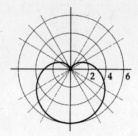

13. $r = 2\sin 4\theta$

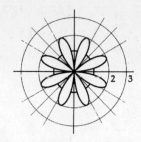

14. $x = r\cos\theta \qquad y = r\sin\theta$

$x = 5\cos\dfrac{7\pi}{3} \qquad y = 5\sin\dfrac{7\pi}{3}$

$x = \dfrac{5}{2} \qquad y = \dfrac{5\sqrt{3}}{2}$

The rectangular coordinates of the point are $(5/2,\ 5\sqrt{3}/2)$.

15. $r - r\cos x = 4$

$\sqrt{x^2+y^2} - x = 4$

$\sqrt{x^2+y^2} = x+4$

$x^2+y^2 = x^2+8x+16$

$y^2 - 8x - 16 = 0$

16. $r = \dfrac{4}{1+\sin\theta}$

$r + r\sin\theta = 4$

$\sqrt{x^2+y^2} + y = 4$

$x^2+y^2 = 16 - 8y + y^2$

$x^2 + 8y - 16 = 0$

17.

$x = t - 3$

$x + 3 = t$

$(x+3)^2 = t^2$

$2(x+3)^2 = 2t^2$

$2(x+3)^2 = y$

$(x+3)^2 = \dfrac{1}{2}y$

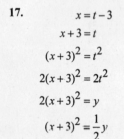

18.

$x = 4\sin\theta \qquad y = \cos\theta + 2$

$\sin\theta = x/4 \qquad \cos\theta = y - 2$

$\sin^2\theta + \cos^2\theta = 1$

$\left(\dfrac{x}{4}\right)^2 + (y-2)^2 = 1$

$\dfrac{x^2}{16} + \dfrac{(y-2)^2}{1} = 1$

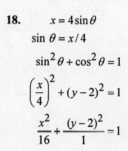

19.

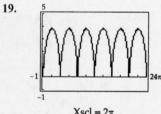

$X\text{scl} = 2\pi$

20.

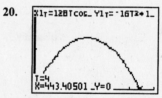

The projectile will travel $256\sqrt{3}$ feet ≈ 443 feet.

SECTION 7.1, Page 386

1. $3^{\sqrt{2}} \approx 4.7288$
3. $10^{\sqrt{7}} \approx 442.3350$
5. $\sqrt{3}^{\sqrt{2}} \approx 2.1746$
7. $e^{5.1} \approx 164.0219$

9. $e^{\sqrt{3}} \approx 5.6522$
11. $e^{-0.031} \approx 0.9695$
13. $f(x) = 3^x$
$f(\sqrt{15}) = 3^{\sqrt{15}} \approx 70.4503$

15. $f(x) = 3^x$
$f(e) = 3^e \approx 19.8130$
17. $g(x) = e^x$
$g(\sqrt{7}) = e^{\sqrt{7}} \approx 14.0940$
19. $g(x) = e^x$
$g(e) = e^e \approx 15.1543$

21. $f[g(x)] = 3^{e^x}$
$f[g(2)] = 3^{e^2} \approx 3353.3255$
23. $g[f(x)] = e^{3^x}$
$g[f(2)] = e^{3^2} \approx 8103.0839$

25.
27.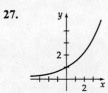
29.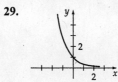
31.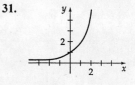

33.
35.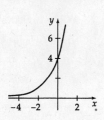
37.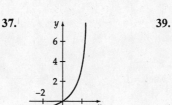
39.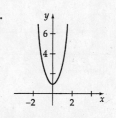

41.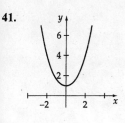
43.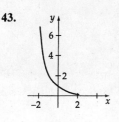
45.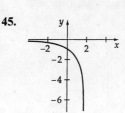
47.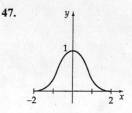

49. $x \approx 1.58$
51. $x \approx 0.69$

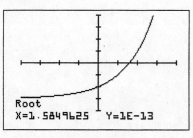

Xmin $= -4$, Xmax $= 4$, Xscl $=1$,
Ymin $= -4$, Ymax $= 4$, Yscl $=1$

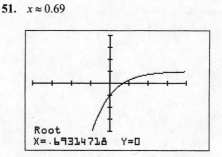

Xmin $= -4$, Xmax $= 4$, Xscl $=1$,
Ymin $= -4$, Ymax $= 4$, Yscl $=1$

228   **Chapter 7/Exponential and Logarithmic Functions**

53. $x \approx 0.79$

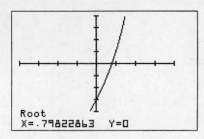

Xmin $= -4$, Xmax $= 4$, Xscl $= 1$,
Ymin $= -4$, Ymax $= 4$, Yscl $= 1$

55. $x \approx 0.80$

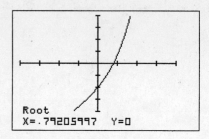

Xmin $= -4$, Xmax $= 4$, Xscl $= 1$,
Ymin $= -4$, Ymax $= 4$, Yscl $= 1$

57. $I(2) = 100e^{-0.95(2)}$
$= 100e^{-1.9}$
$\approx 15.0\%$

59. $f(6) = 1.353(1.9025)^6$
≈ 64
64 million Internet connections

61. Graph $f = x^x$ on $\left(0, 3\right]$.

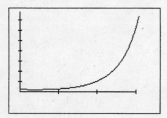

Xmin $= 0$, Xmax $= 3$, Xscl $= 1$,
Ymin $= 0$, Ymax $= 28$, Yscl $= 3.5$

a. The minimum value of f is approximately 0.6922,
which occurs at $x \approx 0.3679$.

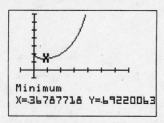

Xmin $= -0.5$, Xmax $= 3.5$, Xscl $= 0.5$,
Ymin $= -0.5$, Ymax $= 3.5$, Yscl $= 0.5$

b. As x approaches 0 from the right, f approaches the
value of 1.

63. $N(t) = 10,000 \left(2^t\right)$ represents the number of bacteria at time t.

a. $N(t) = 10,000(2^t)$
$N(1) = 10,000(2^1) = 20,000$

b. $N(t) = 10,000(2^t)$
$N(2) = 10,000(2^2) = 40,000$

c. $N(t) = 10,000(2^t)$
$N(5) = 10,000(2^5) = 320,000$

65. $\text{PMT} = P\left(\dfrac{i/12}{1-(1+i/12)^{-n}}\right)$ is the formula to calculate the monthly payment (PMT) for a loan amount of P at an annual interest rate of $i\%$.

a. $\text{PMT} = P\left(\dfrac{i/12}{1-(1+i/12)^{-n}}\right)$
$= 9000\left(\dfrac{0.10/12}{1-(1+0.10/12)^{-48}}\right)$
$= \$228.26$

b. Amt Repaid $= 228.26(48)$
$= \$10956.48$
Amt Interest $=$ Amt Repaid $-$ Amt Borrowed
$= 10956.48 - 9000$
$= \$1956.48$

67. $P = \text{PMT}\left(\dfrac{1-(1+i/12)^{-n}}{i/12}\right)$ is the formula to calculate the amount owed on a monthly installment loan.

a. $P = \text{PMT}\left(\dfrac{1-(1+i/12)^{-n}}{i/12}\right)$

$= 258\left(\dfrac{1-(1+0.09/12)^{-60}}{0.09/12}\right)$

$= \$12428.73$

b.   After 12 payments have been made, there are 48 payments remaining to be made.

$P = \text{PMT}\left(\dfrac{1-(1+i/12)^{-n}}{i/12}\right)$

$= 258\left(\dfrac{1-(1+0.09/12)^{-48}}{0.09/12}\right)$

$= \$10367.67$

c.   One way to determine n is to graph

$y = 258\left(\dfrac{1-(1+0.09/12)^{-48}}{0.09/12}\right)$

and determine the value of n for which P is one-half its original value. Using this method, $n \approx 26$. Because n represents the number of remaining payments, 34 payments have been made $(60-26)$.

69.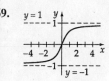

Xmin $= -4.8$, Xmax $= 4.8$, Xscl $= 1$,
Ymin $= -1.5$, Ymax $= 1.5$, Yscl $= 1$

domain: $(-\infty, \infty)$
range: $(-1,1)$
f is an odd function.

$f(x) = \dfrac{e^x - e^{-x}}{e^x + e^{-x}}$

71.

Xmin $= -10$, Xmax $= 10$, Xscl $= 1$,
Ymin $= -1$, Ymax $= 3.5$, Yscl $= 1$

domain: $(-\infty, \infty)$
range: $[0,2.2)$
f is an even function.

$f(x) = \dfrac{4x^2}{e^{|x|}}$

73.

Xmin $= -4$, Xmax $= 6$, Xscl $= 1$,
Ymin $= -1$, Ymax $= 6$, Yscl $= 1$

domain: $(-\infty, \infty)$
range: $(0.5, \infty)$
f is neither even nor odd.

$f(x) = \dfrac{e^{|x|}}{1+e^x}$

75.

Xmin $= -4$, Xmax $= 2$, Xscl $= 1$,
Ymin $= -1.2$, Ymax $= 1.2$, Yscl $= 1$

domain: $(-\infty, \infty]$
range: $[0,1)$
f is neither even nor odd.

$f(x) = \sqrt{1-e^x}$

77. a.

Xmin $= -6$, Xmax $= 6$, Xscl $= 1$,
Ymin $= -2$, Ymax $= 6$, Yscl $= 1$

domain: $(-\infty, \infty)$
range: $[2,\infty)$

$y = (f+g)(x) = e^x + e^{-x}$

b.

Xmin $= -6$, Xmax $= 6$, Xscl $= 1$,
Ymin $= -4$, Ymax $= 4$, Yscl $= 1$

domain: $(-\infty, \infty)$
range: $(-\infty, \infty)$

$y = (f-g)(x) = e^x - e^{-x}$

79. $h(x) = (-4)^x$

   a. $h(1) = (-4)^1$
         $= -4$

   b. $h(2) = (-4)^2$
         $= 16$

   c. $h\left(\dfrac{3}{2}\right) = (-4)^{3/2}$
$$= \left(\sqrt{-4}\right)^3$$
$$= (2i)^3$$
$$= 8i^3$$
$$= 8i(i^2)$$
$$= 8i(-1)$$
$$= -8i$$

   d. h is not a real-valued exponential function because b is not a positive constant.

81.

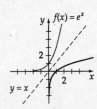

83. $\sinh(x) = \dfrac{e^x - e^{-x}}{2}$ is an odd function. That is, prove $\sinh(-x) = -\sinh(x)$.

   Proof:
$$\sinh(x) = \frac{e^x - e^{-x}}{2}$$
$$\sinh(-x) = \frac{e^{-x} - e^x}{2}$$
$$\sinh(-x) = \frac{-e^{-x} + e^x}{2}$$
$$\sinh(-x) = \frac{\left(e^x - e^{-x}\right)}{2}$$
$$\sinh(-x) = -F(x)$$

85. $e^\pi \approx 23.14069, \pi^e \approx 22.45916$. Therefore, e^π is larger.

87. **a.** VA: None, HA: $y = 0$
     b. None;
     c.

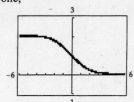

$$y = \frac{2}{1 + e^x}$$

89. **a.** VA: $x = 0$, HA: $y = 0$
     b. $(-1, 0)$
     c.

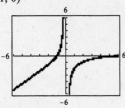

$$y = \frac{x + 1}{1 - 2^x}$$

91. **a.** VA: $x = 0$, HA: $y = 0$
     b. $\left(\dfrac{1}{2}, 0\right)$
     c.

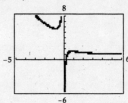

$$y = \frac{1 - 2x}{1 - e^x}$$

SECTION 7.2, Page 396

1. $\log_{10} 100 = 2$
$10^2 = 100$

3. $\log_5 125 = 3$
$5^3 = 125$

5. $\log_3 81 = 4$
$3^4 = 81$

7. $\log_b r = t$
$b^t = r$

9. $-3 = \log_3 \dfrac{1}{27}$
$3^{-3} = \dfrac{1}{27}$

11. $2^4 = 16$
$\log_2 16 = 4$

13. $7^3 = 343$
$\log_7 343 = 3$

15. $10,000 = 10^4$
$\log_{10} 10,000 = 4$

17. $b^k = j$
$\log_b j = k$

19. $b^1 = b$
$\log_b b = 1$

21. $\log_{10} 1,000,000 = n$
$10^n = 1,000,000$
$10^n = 10^6$
$n = 6$

23. $\log_2 32 = n$
$2^n = 32$
$2^n = 2^5$
$n = 5$

25. $\log_{3/2} \dfrac{27}{8} = n$
$\left(\dfrac{3}{2}\right)^n = \dfrac{27}{8}$
$\left(\dfrac{3}{2}\right)^n = \left(\dfrac{3}{2}\right)^3$
$n = 3$

27. $\log_5 \dfrac{1}{25} = n$
$5^n = \dfrac{1}{25}$
$5^n = 5^{-2}$
$n = -2$

29. $\log_b 1 = n$
$b^n = 1 = b^0$
$n = 0$

31. $y = \log_4 x$
$x = 4^y$
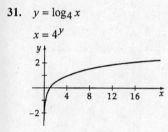

33. $y = \log_{12} x$
$x = 12^y$
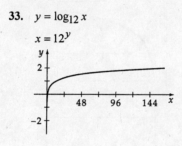

35. $y = \log_{1/2} x$
$x = (1/2)^y$
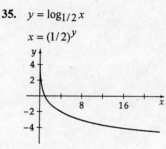

37. $y = \log_{5/2} x$
$x = (5/2)^y$
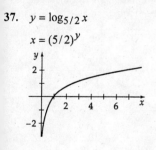

39. $y = -\log_6 x$
$-y = \log_6 x$
$x = 6^{-y}$
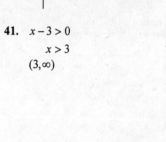

41. $x - 3 > 0$
$x > 3$
$(3, \infty)$
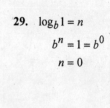

43. $11 - x > 0$
$-x > -11$
$x < 11$
$(-\infty, 11)$

45. $x^2 - 4 > 0$
$(x + 2)(x - 2) > 0$
Critical values are -2 and 2.
Product is positive.
$x < -2$ or $x > 2$
$(-\infty, -2) \cup (2, \infty)$

47. $\dfrac{x^2}{x - 4} > 0$
Critical values are 0 and 4.
Quotient is positive.
$x > 4$
$(4, \infty)$

49. $x^3 - x > 0$

$x(x^2 - 1) > 0$

$x(x + 1)(x - 1) > 0$

Critical values are 0, −1 and 1.
Product is positive.
$-1 < x < 0$ or $x > 1$
$(-1, 0) \cup (1, \infty)$

51. Shift 3 units to the right.

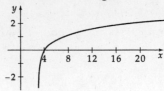

53. Shift 2 units up.

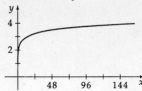

55. Shift 3 units up.

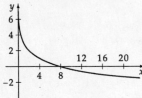

57. Shift 4 units to the right and 1 unit up.

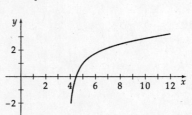

59. Shift 1 unit to the left and 2 units up.

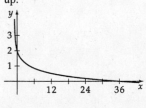

61.

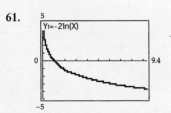

63.

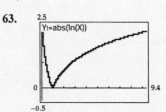

65.

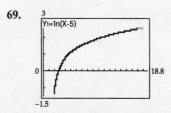

67.

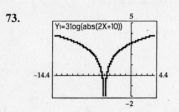

69.

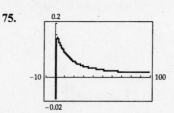

71.

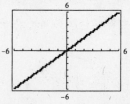

73.

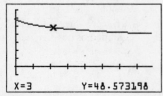

75.

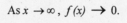

As $x \to \infty$, $f(x) \to 0$.

77. Yes, $\ln e^x = x$ for all values of x since the domain of e^x is every real number. Hence, their graphs are the same.

79. a. $t = 3 \Rightarrow S = 58 - 6.8 \ln(t + 1)$

$= 58 - 6.8 \ln 4$

≈ 49 words per minute

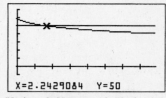

Xmin = 0, Xmax = 12, Xscl = 1.5,
Ymin = −10, Ymax = 70, Yscl = 10

b. Find the value of t for which $S = 50$ using the ZOOM and TRACE features, SOLVER, or the intersection of $S = 58 - 6.8 \ln(t + 1)$ and $S = 50$.
When $S = 50$, $t \approx 2.2$.
Since S is a decreasing function, the speed will fall below 50 after about 2.2 months.

Xmin = 0, Xmax = 12, Xscl = 1.5,
Ymin = −10, Ymax = 70, Yscl = 10

81. a. $BSA = 0.0003207 \cdot 162.56^{0.3} \cdot 49886.6^{(0.7285 - 0.0188 \log 49886.6)} = 1.502963903m^2 \approx 1.50m^2$

b. $BSA = 0.0003207 \cdot 185.42^{0.3} \cdot 81632.7^{(0.7285 - 0.0188 \log 81632.7)} = 2.047681822m^2 \approx 2.05m^2$

c. $BSA = 0.0003207 \cdot 73.66^{0.3} \cdot 18140.6^{(0.7285 - 0.0188 \log 18140.6)} = 0.6722888334m^2 \approx 0.67m^2$

83. a. $b^x = 2^{10} = 1024 \Rightarrow b = 2$ and $x = 10$

$N = \text{int}(x \log b) + 1 = \text{int}(10 \log 2) + 1 = 3 + 1 = 4$ digits

b. $b^x = 3^{200} \Rightarrow b = 3$ and $x = 200$

$N = \text{int}(x \log b) + 1 = \text{int}(200 \log 3) + 1 = 95 + 1 = 96$ digits

c. $b^x = 7^{4005} \Rightarrow b = 7$ and $x = 4005$

$N = \text{int}(x \log b) + 1 = \text{int}(4005 \log 7) + 1 = 3384 + 1 = 3385$ digits

d. $b^x = 2^{6972593} \Rightarrow b = 2$ and $x = 6972593$

$N = \text{int}(x \log b) + 1 = \text{int}(6972593 \log 2) + 1 = 2098959 + 1 = 2098960$ digits in $2^{6972593}$

There are also 2098960 digits in $2^{6972593} - 1$ because no power of 2 is a multiple of 10.

85. $f(x)$ and $g(x)$ are inverse functions

87.

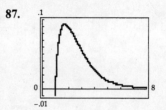

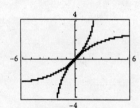

89.

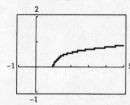

Domain $= \{x \mid x \geq 1\}$

Range $= \{y \mid y \geq 0\}$

91.

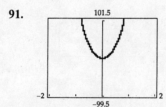

Domain $= \{x \mid -1 < s < 1\}$

Range $= \{y \mid y \geq 100\}$

93.

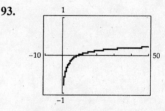

Domain $= \{x \mid x > 1\}$

Range $= \{$all real numbers$)$

SECTION 7.3 Page 408

1. $\log_b xyz = \log_b x + \log_b y + \log_b z$

3. $\log_3 \dfrac{x}{x^4} = \log_3 x - 4\log_3 z$

5. $\log_b \dfrac{\sqrt{x}}{y^3} = \dfrac{1}{2}\log_b x - 3\log_b y$

7. $\log_b x\sqrt[3]{\dfrac{y^2}{z}} = \log_b \dfrac{xy^{2/3}}{z^{1/3}} = \log_b x + \dfrac{2}{3}\log_b y - \dfrac{1}{3}\log_b z$

9. $\log_7 \dfrac{\sqrt{xz}}{y^2} = \log_7 \sqrt{xz} - \log_7(y^2) = \dfrac{1}{2}\log_7(xz) - 2\log_7 y = \dfrac{1}{2}(\log_7 x + \log_7 z) - 2\log_7 y = \dfrac{1}{2}\log_7 x + \dfrac{1}{2}\log_7 z - 2\log_7 y$

11. $\log_{10}(x+5) + 2\log_{10} x = \log_{10}[x^2(x+5)]$

13. $\dfrac{1}{2}(3\log_b(x-y) + \log_b(x+y) - \log_b z) = \dfrac{1}{2}\log_b\left(\dfrac{(x-y)^3(x+y)}{z}\right) = \log_b\sqrt{\dfrac{(x-y)^3(x+y)}{z}}$

15. $\log_8(x^2 - y^2) - \log 8(x - y) = \log_8 \dfrac{x^2 - y^2}{x - y} = \log_8 \dfrac{(x+y)(x-y)}{(x-y)} = \log_8(x+y)$

17. $4\ln(x-3) + 2\ln x = \ln(x-3)^4 x^2 = \ln[x^2(x-3)^4]$

19. $\ln x - \ln y + \ln z = \ln \dfrac{xz}{y}$

21. $\log_7 6 = \log_7[(2)(3)] = \log_7 2 + \log_7 3 \approx 0.3562 + 0.5646 \approx 0.9208$

23. $\log_7 9 = \log_7 3^2 = 2\log_7 3 \approx 2(0.5646) \approx 1.1292$

25. $\log_7 \dfrac{2}{5} = \log_7 2 - \log_7 5 \approx 0.3562 - 0.8271 \approx -0.4709$

27. $\log_7 30 = \log_7[(2)(3)(5)] = \log_7 2 + \log_7 3 + \log_7 5 \approx 0.3562 + 0.5646 + 0.8271 \approx 1.7479$

29. $\log_7 14 = \log_7[(2)(7)] = \log_7 2 + \log_7 7 \approx 0.3562 + 1 \approx 1.3562$

31. $\log_7 20 = \dfrac{\log 20}{\log 7} \approx 1.5395$ **33.** $\log_{11} 8 = \dfrac{\log 8}{\log 11} \approx 0.86719$

35. $\log_6 0.045 = \dfrac{\log 0.045}{\log 6} \approx -1.7308$ **37.** $\log_{0.5} 5 = \dfrac{\log 5}{\log 0.5} \approx -2.3219$ **39.** $\log_\pi e = \dfrac{\log e}{\log \pi} \approx 0.87357$

41. Graph $y = \dfrac{\log x}{\log 4}$.

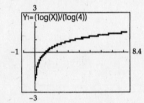

43. Graph $y = \dfrac{\log(x-3)}{\log 8}$.

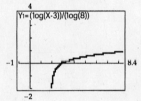

45. Graph $y = \dfrac{2\log(x-3)}{\log 3}$.

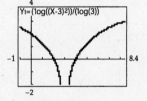

47. Graph $y = \dfrac{-\log(\text{abs}(x-2))}{\log \pi}$.

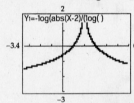

49. Graph $y = \text{abs}\left(\dfrac{\log(2x-1)}{\log 6}\right)$.

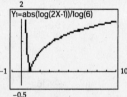

51. Graph $y = \dfrac{-2\log x}{\log 5.5}$.

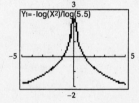

53. Magnitude $M = \log\left(\dfrac{I}{I_0}\right)$

$\qquad = \log\left(\dfrac{100,000 I_0}{I_0}\right)$

$\qquad = \log 100,000$

$\qquad = 5$ on the Richter scale

55. $\log\left(\dfrac{I}{I_0}\right) = 6.5$

$\qquad \dfrac{I}{I_0} = 10^{6.5}$

$\qquad I = 10^{6.5} I_0$

$\qquad I \approx 3,162,277.66 I_0$

57. $\log\left(\dfrac{I_1}{I_0}\right)=5$ and $\log\left(\dfrac{I_2}{I_0}\right)=3$

$\dfrac{I_1}{I_0}=10^5 \qquad \dfrac{I_2}{I_0}=10^3$

$I_1=10^5 I_0 \qquad I_2=10^3 I_0$

To compare the intensities of the earthquakes, compute the ratio I_1/I_2.

$\dfrac{I_1}{I_2}=\dfrac{10^5 I_0}{10^3 I_0}=10^2=100$

The ratio is 100 to 1.

59. $\log\left(\dfrac{I_1}{I_0}\right)=8.9$ and $\log\left(\dfrac{I_2}{I_0}\right)=7.1$

$\dfrac{I_1}{I_0}=10^{8.9} \qquad \dfrac{I_2}{I_0}=10^{7.1}$

$I_1=10^{8.9} I_0 \qquad I_2=10^{7.1} I_0$

$\dfrac{I_1}{I_2}=\dfrac{10^{8.9} I_0}{10^{7.1} I_0}=10^{1.8}\approx 63$

$10^{1.8}$ to 1, or about 63 to 1.

61. $M=\log A+3\log 8t-2.92$

$=\log 18+3\log 8(31)-2.92$

$=\log 18+3\log 248-2.92$

≈ 5.5

63. $\text{pH}=-\log\left[H^+\right]$

$=-\log\left(1.26\times 10^{-12}\right)$

$\approx 11.9;\ \text{base}$

65. $\text{pH}=-\log\left[H^+\right]$

$9.5=-\log\left[H^+\right]$

$\log\left[H^+\right]=-9.5$

$\left[H^+\right]=10^{-9.5}$

$\approx 3.16\times 10^{-10}$

67. $\text{dB}(I)=10\log\left(\dfrac{I}{I_0}\right)$

a. $\text{dB}=10\log\left(\dfrac{1.58\times 10^8\cdot I_0}{I_0}\right)$

$=10\log\left(1.58\times 10^8\right)$

≈ 82.0

b. $\text{dB}=10\log\left(\dfrac{10800\cdot I_0}{I_0}\right)$

$=10\log(10800)$

≈ 40.3

c. $\text{dB}=10\log\left(\dfrac{3.16\times 10^{11}\cdot I_0}{I_0}\right)$

$=10\log\left(3.16\times 10^{11}\right)$

≈ 115.0

d. $\text{dB}=10\log\left(\dfrac{1.58\times 10^{15}\cdot I_0}{I_0}\right)$

$=10\log\left(1.58\times 10^{15}\right)$

≈ 152.0

69. $10\log\left(\dfrac{I_1}{I_0}\right)=120$ and $10\log\left(\dfrac{I_2}{I_0}\right)=110$

$\log\left(\dfrac{I_1}{I_0}\right)=12 \qquad \log\left(\dfrac{I_2}{I_0}\right)=11$

$\dfrac{I_1}{I_0}=10^{12} \qquad \dfrac{I_2}{I_0}=10^{11}$

$I_1=10^{12} I_0 \qquad I_2=10^{11} I_0$

$\dfrac{I_1}{I_2}=\dfrac{10^{12} I_0}{10^{11} I_0}=10$

71. $\log_3 5 \cdot \log_5 7 \cdot \log_7 9 = \dfrac{\log 5}{\log 3} \cdot \dfrac{\log 7}{\log 5} \cdot \dfrac{\log 9}{\log 7}$

$= \dfrac{\log 9}{\log 3}$

$= \dfrac{\log 3^2}{\log 3}$

$= \dfrac{2\log 3}{\log 3}$

$= 2$

73. Definition of a logarithm. (Write in exponential form.)

Multiply each side of $M = b^x$ by the same quantity.
The product property of exponents.
The logarithm-of-each-side property.

The $\log_b b^p = p$ property.
Substitution.

75. $0 \le \log x \le 1000$

$10^0 \le 10^{\log x} \le 10^{1000}$

$1 \le\ \ x\ \ \le 10^{1000}$

$\left[1, 10^{1000} \right]$

77. $e \le \ln x \le e^3$

$e^e \le e^{\ln x} \le e^{(e^3)}$

$e^e \le\ \ x\ \ \le e^{(e^3)}$

$\left[e^e, e^{(e^3)} \right]$

79. $-\log x > 0$ is defined when $x > 0$.

$-\log x > 0$ is defined when $x > 0$.

$\log x < 0$

$10^{\log x} < 10^0$

$x < 1$

$(0, 1)$

81. a.   A line between 40 s on the Time scale and 50 mm on the Amplitude scale intersects the Magnitude scale at about 6.0.
b.   A line between 30 s on the Time scale and 1 mm on the Amplitude scale intersects the Magnitude scale at about 3.8.

Using the amplitude-time-difference formula

$M_1 = \log 50 + 3\log 8(40) - 2.92$

≈ 6.3

$M_2 = \log 1 + 3\log 8(30) - 2.92$

≈ 4.2

The nomogram results are close to the amplitude-time-difference results.

SECTION 7.4, Page 417

1. $2^x = 64$

$2^x = 2^6$

$x = 6$

3. $49^x = \dfrac{1}{343}$

$7^{2x} = 7^{-3}$

$2x = -3$

$x = -\dfrac{3}{2}$

5. $2^{5x+3} = \dfrac{1}{8}$

$2^{5x+3} = 2^{-3}$

$5x + 3 = -3$

$5x = -6$

$x = -\dfrac{6}{5}$

7. $\left(\dfrac{2}{5}\right)^x = \dfrac{8}{125}$

$\left(\dfrac{2}{5}\right)^x = \left(\dfrac{2}{5}\right)^3$

$x = 3$

9. $5^x = 70$

$\log(5^x) = \log 70$

$x \log 5 = \log 70$

$x = \dfrac{\log 70}{\log 5}$

11. $3^{-x} = 120$

$\log(3^{-x}) = \log 120$

$-x \log 3 = \log 120$

$-x = \dfrac{\log 120}{\log 3}$

$x = -\dfrac{\log 120}{\log 3}$

13. $10^{2x+3} = 315$

$\log 10^{2x+3} = \log 315$

$(2x + 3)\log 10 = \log 315$

$2x + 3 = \log 315$

$x = \dfrac{\log 315 - 3}{2}$

15. $e^x = 10$

$\ln e^x = \ln 10$

$x = \ln 10$

17.
$$2^{1-x} = 3^{x+1}$$
$$\log 2^{1-x} = \log 3^{x+1}$$
$$(1-x)\log 2 = (x+1)\log 3$$
$$\log 2 - x\log 2 = x\log 3 + \log 3$$
$$\log 2 - x\log 2 - x\log 3 = \log 3$$
$$-x\log 2 - x\log 3 = \log 3 - \log 2$$
$$-x(\log 2 + \log 3) = \log 3 - \log 2$$
$$x = -\frac{(\log 3 - \log 2)}{(\log 2 + \log 3)}$$
$$x = \frac{\log 2 - \log 3}{\log 2 + \log 3} \text{ or } \frac{\log 2 - \log 3}{\log 6}$$

19.
$$2^{2x-3} = 5^{-x-1}$$
$$\log 2^{2x-3} = \log 5^{-x-1}$$
$$(2x-3)\log 2 = (-x-1)\log 5$$
$$2x\log 2 - 3\log 2 = -x\log 5 - \log 5$$
$$2x\log 2 + x\log 5 - 3\log 2 = -\log 5$$
$$2x\log 2 + x\log 5 = 3\log 2 - \log 5$$
$$x(2\log 2 + \log 5) = 3\log 2 - \log 5$$
$$x = \frac{3\log 2 - \log 5}{2\log 2 + \log 5}$$

21. $\log(4x-18) = 1$
$$4x-18 = 10^1$$
$$4x-18 = 10$$
$$4x = 28$$
$$x = 7$$

23. $\ln(x^2-12) = \ln x$
$$x^2-12 = x$$
$$x^2-x-12 = 0$$
$$(x-4)(x+3) = 0$$
$$x = 4 \text{ or } x = -3 \text{ (No; not in domain.)}$$
$$x = 4$$

25. $\log_2 + \log_2(x-4) = 2$
$$\log_2 x(x-4) = 2$$
$$\log_2(x^2-4x) = 2$$
$$2^2 = x^2-4x$$
$$0 = x^2-4x-4$$
$$x = \frac{4 \pm \sqrt{16-4(1)(-4)}}{2} = \frac{4 \pm 4\sqrt{2}^2}{2}$$
$$x = 2 \pm 2\sqrt{2}$$
$2-2\sqrt{2}$ is not a solution because the logarithm of a negative number is not defined. The solution is $x = 2 + 2\sqrt{2}$.

27.
$$\log(5x-1) = 2 + \log(x-2)$$
$$\log(5x-1) - \log(x-2) = 2$$
$$\log\frac{(5x-1)}{(x-2)} = 2$$
$$10^2 = \frac{(5x-1)}{(x-2)}$$
$$100(x-2) = 5x-1$$
$$100x-200 = 5x-1$$
$$95x = 199$$
$$x = \frac{199}{95}$$

29. $\ln(1-x) + \ln(3-x) = \ln 8$
$$\ln[(1-x)(3-x)] = \ln 8$$
$$(1-x)(3-x) = 8$$
$$3-4x+x^2 = 8$$
$$x^2-4x-5 = 0$$
$$(x+1)(x-5) = 0$$
$x = -1$ or $x = 5$ (No; not in domain.)
The solution is $x = -1$.

31. $\log\sqrt{x^3-17} = \frac{1}{2}$
$$\frac{1}{2}\log(x^3-17) = \frac{1}{2}$$
$$10^1 = x^3-17$$
$$27 = x^3$$
$$\sqrt[3]{27} = \sqrt[3]{x^3}$$
$$3 = x$$
The solution is $x = 3$.

33. $\log(\log x) = 1$
$$10^1 = \log x$$
$$10^{10} = x$$

35. $\ln(e^{3x}) = 6$
$$3x\ln e = 6$$
$$3x(1) = 6$$
$$3x = 6$$
$$x = 2$$

37. $e^{\ln(x-1)} = 4$
$$\ln e^{\ln(x-1)} = \ln 4$$
$$\ln(x-1)\ln e = \ln 4$$
$$\ln(x-1)(1) = \ln 4$$
$$(x-1) = 4$$
$$x = 5$$

39.

$$\frac{10^x - 10^{-x}}{2} = 20$$

$$10^x\left(10^x - 10^{-x}\right) = 40\left(10^x\right)$$

$$10^{2x} - 1 = 40\left(10^x\right)$$

$$10^{2x} - 40(10)^x - 1 = 0$$

Let $u = 10^x$.

$$u^2 - 40u - 1 = 0$$

$$u = \frac{40 \pm \sqrt{40^2 - 4(1)(-1)}}{2}$$

$$= \frac{40 \pm \sqrt{1600 + 4}}{2}$$

$$= \frac{40 \pm \sqrt{1604}}{2}$$

$$= \frac{40 \pm 2\sqrt{401}}{2}$$

$$= 20 \pm \sqrt{401}$$

$$10^x = 20 + \sqrt{401}$$

$$\log 10^x = \log\left(20 + \sqrt{401}\right)$$

$$x = \log\left(20 + \sqrt{401}\right)$$

41.

$$\frac{10^x + 10^{-x}}{10^x - 10^{-x}} = 5$$

$$10^x + 10^{-x} = 5\left(10^x - 10^{-x}\right)$$

$$10^x\left(10^x + 10^{-x}\right) = 5\left(10^x - 10^{-x}\right)10^x$$

$$10^{2x} + 1 = 5\left(10^{2x} - 1\right)$$

$$4\left(10^{2x}\right) = 6$$

$$2\left(10^{2x}\right) = 3$$

$$\left(10^x\right)^2 = \frac{3}{2}$$

$$10^x = \sqrt{\frac{3}{2}}$$

$$x \log 10 = \log\sqrt{\frac{3}{2}}$$

$$x = \log\sqrt{\frac{3}{2}}$$

$$x = \frac{1}{2}\log\left(\frac{3}{2}\right)$$

43.

$$\frac{e^x + e^{-x}}{2} = 15$$

$$e^x\left(e^x + e^{-x}\right) = (30)e^x$$

$$e^{2x} + 1 = e^x(30)$$

$$e^{2x} - 30e^x + 1 = 0$$

Let $u = e^x$.

$$u^2 - 30u + 1 = 0$$

$$u = \frac{30 \pm \sqrt{900 - 4}}{2}$$

$$u = \frac{30 \pm \sqrt{896}}{2}$$

$$u = \frac{30 \pm 8\sqrt{14}}{2}$$

$$u = 15 \pm 4\sqrt{14}$$

$$e^x = 15 \pm 4\sqrt{14}$$

$$x \ln e = \ln\left(15 \pm 4\sqrt{14}\right)$$

$$x = \ln\left(15 \pm 4\sqrt{14}\right)$$

45.

$$\frac{1}{e^x - e^{-x}} = 4$$

$$1 = 4(e^x - e^{-x})$$

$$1(e^x) = 4(e^x)(e^x - e^{-x})$$

$$e^x = 4(e^{2x} - 1)$$

$$e^x = 4e^{2x} - 4$$

$$0 = 4e^{2x} - e^x - 4$$

Let $u = e^x$.

$$0 = 4u^2 - u - 4$$

$$u = \frac{1 \pm \sqrt{1 - 4(4)(-4)}}{8}$$

$$u = \frac{1 \pm \sqrt{65}}{8}$$

$$e^x = \frac{1 + \sqrt{65}}{8}$$

$$x \ln e = \ln\left(\frac{1 + \sqrt{65}}{8}\right)$$

$$x = \ln\left(1 + \sqrt{65}\right) - \ln 8$$

47. $2^{-x+3} = x+1$

Graph $f = 2^{-x+3} - (x+1)$.
Its x-intercept is the solution.
$x \approx 1.61$

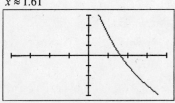

Xmin = −4, Xmax = 4, Xscl = 1,
Ymin = −4, Ymax = 4, Yscl = 1

49. $e^{3-2x} - 2x = 1$

Graph $f = e^{3-2x} - 2x - 1$.
Its x-intercept is the solution.
$x \approx 0.96$

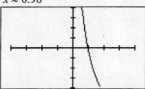

Xmin = −4, Xmax = 4, Xscl = 1,
Ymin = −4, Ymax = 4, Yscl = 1

51. $3\log_2(x-1) = x+3$

Graph $f = \dfrac{3\log(x-1)}{\log 2} + x - 3$.
Its x-intercept is the solution.
$x \approx 2.20$

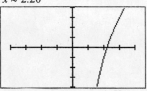

Xmin = −4, Xmax = 4, Xscl = 1,
Ymin = −4, Ymax = 4, Yscl = 1

53. $\ln(2x+4) + \dfrac{1}{2}x = -3$

Graph $f = \ln(2x+4) + \dfrac{1}{2}x + 3$.
Its x-intercept is the solution.
$x \approx -1.93$

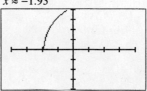

Xmin = −4, Xmax = 4, Xscl = 1,
Ymin = −4, Ymax = 4, Yscl = 1

55. $2^{x+1} = x^2 - 1$

Graph $f = 2^{x+1} - x^2 + 1$.
Its x-intercept is the solution.
$x \approx -1.34$

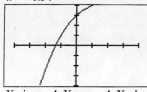

Xmin = −4, Xmax = 4, Xscl = 1,
Ymin = −4, Ymax = 4, Yscl = 1

57. a. $P(0) = 8500(1.1)^0 = 8500(1) = 8500$

$P(2) = 8500(1.1)^2 = 10,285$

b.
$$15,000 = 8500(1.1)^t$$
$$\ln 15,000 = 8500(1.1)^t$$
$$\ln 51,000 = \ln 8500 + t\ln(1.1)$$
$$\frac{\ln 15,000 - \ln 8500}{\ln(1.1)} = t$$
$$6 \approx t$$

The population will reach 15,000 in 6 years.

59. a. $T(10) = 36 + 43e^{-0.058(10)} = 36 + 43e^{-0.58}$

$T \approx 60^\circ F$

b.
$$45 = 36 + 43e^{-0.058t}$$
$$\ln(45-36) = \ln 43 - 0.058t \ln e$$
$$\frac{\ln(45-36) - \ln 43}{-0.058} = t$$
$$t \approx 27 \text{ minutes}$$

61. a.

b.   48 hours

c. $P = 100$

d.   As the number of hours of training increases, the test scores approach 100%.

63. a.

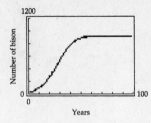

    b. in 27 years or 2026

    c. $B = 1000$

    d. As the number of years increases, the bison population approaches but never exceeds 1000.

67. a.

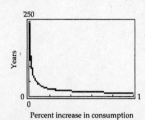

    b. When $r = 3\%$, or 0.03, $T \approx 78$ years.

    c. When $T = 100$, $r \approx 0.019$, or 1.9%.

    d. More. For $r > 0$, $r > \ln(1+r)$. Therefore,

$$\frac{\ln(300r+1)}{r} < \frac{\ln(300r+1)}{\ln(1+r)}$$

65. a.

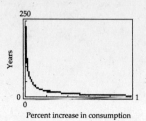

    b. When $r = 3\%$, or 0.03, $T \approx 77$ years

    c. When $T = 100$, $r \approx 0.019$, or 1.9%

69. a.

$$t = -\frac{175}{32}\ln\left(1-\frac{v}{175}\right)$$

$$10 = -\frac{175}{32}\ln\left(1-\frac{v}{175}\right)$$

$$\frac{-32(10)}{175} = \ln\left(1-\frac{v}{175}\right)$$

$$-1.8285714 = \ln\left(1-\frac{v}{175}\right)$$

$$e^{-1.8285714} = 1-\frac{v}{175}$$

$$e^{-1.8285714} - 1 = \frac{v}{175}$$

$$v = -175\left(e^{-1.8285714} - 1\right)$$

$$v \approx 146.9$$

The velocity after 10 seconds is approximately 146.9 feet per second.

    b. The vertical asymptote occurs when

$$1-\frac{v}{175} = 0, \text{ or when } v = 175.$$

    c. The vertical asymptote means that the velocity of the skydiver cannot exceed 175 feet per second.

71. a.

$$v = 100\left(\frac{e^{0.64t} - 1}{e^{0.64t} + 1}\right)$$

$$50 = 100\left(\frac{e^{0.64t} - 1}{e^{0.64t} + 1}\right)$$

$$\frac{50}{100} = \frac{e^{0.64t} - 1}{e^{0.64t} + 1}$$

$$0.5 = \frac{e^{0.64t} - 1}{e^{0.64t} + 1}$$

$$0.5(e^{0.64t} + 1) = e^{0.64t} - 1$$

$$0.5e0^{1.64t} + 0.5 = e^{0.64t} - 1$$

$$0.5e^{0.64t} - e^{0.64t} = -1.5$$

$$-0.5e^{0.64t} = -1.5$$

$$e^{0.64t} = 3$$

$$0.64t = \ln 3$$

$$t = \frac{\ln 3}{0.64}$$

$$t \approx 1.72$$

In approximately 1.72 seconds, the velocity will be 50 feet per second.

b.   The horizontal asymptote is the value of

$$100\left[\frac{e^{0.64t} - 1}{e^{0.64t} + 1}\right] \text{ as } t \to \infty. \text{ Therefore, the horizontal}$$

asymptote is $v = 100$ feet per second.

c.   The object cannot fall faster than 100 feet per second.

75. a. $x = 0, 1, 2, \ldots$
The domain is the number of months after February 1, 1999 until the account is depleted, or the whole numbers 0, 1, 2, 3, ..., 196.

b.   Graph $V = 400{,}000 - 150{,}000(1.005)^x$ and $V = 100.000$.
They intersect when $x \approx 138.97$.
After 138 withdrawals, the account has $101,456.39.
After 139 withdrawals, the account has $99,963.67.
The designer can make at most 138 withdrawals and still have $100,000.

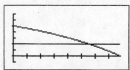

Xmin = 0, Xmax = 200, Xscl = 25
Ymin = −50000, Ymax = 350000, Yscl = 50000

73. a.

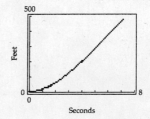

b.   When $s = 100$, $t \approx 2.6$ seconds.

77.   The second step because log $0.5 < 0$. Thus the inequality sign must be reversed.

79. $\log(x+y) = \log x + \log y$

$\log(x+y) = \log xy$

Therefore $x + y = xy$

$x - xy = -y$

$x(1-y) = -y$

$x = \dfrac{-y}{1-y}$

$x = \dfrac{y}{y-1}$

81. Since $e^{0.336} \approx 1.4$,

$F(x) = (1.4)^x \approx (e^{0.336})^x = e^{0.336x} = G(x)$

83. $e^{1/x} > 2$

$\dfrac{1}{x}\ln e > \ln 2$

$\dfrac{1}{x} > \ln 2$ $\left(x > 0,\text{ since}\dfrac{1}{x} \not{>} \ln 2 \text{ when } x < 0\right).$

$1 > x\ln 2$

$\dfrac{1}{\ln 2} > x$

The solution is $\left(0, \dfrac{1}{\ln 2}\right).$

SECTION 7.5, Page 430

1. **a.** $P = 8000, r = 0.05, t = 4, n = 1$

$B = 8000\left(1 + \dfrac{0.05}{1}\right)^4 \approx \$9724.05$

b. $t = 7, B = 8000\left(1 + \dfrac{0.05}{1}\right)^7 \approx \$11,256.80$

3. **a.** $P = 38,000, r = 0.065, t = 4, n = 1$

$B = 38,000\left(1 + \dfrac{0.065}{1}\right)^4 \approx \$48,885.72$

b. $n = 365, B = 38,000\left(1 + \dfrac{0.065}{365}\right)^{4(365)} \approx \$49,282.20$

c. $n = 8760, B = 38,000\left(1 + \dfrac{0.065}{8760}\right)^{4(8760)} \approx \$49,283.30$

5. $P = 15,000,\quad r = 0.1, t = 5$

$B = 15,000e^{5(0.1)} \approx \$24,730.82$

7. $t = \dfrac{\ln 2}{r}$

$t = \dfrac{\ln 2}{0.0784}$

$t \approx 8.8$ years

$r = 0.0784$

9. $B = Pe^{rt}$ Let $B = 3P$

$3P = Pe^{rt}$

$3 = e^{rt}$

$\ln 3 = rt\ln e$

$t = \dfrac{\ln 3}{r}$

11. $t = \dfrac{\ln 3}{r}$ $r = 0.076$

$t = \dfrac{\ln 3}{0.076} \approx 14$ years

13. **a.** $t = 0$ hours, $N(0) = 2200(2)^0 = 2200$ bacteria

b. $t = 3$ hours, $N(3) = 2200(2)^3 = 17,600$ bacteria

15. **a.** $N(t) = N_0 e^{kt}$ where $N_0 = 24600$

$N(5) = 22600e^{k(5)}$

$24200 = 22600e^{5k}$

$\dfrac{24200}{22600} = e^{5k}$

$\ln\left(\dfrac{24200}{22600}\right) = \ln(e^{5k}) = 5k$

$k = \dfrac{1}{5}\left[\ln\dfrac{24200}{22600}\right] \approx 0.01368$

$N(t) = 22600e^{0.01368t}$

b. $t = 15$

$N(15) = 22600e^{0.01368(15)}$

$= 22600e^{0.2052}$

$\approx 27,700$

17. a. $P = 10{,}130(1.005)^t$ where $t = 12$

$= 10.130(1.005)^{12}$

$= 110.130(1.061677812)$

$= 10{,}754.79623$ thousand

$P \approx 10{,}775{,}000$

b. $13{,}000 = 10{,}130(1.005)^t$

$$\frac{13{,}000}{10{,}130} = 1.005^t$$

$$\log\left(\frac{13{,}000}{10{,}130}\right) = \log 1.005^t$$

$$\log\left(\frac{13{,}000}{10{,}130}\right) = t\log 1.005$$

$$\frac{\log\left(\frac{13{,}000}{10{,}130}\right)}{\log 1.005} = t \approx 50$$

in 50 years, or in 2042

19. a.

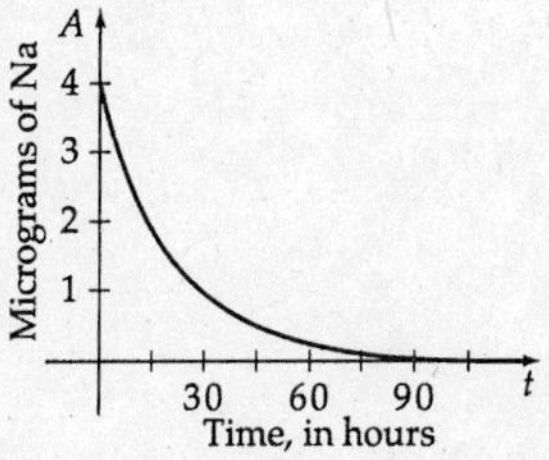

b. $A(5) = 4e^{-0.23} \approx 3.18$ micrograms

c. Since $A = 4$ micrograms are present when $t = 0$, find the time t at which half remains—that is when $A = 2$.

$$2 = 4e^{-0.046t}$$

$$\frac{1}{2} = e^{-0.046t}$$

$$\ln\left(\frac{1}{2}\right) = -0.046t$$

$$\frac{\ln\left(\frac{1}{2}\right)}{-0.046} = t$$

$$15.07 \approx t$$

The half-life of sodium-24 is about 15.07 hours.

d. $1 = 4e^{-0.046t}$

$$\frac{1}{4} = e^{-0.046t}$$

$$\ln\left(\frac{1}{4}\right) = -0.046t$$

$$\frac{\ln\left(\frac{1}{4}\right)}{-0.046t} = t$$

$$30.14 \approx t$$

The amount of sodium-24 will be 1 microgram after 30.14 hours.

21.

$$N(t) = N_0(0.5)^{t/5730}$$

$$N(t) = 0.45N_0$$

$$0.45N_0 = N_0(0.5)^{t/5730}$$

$$\ln(0.45) = \frac{t}{5730}\ln 0.5$$

$$5730\frac{\ln 0.45}{\ln 0.5} = t$$

$$6601 \approx t$$

The bone is about 6601 years old.

23.

$$N(t) = N_0(0.5)^{t/5730}$$

$$N(t) = 0.75N_0$$

$$0.75N_0 = N_0(0.5)^{t/5730}$$

$$\ln 0.75 = \frac{t}{5730}\ln 0.5$$

$$5730\frac{\ln 0.75}{\ln 0.5} = t$$

$$2378 \approx t$$

The Rhind papyrus is about 2378 years old

25. a. $A = 34^\circ\text{F}$, $T_0 = 75^\circ\text{F}$, $T_t = 65^\circ\text{F}$, $t = 5$. Find k.

$$65 = 34 + (75 - 34)e^{-5k}$$

$$31 = 41e^{-5k}$$

$$\frac{31}{41} = e^{-5k}$$

$$\ln\left(\frac{31}{41}\right) = -5k$$

$$k = -\frac{1}{5}\ln\left(\frac{31}{41}\right)$$

$$k \approx 0.056$$

b. $A = 34^\circ\text{F}$, $k = 0.056$, $T_0 = 75^\circ\text{F}$, $t = 30$

$$T_t = 34 + (75 - 34)e^{-30(0.056)}$$

$$T_t = 34 + (41)e^{-1.68}$$

$$T_t \approx 42^\circ\text{F}$$

c. $T_t = 36^\circ\text{F}$, $k = 0.056$, $T_t = 75^\circ\text{F}$, $A = 34^\circ\text{F}$

$$36 = 34 + (75 - 34)e^{-0.056t}$$

$$2 = 41e^{-0.056t}$$

$$t \approx 54 \text{ minutes}$$

244   **Chapter 7/Exponential and Logarithmic Functions**

27. a.   10% of 80,000 is 8000.

$$8000 = 80,000\left(1 - e^{-0.0005t}\right)$$

$$0.1 = 1 - e^{-0.0005t}$$

$$-0.9 = -e^{-0.0005t}$$

$$0.9 = e^{-0.0005t}$$

$$\ln 0.9 = -0.0005t \ln e$$

$$\ln 0.9 = -0.0005t$$

$$\frac{\ln 0.9}{-0.0005} = t$$

$$211 \approx t$$

b.   50% of 80,000 is 40,000.

$$40,000 = 80,000\left(1 - e^{-0.0005t}\right)$$

$$0.5 = 1 - e^{-0.0005t}$$

$$-0.5 = -e^{-0.0005t}$$

$$0.5 = e^{-0.0005t}$$

$$\ln(0.5) = \ln\left(e^{-0.0005t}\right)$$

$$\ln(0.5) = -0.0005t$$

$$\frac{\ln(0.5)}{-0.0005} = t$$

$$1386 \approx t$$

29. $V(t) = V_0(1 - r)^t$

$$0.5V_0 = V_0(1 - 0.20)^t$$

$$0.5 = (1 - 0.20)^t$$

$$0.5 = 0.8^t$$

$$\ln 0.5 = \ln 0.8^t$$

$$\ln 0.5 = t \ln 0.8$$

$$\frac{\ln 0.5}{\ln 0.8} = t$$

$$3.1 \text{ years} \approx t$$

31. a.
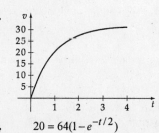

b. $$20 = 64(1 - e^{-t/2})$$

$$0.625 = 1 - e^{-t/2}$$

$$e^{-t/2} = 0.375$$

$$-t/2 = \ln 0.375$$

$$t \approx 0.98 \text{ seconds}$$

c.   The horizontal asymptote is $v = 32$.

d.   As time increases, the object's velocity approaches but never exceeds 32 ft/sec.

33. a.
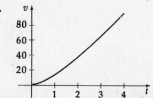

b.   The graphs of $s = 32t + 32(e^{-t} - 1)$ and $s = 50$ intersect when $t \approx 2.5$ seconds.

c.   The slope m of the secant line containing $(1, s(1))$ and $(2, s(2))$ is $m = \frac{s(2) - s(1)}{2 - 1} \approx 24.56$ ft/sec

d.   The average speed of the object was 24.56 feet per second between $t = 1$ and $t = 2$.

35. a. $P_0 = 1500$, $m = 16{,}500$, $t = 2$, $P(2) = 2500$

$$P(T) = \frac{mP_0}{P_0 + (m - P_0)e^{-kt}}$$

$$2500 = \frac{(16{,}500)(1500)}{1500 + (15{,}000)e^{-2k}}$$

$$2500 = \frac{24{,}750{,}000}{1500 + (15{,}000)e^{-2k}}$$

$$2500(1500 + (15{,}000)e^{-2k}) = 24{,}750{,}000$$

$$3{,}750{,}000 + 37{,}500{,}000e^{-2k}) = 24{,}750{,}000$$

$$37{,}500{,}000e^{-2k} = 20{,}999{,}500$$

$$e^{-2k} = 0.56$$

$$-2k = \ln 0.56$$

$$k = -\tfrac{1}{2}\ln 0.56 \approx 0.29$$

b.   Graph $P = \dfrac{24{,}750{,}000}{1500 + 15{,}000e^{-0.29t}}$

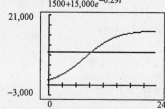

When $P = 10{,}000$, $t \approx 9.4$.

Thus the year is $1995 + 9 = 2004$.

37. a. Graph: $R = \dfrac{47850e^{0.29t}}{(10+e^{0.29t})^2}$

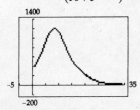

b. The maximum value of R is 1196 squirrels per year.
c. $t \approx 7.94$ years
d. The squirrel population increases the fastest when $t \approx 7.94$ years at a rate of 1196 squirrels per year.

39. a. Graph: $R = \dfrac{3810e^{0.127t}}{(1+30e^{0.127t})^2}$

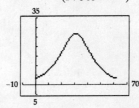

b. The maximum value of R is about 32 bison per year.
c. $t \approx 26.78$ years
d. The bison population increases the fastest when $t \approx 26.78$ years at a rate of about 32 bison per year.

41.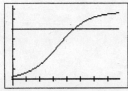

Xmin = 0, Xmax = 80, Xscl = 10,
Ymin = −10, Ymax = 110, Yscl = 15

When $P = 75\%$, $t \approx 45$ hours.

43. a.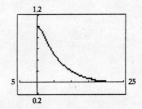

b. When $x = 5$, $P \approx 0.504$.
c. When $P = 0.25$, $x \approx 8.08$ million liters.
d. As the amount of water required per day becomes infinitely large, it becomes nearly impossible to meet the demand.

45. a.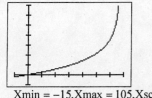

Xmin = −15, Xmax = 105, Xscl = 15,
Ymin = −2, Ymax = 14, Yscl = 2

When $v = 50$ ft/sec, $t \approx 2.2$ seconds.
b. The velocity approaches but never exceeds 100 ft/sec.

47. a. $A(1) = 0.5^{1/2}$
≈ 0.71 gram
b. $A(4) = 0.5^{4/2} + 0.5^{(4-3)/2}$
$= 0.5^2 + 0.5^{1/2}$
≈ 0.96 gram
c. $A(9) = 0.5^{9/2} + 0.5^{(9-3)/2} + 0.5^{(9-6)/2}$
$= 0.5^{4.5} + 0.5^3 + 0.5^{1.5}$
≈ 0.52 gram

49. a. $P(0) = \dfrac{4.1^0 e^{-4.1}}{0!} = \dfrac{1 \cdot e^{-4.1}}{1} \approx 0.017 = 1.7\%$

b. $P(2) = \dfrac{4.1^2 e^{-4.1}}{2!} = \dfrac{16.81e^{-4.1}}{2} \approx 0.139 = 13.9\%$

c. $P(3) = \dfrac{4.1^3 e^{-4.1}}{3!} = \dfrac{68.921e^{-4.1}}{6} \approx 0.190 = 19.0\%$

d. $P(4) = \dfrac{4.1^4 e^{-4.1}}{4!} = \dfrac{282.5761e^{-4.1}}{24} \approx 0.195 = 19.5\%$

e. $P(9) = \dfrac{4.1^9 e^{-4.1}}{9!} = \dfrac{327381.9344e^{-4.1}}{362880} \approx 0.015 = 1.5\%$

As $x \to \infty$, $P \to 0$.

51. $\dfrac{3^8 2^{12} 8! 12!}{2 \cdot 3 \cdot 2} = 3^7 2^{10} 8! 12!$

time $= \left(3^7 2^{10} 8! 12! \text{ arrangements} \right) \left(\dfrac{1 \text{ second}}{\text{arrangement}} \right) \left(\dfrac{1 \text{ minute}}{60 \text{ seconds}} \right) \left(\dfrac{1 \text{ hour}}{60 \text{ minutes}} \right) \left(\dfrac{1 \text{ day}}{24 \text{ hours}} \right) \left(\dfrac{1 \text{ year}}{365 \text{ days}} \right) \left(\dfrac{1 \text{ century}}{100 \text{ years}} \right)$

$\qquad = 3^7 2^{10} 8! 12! \left(\dfrac{1}{60^2 \cdot 24 \cdot 365 \cdot 100} \right)$

$\qquad = \dfrac{(2187)(1024)(40320)(479001600)}{(3600)(24)(365)(100)}$

$\qquad = 13,715,120,270 \text{ centuries}$

53. a. $V(3) = 350,000 \left(\dfrac{7}{8} \right)^{3/2} \approx 286,471 \text{ gallons}$

b. $V(5) = 350,000 \left(\dfrac{7}{8} \right)^{5/2} \approx 250,662 \text{ gallons}$

c. $0.10(350,000) = 350,000(.875)^{t/2}$

$\qquad 0.10 = 0.875^{t/2}$

$\qquad \ln 0.10 = \dfrac{t}{2} \ln 0.875$

$\qquad t = 2 \dfrac{\ln 0.10}{\ln 0.875} \approx 34$

$\qquad t \approx 34 \text{ hours}$

55. a. $C = \left(\dfrac{4}{5 \left[5000(0.5)^2 + 350(0.5) + 6 \right] \pi} \right) e^{\left(-1.4^2 / \; [50(0.5) + 2]^2 \right)}$

$\qquad = \dfrac{4}{5(1431)\pi} e^{(-1.4^2 / 27^2)}$

$\qquad \approx 1.7747 \times 10^{-4} \text{ g/m}^3$

b. $C = \left(\dfrac{4}{5 \left[5000(1)^2 + 350(1) + 6 \right] \pi} \right) e^{\left(-1.4^2 / \; [50(1) + 2]^2 \right)}$

$\qquad = \dfrac{4}{5(5356)\pi} e^{(-1.4^2 / 52^2)}$

$\qquad \approx 4.7510 \times 10^{-5} \text{ g/m}^3$

c. $C = \left(\dfrac{4}{5 \left[5000(1.5)^2 + 350(1.5) + 6 \right] \pi} \right) e^{(-1.4^2 / [50(1.5) + 2]^2)}$

$\qquad = \dfrac{4}{5(11781)\pi} e^{(-1.4^2 / 77^2)}$

$\qquad \approx 2.1608 \times 10^{-5} \text{ g/m}^3$

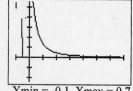

Xmin = -0.1, Xmax = 0.7, Xscl = 0.1,
Ymin = -0.001, Ymax = 0.007, Yscl = 0.001

Use the TRACE feature and the graph of

$y = \left(\dfrac{4}{5(5000x^2 + 350x + 6)\pi} \right) e^{(-1.4^2 / (50x+2)^2)}$ to

determine that $y = 0.001$ gram per cubic meter when $x \approx 0.19$ km.

SECTION 7.6, Page 444

1.

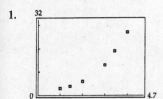

Quadratic function; increasing exponential function

3.

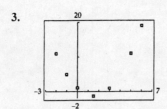

Quadratic function

5.

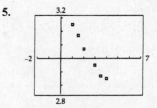

Linear function; quadratic function; decreasing logarithmic function; decreasing exponential function

7.

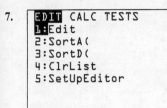

EDIT CALC TESTS
1:Edit
2:SortA(
3:SortD(
4:ClrList
5:SetUpEditor

L1	L2	L3 3
10	6.8	------
12	6.9	
14	15	
16	16.1	
18	50	
19	20	
L3(1)=		

EDIT **CALC** TESTS
4↑LinReg(ax+b)
5:QuadReg
6:CubicReg
7:QuartReg
8:LinReg(a+bx)
9:LnReg
0↓ExpReg

ExpReg
y=a*b^x
a=.9962772586
b=1.200523643
r²=.7345321476
r=.8570485095

$y = 0.9962772586(1.200523643)^x$

$r = 0.8570485095$

9.

EDIT CALC TESTS
1:Edit
2:SortA(
3:SortD(
4:ClrList
5:SetUpEditor

L1	L2	L3
0	1.83	------
1	.92	
2	.51	
3	.25	
4	.13	
5	.07	
L3(1)=		

EDIT **CALC** TESTS
4↑LinReg(ax+b)
5:QuadReg
6:CubicReg
7:QuartReg
8:LinReg(a+bx)
9:LnReg
0↓ExpReg

ExpReg
y=a*b^x
a=1.815049907
b=.519793213
r²=.9995676351
r=-.9997837942

$y = 1.815049907(0.519793213)^x$

$r = -0.9997837942$

11.

EDIT CALC TESTS
1:Edit
2:SortA(
3:SortD(
4:ClrList
5:SetUpEditor

L1	L2	L3 3
5	2.7	------
6	2.5	
7.2	2.2	
9.3	1.9	
11.4	1.6	
14.2	1.3	
L3(1)=		

EDIT **CALC** TESTS
4↑LinReg(ax+b)
5:QuadReg
6:CubicReg
7:QuartReg
8:LinReg(a+bx)
9:LnReg
0↓ExpReg

LnReg
y=a+blnx
a=4.890602565
b=-1.350726072
r²=.9984239966
r=-.9992116876

$y = 4.890602565 - 1.350726072 \ln x$

$r = -0.9992116876$

13.

EDIT CALC TESTS
1:Edit
2:SortA(
3:SortD(
4:ClrList
5:SetUpEditor

L1	L2	L3 3
3	16	------
4	16.5	
5	16.9	
7	17.5	
8	17.7	
9.8	18.1	
L3(1)=		

EDIT **CALC** TESTS
4↑LinReg(ax+b)
5:QuadReg
6:CubicReg
7:QuartReg
8:LinReg(a+bx)
9:LnReg
0↓ExpReg

LnReg
y=a+blnx
a=14.05858424
b=1.76392577
r²=.9996613775
r=-.9998306744

$y = 14.05858424 + 1.76392577 \ln x$

$r = 0.9998306744$

15.

```
EDIT CALC TESTS
1:Edit
2:SortA(
3:SortD(
4:ClrList
5:SetUpEditor
```

```
L1    L2     L3  3
0     6.5    -----
.6    9.1
1     11.1
1.8   15.2
2.2   17.1
3     20.4
L3(1)=
```

```
EDIT CALC TESTS
6↑CubicReg
7:QuartReg
8:LinReg(a+bx)
9:LnReg
0:ExpReg
A:PwrReg
B↓Logistic
```

```
Logistic
y=c/(1+ae^(-bx))
a=2.993715983
b=.7965003627
c=26.01827446
```

$$y = \frac{26.01827446}{1 + 2.993715983e^{-0.7965003627x}}$$

17.

```
EDIT CALC TESTS
1:Edit
2:SortA(
3:SortD(
4:ClrList
5:SetUpEditor
```

```
L1    L2      L3  3
1.6   151.2   -----
2     191.9
2.5   251.8
3.4   382.5
4     468.6
6.1   700.2
L3(1)=
```

```
EDIT CALC TESTS
6↑CubicReg
7:QuartReg
8:LinReg(a+bx)
9:LnReg
0:ExpReg
A:PwrReg
B↓Logistic
```

```
Logistic
y=c/(1+ae^(-bx))
a=14.33048267
b=.7547088681
c=799.7677348
```

$$y = \frac{799.7677348}{1 + 14.33048267e^{-0.7547088681x}}$$

19. a.

```
EDIT CALC TESTS
1:Edit
2:SortA(
3:SortD(
4:ClrList
5:SetUpEditor
```

```
L1   L2      L3  3
68   71.75   -----
72   75.625
76   76
80   77.5
84   79.5
88   80
L3(1)=
```

```
EDIT CALC TESTS
4↑LinReg(ax+b)
5:QuadReg
6:CubicReg
7:QuartReg
8:LinReg(a+bx)
9:LnReg
0↓ExpReg
```

```
LinReg
y=ax+b
a=.2244791667
b=58.87986111
r²=.7398085705
r=.8601212534
```

```
EDIT CALC TESTS
4↑LinReg(ax+b)
5:QuadReg
6:CubicReg
7:QuartReg
8:LinReg(a+bx)
9:LnReg
0↓ExpReg
```

```
LnReg
y=a+blnx
a=-7.07160205
b=19.17357961
r²=.7812119302
r=.8838619407
```

linear: height $= 0.2244791667t + 58.87986111$

logarithmic: height $= -7.07160205 + 19.17357961\ln t$

b.   The logarithmic model is a better fit for the data; $|r|$ for the logarithmic model is closer to 1.

c. $h = -7.07160205 + 19.17357961\ln 112$

   ≈ 83.4 inches $= 6$ ft 11.4 inches

21. a.

```
EDIT CALC TESTS
1:Edit
2:SortA(
3:SortD(
4:ClrList
5:SetUpEditor
```

L1	L2	L3	3
1	153	------	
2	154		
3	156		
4	155		
5	160		
6	164		

L3(1)=

```
EDIT CALC TESTS
4↑LinReg(ax+b)
5:QuadReg
6:CubicReg
7:QuartReg
8:LinReg(a+bx)
9:LnReg
0↓ExpReg
```

```
LinReg
y=ax+b
a=1.040350877
b=154.9649123
r²=.7824855363
r=.8845821252
```

```
EDIT CALC TESTS
4↑LinReg(ax+b)
5:QuadReg
6:CubicReg
7:QuartReg
8:LinReg(a+bx)
9:LnReg
0↓ExpReg
```

```
LnReg
y=a+blnx
a=149.5687609
b=7.630768331
r²=.8667828751
r=.9310117481
```

linear: time $= 1.040350877x + 154.9649123$
logarithmic: time $= 149.5687609 + 7.630768331 \ln x$

b. The logarithmic model provides a better fit for the data; its correlation coefficient is closer to 1.

c. Use the logarithmic model with time = 175

$$175 = 149.5687609 + 7.630768331 \ln x$$

$$25.4312391 = 7.630768331 \ln x$$

$$3.332723259 = \ln x$$

$$e^{3.332723259} = x$$

$$28 \approx x$$

28 years after 1980, or 2008

23. a.

```
EDIT CALC TESTS
1:Edit
2:SortA(
3:SortD(
4:ClrList
5:SetUpEditor
```

L1	L2	L3	3
0	95	------	
5	70		
10	51		
15	37		
20	27		
25	19		

L3(1)=

```
EDIT CALC TESTS
6↑CubicReg
7:QuartReg
8:LinReg(a+bx)
9:LnReg
0:ExpReg
A:PwrReg
B↓Logistic
```

```
ExpReg
y=a*b^x
a=96.16776667
b=.9378652611
r²=.9995866444
r=.9997933008
```

difference $= 96.16776667(0.9378652611)^t$

b. When the temperature is 80° F, the difference is 10° F.

$$10 = 96.16776667(0.9378652611)^t$$

$$\frac{10}{96.16776667} = 0.9378652611^t$$

$$0.1039849457 = 0.9378652611^t$$

$$\ln 0.1039849457 = \ln 0.9378652611^t$$

$$\ln 0.1039849457 = t \ln 0.9378652611$$

$$\frac{\ln 0.1039849457}{\ln 0.9378652611} = t$$

$$35 \approx t$$

35 minutes

25. a.

```
EDIT CALC TESTS
1:Edit
2:SortA(
3:SortD(
4:ClrList
5:SetUpEditor
```

```
 L1    L2    L3   3
 60   2.66  ------
 70   3.27
 80   3.61
 90   4
100   4.3
-----  -----
L3(1)=
```

```
EDIT CALC TESTS
4↑LinReg(ax+b)
5:QuadReg
6:CubicReg
7:QuartReg
8:LinReg(a+bx)
9:LnReg
0↓ExpReg
```

```
LinReg
y=ax+b
a=.0401
b=.36
r²=.9820028336
r=.9909605611
```

```
EDIT CALC TESTS
4↑LinReg(ax+b)
5:QuadReg
6:CubicReg
7:QuartReg
8:LinReg(a+bx)
9:LnReg
0↓ExpReg
```

```
LnReg
y=a+blnx
a=-10.23519266
b=3.161541421
r²=.9947750419
r=.9973840995
```

linear: $p = 0.0401t + 0.36$

logarithmic: $p = -10.23519266 + 3.161541421\ln t$

b. The logarithmic model is a better fit for the data; $|r|$ for the logarithmic model is closer to 1.

c. $p = -10.23519266 + 3.161541421\ln 105$

≈ 4.48 pounds

27. a.

```
EDIT CALC TESTS
1:Edit
2:SortA(
3:SortD(
4:ClrList
5:SetUpEditor
```

```
 L1    L2    L3   3
 60    3   ------
 74    4
 87    5
 99    6
-----  -----
L3(1)=
```

```
EDIT CALC TESTS
4↑LinReg(ax+b)
5:QuadReg
6:CubicReg
7:QuartReg
8:LinReg(a+bx)
9:LnReg
0↓ExpReg
```

```
ExpReg
y=a*b^x
a=1.052579579
b=1.017912057
r²=.995336125
r=.9976653372
```

$$p = 1.052579579(1.017912057)^t$$

$$12 = 1.052579579(1.017912057)^t$$

$$\frac{12}{1.052579579} = 1.017912057^t$$

$$\ln\left(\frac{12}{1.052579579}\right) = \ln(1.017912057^t)$$

$$\ln\left(\frac{12}{1.052579579}\right) = t\ln 1.017912057$$

$$\frac{\ln\left(\dfrac{12}{1.052579579}\right)}{\ln 1.017912057} = t$$

$$137 \approx t$$

137 years after 1900, or 2037

b.

```
EDIT CALC TESTS
1:Edit
2:SortA(
3:SortD(
4:ClrList
5:SetUpEditor
```

```
 L1   L2  L3   3
 60   3  ------
 74   4
 87   5
 99   6
-----
L3(1)=
```

```
EDIT CALC TESTS
6↑CubicReg
7:QuartReg
8:LinReg(a+bx)
9:LnReg
0:ExpReg
A:PwrReg
B↓Logistic
```

```
Logistic
y=c/(1+ae^(-bx))
a=15.89101611
b=.0292380039
c=11.26828438
```

$$y = \frac{11.26828438}{1+15.89101611e^{-0.0292380039t}}$$

As $t \to \infty, e^{-0.0292380039t} \to 0$ and thus $y \to \dfrac{11.26828438}{1+15.89101611(0)} = \dfrac{11.26828438}{1} = 11.26828438$ billion

c. The logistic model is more realistic. The world's future population is more likely to level off due to finite resources than to increase without bound.

29. a.

```
 EDIT  CALC TESTS
 1:Edit
 2:SortA(
 3:SortD(
 4:ClrList
 5:SetUpEditor
```

```
 L1    L2    L3   3
 0     3200  ------
 20    590
 -----  -----


 L3(1)=
```

```
 EDIT  CALC  TESTS
 4↑LinReg(ax+b)
 5:QuadReg
 6:CubicReg
 7:QuartReg
 8:LinReg(a+bx)
 9:LnReg
 0↓ExpReg
```

```
 ExpReg
 y=a*b^x
 a=3200
 b=.918935653
 r²=1
 r=-1
```

$$p = 3200(0.918935653)^t$$

$$200 = 3200(0.918935653)^t$$

$$\frac{200}{3200} = 0.918935653^t$$

$$0.0625 = 0.918935653^t$$

$$\ln 0.0625 = \ln 0.918935653^t$$

$$\ln 0.0625 = t \ln 0.918935653$$

$$\frac{\ln 0.0625}{\ln 0.918935653} = t$$

$$32.8 \approx t$$

32 years after 1980, or 2012

b. No. The model fits the data perfectly because there are only two data points.

31.

```
 EDIT  CALC TESTS
 1:Edit
 2:SortA(
 3:SortD(
 4:ClrList
 5:SetUpEditor
```

```
 L1    L2    L3   3
 0     6.5   ------
 .6    9.1
 1     11.1
 1.8   15.2
 2.2   17.1
 3     20.4
 L3(1)=
```

```
 EDIT  CALC  TESTS
 6↑CubicReg
 7:QuartReg
 8:LinReg(a+bx)
 9:LnReg
 0:ExpReg
 A:PwrReg
 B↓Logistic
```

```
 Logistic
 y=c/(1+ae^(-bx))
 a=2.993715983
 b=.7965003627
 c=26.01827446
```

$$P = \left(\frac{\ln 2.993715983}{0.7965003627}, \frac{26.01827446}{2} \right) \approx (1.4, \ 13.0)$$

33. $P = \left(\dfrac{\ln 11.44466821}{0.3115234553}, \dfrac{3965.337214}{2} \right) \approx (7.8, \ 1982.7)$

35. $P(x) = \dfrac{mP_0}{P_0 + (m - P_0)e^{-kx}} = \dfrac{\dfrac{mP_0}{P_0}}{\dfrac{P_0 + (m - P_0)e^{-kx}}{P_0}} = \dfrac{m}{1 + \dfrac{m - P_0}{P_0}e^{-kx}}$ Compare this with $\dfrac{c}{1 + ae^{-bx}}$.

a. $c = m$

b. $b = k$

c. $a = \dfrac{m - P_0}{P_0}$

37. a. The x-coordinate of the first ordered pair is 0, and 0 is not in the domain of $y = \ln x$.

    b. Use a horizontal translation of the data. For instance, add 1 to each of the x-coordinates. Find the logarithmic regression function for this new data set. Remember that each x-value in the regression represents $x - 1$ in the original data set.

39. a.

```
EDIT CALC TESTS
1:Edit
2:SortA(
3:SortD(
4:ClrList
5:SetUpEditor
```

```
L1   L2    L3   3
 1   2.1   ------
 2   5.5
 3   9.8
 4   14.6
 5   20.1
 6   25.8
L3(1)=
```

```
EDIT CALC TESTS
4↑LinReg(ax+b)
5:QuadReg
6:CubicReg
7:QuartReg
8:LinReg(a+bx)
9:LnReg
0↓ExpReg
```

```
ExpReg
y=a*b^x
a=1.811200596
b=1.617401977
r²=.9368807058
r=.9679259816
```

```
EDIT CALC TESTS
5↑QuadReg
6:CubicReg
7:QuartReg
8:LinReg(a+bx)
9:LnReg
0:ExpReg
A↓PwrReg
```

```
PwrReg
y=a*x^b
a=2.093851108
b=1.402462579
r²=.9999806043
r=.9999903021
```

exponential: $y = 1.811200596(1.617401977^x)$

power: $y = 2.093851108x^{1.402462579}$

b. The power regression is a better fit for the data; $|r|$ for the power model is closer to 1.

41. a.

```
EDIT CALC TESTS
1:Edit
2:SortA(
3:SortD(
4:ClrList
5:SetUpEditor
```

```
L1   L2    L3   3
 1   1.11  ------
 2   1.57
 3   1.92
 4   2.25
 6   2.72
 8   3.14
L3(1)=
```

```
EDIT CALC TESTS
5↑QuadReg
6:CubicReg
7:QuartReg
8:LinReg(a+bx)
9:LnReg
0:ExpReg
A↓PwrReg
```

```
PwrReg
y=a*x^b
a=1.110883155
b=.5011329269
r²=.9997894216
r=.9998947053
```

```
EDIT CALC TESTS
4↑LinReg(ax+b)
5:QuadReg
6:CubicReg
7:QuartReg
8:LinReg(a+bx)
9:LnReg
0↓ExpReg
```

```
ExpReg
y=a*b^x
a=1.14997043
b=1.148603681
r²=.9079306949
r=.9528539735
```

```
EDIT CALC TESTS
4↑LinReg(ax+b)
5:QuadReg
6:CubicReg
7:QuartReg
8:LinReg(a+bx)
9:LnReg
0↓ExpReg
```

```
LnReg
y=a+blnx
a=.9740824917
b=.9739334447
r²=.9764616924
r=.9881607624
```

The power regression is the best fit for the data; $|r|$ for the power model is closest to 1.
The power regression equation is $t = 1.110883155l^{0.5011329269}$

b.

$$1.110883155l^{0.5011329269} = 12$$

$$l^{0.5011329269} = \frac{12}{1.110883155}$$

$$\left(l^{0.5011329269}\right)^{1/0.5011329269} = \left(\frac{12}{1.110883155}\right)^{1/0.5011329269}$$

$$l \approx 115.4 \text{ feet}$$

EXPLORING CONCEPTS WITH TECHNOLOGY, Page 448

1. a.

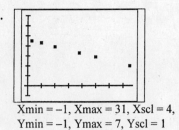

Xmin = −1, Xmax = 31, Xscl = 4,
Ymin = −1, Ymax = 7, Yscl = 1

b. $m = \frac{4.3-3.3}{4-15} \approx -0.0909$

c. $\ln A - 4.3 = -0.0909(t-4)$
$\ln A = -0.0909t + 4.664$

d. $e^{\ln A} = e^{-0.0909\,t+4.664}$
$A = e^{-0.0909\,t}e^{4.664}$
$A = 106e^{-0.0909\,t}$

e.

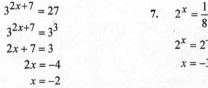

Xmin = −5, Xmax = 35, Xscl = 5,
Ymin = −10, Ymax = 110, Yscl = 15

f. $A(t) = 106e^{-0.0909t}$
At $t = 0$ there is $A = 106$ mg present
We must find t where $A = \frac{1}{2}(106) = 53$ mg.

$$53 = 106e^{-0.0909\,t}$$
$$\frac{53}{106} = e^{-0.0909\,t}$$
$$\frac{1}{2} = e^{-0.0909\,t}$$
$$\ln 0.5 = -0.0909t$$
$$t = \frac{\ln 0.5}{-0.0909} \approx 7.6 \text{ days}$$

CHAPTER 7 TRUE/FALSE EXERCISES, Page 451

1. True

3. True

5. False; $h(x)$ is not an increasing function for $0 < b < 1$.

7. True

9. True, because $f(-x) = \frac{2^{-x}+2^{-(-x)}}{2} = \frac{2^{-x}+2^{x}}{2} = f(x)$

11. False, $\log x + \log y = \log xy$.

13. True

CHAPTER 7 REVIEW EXERCISES, Page 452

1. $\log_5 25 = x$
$5^x = 25$
$5^x = 5^2$
$x = 2$

3. $\ln e^3 = x$
$e^x = e^3$
$x = 3$

5. $3^{2x+7} = 27$
$3^{2x+7} = 3^3$
$2x + 7 = 3$
$2x = -4$
$x = -2$

7. $2^x = \frac{1}{8}$
$2^x = 2^{-3}$
$x = -3$

9. $\log x^2 = 6$
$10^6 = x^2$
$1{,}000{,}000 = x^2$
$\pm\sqrt{1{,}000{,}000} = x$
$\pm 1000 = x$

11. $10^{\log 2x} = 14$
$2x = 14$
$x = 7$

13.

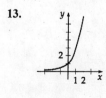

15.

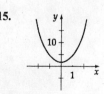

17.

19.

21.

23.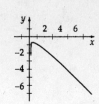

25.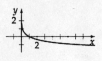

27. $\log_4 64 = 3$

   $4^3 = 64$

29. $\log_{\sqrt{2}} 4 = 4$

   $\left(\sqrt{2}\right)^4 = 4$

31. $5^3 = 125$

   $\log_5 125 = 3$

33. $10^0 = 1$

   $\log_{10} 1 = 0$

35. $\log_b \dfrac{x^2 y^3}{z} = 2\log_b x + 3\log_b y - \log_b z$

37. $\ln xy^3 = \ln x + 3\ln y$

39. $2\log x + \dfrac{1}{3}\log(x+1) = \log\left(x^2 \sqrt[3]{x+1}\right)$

41. $\dfrac{1}{2}\ln 2xy - 3\ln z = \ln \dfrac{\sqrt{2xy}}{z^3}$

43. $\log_5 101 = \dfrac{\log 101}{\log 5} \approx 2.86754$

45. $\log_4 0.85 = \dfrac{\log 0.85}{\log 4} \approx -0.117233$

47. $4^x = 30$

   $\log 4^x = \log 30$

   $x\log 4 = \log 30$

   $x = \dfrac{\log 30}{\log 4}$

49. $\ln(3x) - \ln(x-1) = \ln 4$

   $\ln \dfrac{3x}{x-1} = \ln 4$

   $\dfrac{3x}{x-1} = 4$

   $3x = 4(x-1)$

   $3x = 4x - 4$

   $4 = x$

51. $e^{\ln(x+2)} = 6$

   $(x+2) = 6$

   $x + 2 = 6$

   $x = 4$

53. $\dfrac{4^x + 4^{-x}}{4^x - 4^{-x}} = 2$

   $4^x\left(4^x + 4^{-x}\right) = 2\left(4^x - 4^{-x}\right)4^x$

   $4^{2x} + 1 = 2\left(4^{2x} - 1\right)$

   $4^{2x} + 1 = 2 \cdot 4^{2x} - 2$

   $4^{2x} - 2 \cdot 4^{2x} + 3 = 0$

   $4^{2x} = 3$

   $2^x \ln 4 = \ln 3$

   $x = \dfrac{\ln 3}{2\ln 4}$

55. $\log \sqrt{x-5} = 3$

   $10^3 = \sqrt{x-5}$

   $10^6 = x - 5$

   $10^6 + 5 = x$

   $x = 1{,}000{,}005$

57. $\log(\log x) = 3$

   $10^3 = \log x$

   $10^{(10^3)} = x$

   $10^{1000} = x$

59. $\log_4(\log_3 x) = 1$

   $4 = \log_3 x$

   $3^4 = x$

   $81 = x$

61. $\log_5 x^3 = \log_5 16x$

   $x^3 = 16x$

   $x^2 = 16$

   $x = 4$

63. $m = \log\left(\dfrac{I}{I_0}\right)$

   $= \log\left(\dfrac{51{,}782{,}000 I_0}{I_0}\right)$

   $= \log 51{,}782{,}000$

   ≈ 7.7

65. $\log\left(\dfrac{I_1}{I_0}\right) = 7.2$ and $\log\left(\dfrac{I_2}{I_0}\right) = 3.7$

$\dfrac{I_1}{I_0} = 10^{7.2}$ $\dfrac{I_2}{I_0} = 10^{3.7}$

$I_1 = 10^{7.2}I_0$ $I_2 = 10^{3.7}I_0$

$\dfrac{I_1}{I_2} = \dfrac{10^{7.2}I_0}{10^{3.7}I_0} = \dfrac{10^{3.5}}{1} \approx \dfrac{3162}{1}$

3162 to 1

67. $\text{pH} = -\log\left[H_3O^+\right]$

$= -\log\left[6.28 \times 10^{-5}\right]$

≈ 4.2

69. $P = 16{,}000, r = 0.08, t = 3$

a. $B = 16{,}000\left(1 + \dfrac{0.08}{12}\right)^{36} \approx \$20{,}323.79$

b. $B = 16{,}000e^{0.08(3)}$

$B = 16{,}000e^{0.24} \approx \$20{,}339.99$

71. $S(n) = P(1 - r)^n, P = 12{,}400, r = 0.29, t = 3$

$S(n) = 12{,}400(1 - 0.29)^3 \approx \$4438.10$

73. $N(0) = 1$ $N(2) = 5$

$1 = N_0 e^{k(0)}$ $5 = e^{2k}$

$1 = N_0$ $\ln 5 = 2k$

$k = \dfrac{\ln 5}{2} \approx 0.8047$

Thus $N(t) = e^{0.8047t}$

75. $4 = N(1) = N_0 e^k$ and thus $\dfrac{4}{N_0} = e^k$. Now, we also

have $N(5) = 5 = N_0 e^{5k} = N_0\left(\dfrac{4}{N_0}\right)^5 = \dfrac{1024}{N_0^4}$.

$N_0 = \sqrt[4]{\dfrac{1024}{5}} \approx 3.783$

Thus $4 = 3.783e^k$.

$\ln\left(\dfrac{4}{3.783}\right) = k$

$k \approx 0.0558$

Thus $N_0 = 3.783e^{0.0558t}$.

77. a. $N(1) = 25200e^{k(1)} = 26800$

$e^k = \dfrac{26800}{25200}$

$\ln e^k = \ln\left(\dfrac{26800}{25200}\right)$

$k \approx 0.061557893$

$N(t) = 25200e^{0.061557893\,t}$

b. $N(7) = 25200e^{0.061557893(7)}$

$= 25200e^{0.430905251}$

≈ 38800

79. a.

L1	L2	L3 2
90	2.04E6	------
91	1.99E6	
92	1.88E6	
93	1.71E6	
94	1.61E6	
95	1.52E6	
L2(1)=2043705		

```
LinReg
y=ax+b
a=-76740
b=8888642.4
r²=.9322468826
r=-.9655293277
```

```
ExpReg
y=a*b^x
a=125169617.3
b=.9550487932
r²=.9526445023
r=-.9760350928
```

```
LnReg
y=a+blnx
a=34707491.3
b=-7271279.418
r²=.9384231477
r=-.9687224307
```

linear: $P = -76740t + 8888642.4$

exponential: $P = 125169617.3\left(0.9550487932^t\right)$

logarithmic: $P = 34707491.3 - 7271279.418\ln t$

b. The exponential model provides the best fit to the data; its value of $|r|$ is closest to 1.

c. $P = 125169617.3(0.9550487932)^{104}$

   $\approx 1,050,000$

81. a.

$$P(t) = \frac{mP_0}{P_0 + (m - P_0)e^{-kt}}$$

$$P(3) = 360 = \frac{1400(210)}{210 + (1400 - 210)e^{-k(3)}}$$

$$360 = \frac{294000}{210 + 1190e^{-3k}}$$

$$360\left(210 + 1190e^{-3k}\right) = 294000$$

$$210 + 1190e^{-3k} = \frac{294000}{360}$$

$$1190e^{-3k} = \frac{29400}{36} - 210$$

$$e^{-3k} = \frac{29400/36 - 210}{1190}$$

$$\ln e^{-3k} = \ln\left(\frac{29400/36 - 210}{1190}\right)$$

$$-3k = \ln\left(\frac{29400/36 - 210}{1190}\right)$$

$$k = -\frac{1}{3}\ln\left(\frac{29400/36 - 210}{1190}\right)$$

$$k \approx 0.2245763649$$

$$P(t) = \frac{294000}{210 + 1190e^{-0.2245763649t}}$$

b.

$$P(13) = \frac{294000}{210 + 1190e^{-0.2245763649(13)}}$$

$$= \frac{294000}{210 + 1190e^{-2.919492744}}$$

$$\approx 1070$$

CHAPTER 7 TEST, Page 454

1.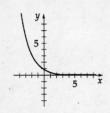

2.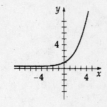

3. $\log_b(5x - 3) = c$

   $b^c = 5x - 3$

4. $3^{x/2} = y$

   $\log_3 y = \dfrac{x}{2}$

5. $\log_b \dfrac{z^2}{y^3\sqrt{x}} = \log_b z^2 - \log_b y^3 - \log_b x^{1/2}$

   $= 2\log_b z - 3\log_b y - \dfrac{1}{2}\log_b x$

6. $\log_{10}(2x + 3) - 3\log_{10}(x - 2) = \log_{10}(2x + 3) - \log_{10}(x - 2)^3$

   $= \log_{10}\dfrac{2x + 3}{(x - 2)^3}$

7. $\log_4 12 = \dfrac{\log 12}{\log 4}$

          ≈ 1.7925

8.

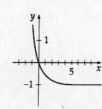

9. $5^x = 22$

        $x \log 5 = \log 22$

           $x = \dfrac{\log 22}{\log 5}$

           $x \approx 1.9206$

10. $\dfrac{3^x + 3^{-x}}{2} = 21$

           $3^x + 3^{-x} = 42$

             $3^{2x} + 1 = 42(3^x)$

       $3^{2x} - 42(3^x) + 1 = 0$

     $(3^x)^2 - 42(3^x) + 1 = 0$

          $u^2 - 42u + 1 = 0$

$$u = \frac{-(-42) \pm \sqrt{(-42)^2 - 4(1)(1)}}{2(1)}$$

$$u = \frac{42 \pm \sqrt{1764 - 4}}{2}$$

$$u = \frac{42 \pm \sqrt{1760}}{2}$$

$$u = \frac{42 \pm 4\sqrt{110}}{2}$$

$$u = 21 \pm 2\sqrt{110}$$

$$3^x = 21 \pm 2\sqrt{110}$$

$$\ln 3^x = \ln\left(21 \pm 2\sqrt{110}\right)$$

$$x \ln 3 = \ln\left(21 \pm 2\sqrt{110}\right)$$

$$x = \frac{\ln\left(21 \pm 2\sqrt{110}\right)}{\ln 3}$$

11. $\log(x + 99) - \log(3x - 2) = 2$

            $\log \dfrac{x + 99}{3x - 2} = 2$

               $\dfrac{x + 99}{3x - 2} = 10^2$

          $x + 99 = 100(3x - 2)$

          $x + 99 = 300x - 200$

        $-299x = -299$

              $x = 1$

12. $\ln(2 - x) + \ln(5 - x) = \ln(37 - x)$

         $\ln(2 - x)(5 - x) = \ln(37 - x)$

           $(2 - x)(5 - x) = (37 - x)$

          $10 - 7x + x^2 = 37 - x$

            $x^2 - 6x - 27 = 0$

           $(x - 9)(x + 3) = 0$

   $x = 9$ (not in domain)   or $x = -3$

$x = -3$

13.   **a.** $A = P\left(1 + \dfrac{r}{n}\right)^{nt}$

            $= 20{,}000\left(1 + \dfrac{0.078}{12}\right)^{12(5)}$

            $= 20{,}000(1.0065)^{60}$

            $= \$29{,}502.36$

   b. $A = Pe^{rt}$

            $= 20{,}000e^{0.078(5)}$

            $= 20{,}000e^{0.39}$

            $= \$29{,}539.62$

14. a. $M = \log\left(\dfrac{I}{I_0}\right)$

$= \log\left(\dfrac{42{,}304{,}000 I_0}{I_0}\right)$

$= \log 42{,}304{,}000$

≈ 7.6

b. $\log\left(\dfrac{I_1}{I_0}\right) = 6.3$ and $\log\left(\dfrac{I_2}{I_0}\right) = 4.5$

$\dfrac{I_1}{I_0} = 10^{6.3}$ $\dfrac{I_2}{I_0} = 10^{4.5}$

$I_1 = 10^{6.3} I_0$ $I_2 = 10^{4.5} I_0$

$\dfrac{I_1}{I_2} = \dfrac{10^{6.3} I_0}{10^{4.5} I_0} = \dfrac{10^{1.8}}{1} \approx \dfrac{63}{1}$

Therefore the ratio is 63 to 1.

15. a. $N(3) = 34600 e^{k(3)} = 39800$

$34600 e^{3k} = 39800$

$e^{3k} = \dfrac{39800}{34600}$

$\ln e^{3k} = \ln\left(\dfrac{398}{346}\right)$

$3k = \ln\left(\dfrac{398}{346}\right)$

$k = \dfrac{1}{3}\ln\left(\dfrac{398}{346}\right)$

$k \approx 0.0466710767$

$N(t) = 34600 e^{0.0466710767\,t}$

b. $N(10) = 34600 e^{0.0466710767(10)}$

$= 34600 e^{0.466710767}$

$\approx 55{,}000$

16. $P(t) = 0.5^{\,t/5730} = 0.92$

$\log 0.5^{\,t/5730} = \log 0.92$

$\dfrac{t}{5730}\log 0.5 = \log 0.92$

$\dfrac{t}{5730} = \dfrac{\log 0.92}{\log 0.5}$

$t = 5730\left(\dfrac{\log 0.92}{\log 0.5}\right)$

$t \approx 690$ years

17. a.

```
EDIT CALC TESTS
1:Edit
2:SortA(
3:SortD(
4:ClrList
5:SetUpEditor
```

```
L1    L2    L3    3
2.5   16    -----
3.7   48
5     155
6.5   571
6.9   896
----- -----
L3(1)=
```

```
EDIT CALC TESTS
4↑LinReg(ax+b)
5:QuadReg
6:CubicReg
7:QuartReg
8:LinReg(a+bx)
9:LnReg
0↓ExpReg
```

```
ExpReg
y=a*b^x
a=1.671991998
b=2.471878247
r²=.9996384751
r=.9998192212
```

$y = 1.671991998(2.471878247)^x$

b. $y = 1.671991998(2.471878247)^{7.8}$

≈ 1945

18. a.

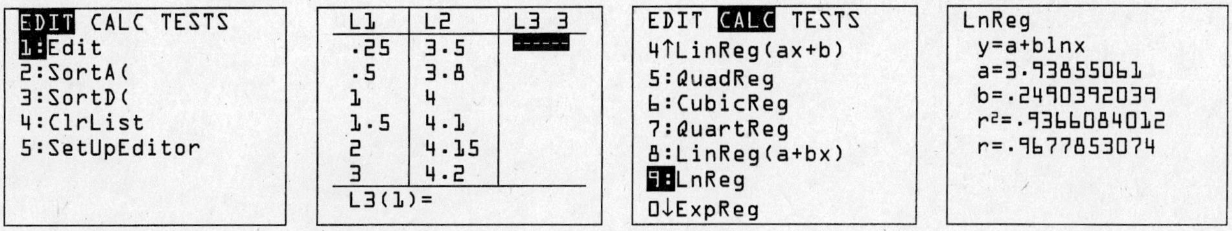

```
EDIT CALC TESTS
1:Edit
2:SortA(
3:SortD(
4:ClrList
5:SetUpEditor
```

L1	L2	L3 3
.25	3.5	-----
.5	3.8	
1	4	
1.5	4.1	
2	4.15	
3	4.2	
L3(1)=		

```
EDIT CALC TESTS
4↑LinReg(ax+b)
5:QuadReg
6:CubicReg
7:QuartReg
8:LinReg(a+bx)
9:LnReg
0↓ExpReg
```

```
LnReg
y=a+blnx
a=3.93855061
b=.2490392039
r²=.9366084012
r=.9677853074
```

$R(t) = 3.93855061 + 0.2490392039 \ln t$

$R(3.5) = 3.93855061 + 0.2490392039 \ln 3.5$

≈ 4.25

The predicted interest rate for 3.5 years is 4.25%.

b. $3.93855061 + 0.2490392039 \ln t = 4.4$

$0.2490392039 \ln t = 0.46144939$

$\ln t = \dfrac{0.46144939}{0.2490392039}$

$e^{\ln t} = e^{0.46144939/0.2490392039}$

$t = e^{0.46144939/0.2490392039}$

$t \approx 6.4 \text{ years}$